Topics in Applied Physics   Volume 58

Topics in Applied Physics Founded by Helmut K. V. Lotsch

Volumes 1–56 are listed on the back inside cover

Hot-Electron Transport in Semiconductors

Edited by L. Reggiani

With Contributions by
M. Asche   C. Canali   E. Constant   K. Hess
G. J. Iafrate   S. Komiyama   T. Kurosawa   T. Masumi
F. Nava   Y. K. Pozhela   L. Reggiani

With 152 Figures

Springer-Verlag Berlin Heidelberg GmbH

Professor *Lino Reggiani*

Dipartimento di Fisica, Università di Modena, Via Campi 213/A,
I-41100 Modena, Italy

ISBN 978-3-662-30935-3   ISBN 978-3-540-38849-4 (eBook)
DOI 10.1007/978-3-540-38849-4

Library of Congress Cataloging in Publication Data. Main entry under title: Hot-electron transport in semiconductors. (Topics in applied physics; v. 58) Includes index. 1. Hot carriers. 2. Semiconductors. I. Reggiani, L. (Lino), 1941–. II. Asche, M. (Marion) III. Series. QC611.6.H67H68 1985   537.6'22   84-19331

© Springer-Verlag Berlin Heidelberg 1985
Originally published by Springer-Verlag Berlin Heidelberg New York in 1985
Softcover reprint of the hardcover 1st edition 1985

Monophoto typesetting, offset printing and bookbinding: Brühlsche Universitätsdruckerei, Giessen
2153/3130-543210

To Annalita, Rossana, and Lorenzo
my wife and my children

Preface

In a heuristic approach the concept of hot electrons is associated with a temperature of the electron gas which is higher than that of the host lattice. This is usually realized by applying electric fields of sufficiently high strength, and the study of charge-carrier dynamics under such a condition is commonly called hot-electron transport.

Owing to the growing importance which semiconductor devices are assuming in the technology of computers and telecommunications, hot-electron transport in semiconductors is rapidly developing as a research subject. Indeed, modern microelectronics has now entered the submicrometer scale of miniaturization, and it is easy to understand that even a few volts in the applied voltage can lead to very high electric fields of the order of 10,000 V/cm. These high fields, by leading to values of the carrier drift velocity of the order of 10^7 cm/s, are also at the basis of devices operating at frequencies as high as 100 GHz.

Physical understanding of most of the microscopic processes which underlie the performances of semiconductor devices at high electric fields is provided by research into hot-electron phenomena. A first survey, describing theoretical and experimental findings up to 1965, is given in Conwell's well-known book *High Field Transport in Semiconductors*, issued as supplement in the Solid State Physics Series (Academic, New York 1967). Since then, a notable amount of work has been done. New experimental techniques have been used and different materials have been characterized. Also new theoretical methods have been introduced, enabling exact numerical solutions of the Boltzmann equation.

The purpose of this volume is to give a unifying physical interpretation of the main results which have appeared in the literature of the past 20 years on hot-electron transport in bulk semiconductors. This aim is pursued by a combination of tutorial and educational background material, and up-to-date applications to problem areas of current interest.

Many authors with international recognition have contributed to make this book highly beneficial for the reader.

Together with the description of different theoretical and experimental techniques, a great effort has been devoted to the collection and display of the most reliable data, not otherwise available in a single textbook, on the drift velocity, the diffusion coefficient and the equivalent noise temperature associated with velocity fluctuations of the best known semiconductors. Furthermore, the microscopic models pertaining to different materials have been widely discussed and summarized in useful tables. The content of the book should therefore satisfy the basic requirement of offering an up-to-date microscopic description of hot-electron phenomena, and could be destined to become a helpful standard of reference. It may be of particular interest to researchers and graduate students in the field of microelectronics, VLSI, and device modeling, in particular. It may thus be recommended as a textbook for graduate courses in physics of electronics, and electrical engineering departments.

Finally, the editor would like to express his thanks to Professor M. Cardona for having solicited this effort, to Dr. H. Lotsch for his cooperation in the editing procedure, to the authors who have contributed to this volume and to the colleagues in the Physics Department for having provided a critical scientific environment of high standard. Also acknowledged are the Computer Center of the Modena University, the Ministero della Pubblica Istruzione (MPI) and the European Research Office (ERO) for their financial support.

Modena, October 1984   *Lino Reggiani*

Contents

Contributors

Asche, Marion
Zentralinstitut für Elektronenphysik der AdW der DDR,
Hausvogteiplatz 5–7, DDR-1086 Berlin

Canali, Claudio
Istituto di Elettrotecnica ed Elettronica, Via Gradenigo 6/a,
I-35131 Padova, Italy

Constant, Eugene
Centre Hyperfrequences et Semiconducteurs, La CNRS N° 287, Bat. P3,
F-59655 Villeneuve d'Ascq Cedex Lille, France

Hess, Karl
Department of Electrical Engineering, University of Illinois,
Urbana-Champaign, Urbana, IL 61801, USA

Iafrate, Gerald J.
U.S. Army Electronics Technology and Devices Laboratory,
Forth Monmouth, NJ 07703, USA

Komiyama, Susumu
Department of Pure and Applied Sciences, University of Tokyo, Komaba,
3-8-1, Komaba, Meguro-ku, Tokyo 153, Japan

Kurosawa, Tatsumi
Department of Physics, Chuo University, 1-13-27 Kasuga,
Bunkyo-ku, Tokyo 112, Japan

Masumi, Taizo
Department of Pure and Applied Sciences, University of Tokyo, Komaba,
3-8-1, Komaba, Meguro-ku, Tokyo 153, Japan

Nava, Filippo

Dipartimento di Fisica, Università di Modena, Via Campi 213/A,
I-41100 Modena, Italy

Pozhela, Yuras K.

Institute of Semiconductor Physics, Academy of Sciences
of the Lithuanian SSR, Lenino 3, SU-232600 Vilnius, USSR

Reggiani, Lino

Dipartimento di Fisica, Università di Modena, Via Campi 213/A,
I-41100 Modena, Italy

1. Introduction

Lino Reggiani

1.1 Historical Survey and Scientific Motivations

For about 40 years now hot electron transport has been a fruitful subject in the field of solid-state physics both for theory and experiments. This is easily understandable in view of the determinant role that semiconductor devices are continuously playing in the developing fields of computers and telecommunications. Modern microelectronics has by now entered the submicrometer scale of miniaturization, and it can easily be seen that even a few volts in the applied voltage can lead to very high electric fields of the order of 10,000 V/cm. These high fields, by leading to values of the carrier drift velocity v_d of the order of 10^7 cm/s, are also at the basis of devices operating at frequencies as high as 100 GHz.

One should emphasize that this subject has taken advantage of most of the knowledge in the parallel field of transport in an ionized gas [1.1,2], in many cases offering valuable testing of theoretical models. Indeed, the electron gas can be confined within very small volumes (typically a few cubic millimeters) of the host crystal, which in turn can be shaped appropriately, by making use of the sophisticated technology borrowed from electrical engineering.

This has enabled scientists to devise and develop a wide series of experiments which hitherto seems to be limited only by the ingenious imagination of researchers.

From an historical point of view, the beginning of systematic analysis in hot-electron problems could be dated to the end of the 40's. At that time the scientific motivation was related to the study of dielectric breakdown in insulators, from which the concept of hot electron was originally introduced [1.3,4]. Subsequently, taking impetus from the discovery of the transistor, the study of the nonlinear behaviour of current-voltage characteristics (deviation from Ohm's law) [1.5] and of instability phenomena (the Gunn effect) [1.6] was carried out on a few semiconductors, Ge and GaAs in particular. Even if these pioneering measurements were somewhat incomplete (for instance, diffusion data were not available and the range of electric field strengths and temperatures was too limited) and the theoretical interpretation suffered from too rough analytical approximations, the agreement between theory and experiments appeared to be reasonably satisfactory. The possibility of superimposing on a static electric field other fields such as magnetic, strain, etc. enlarged the subject, which received a first general survey up to about 1965 in *Conwell*'s book [1.7].

In the following years, 1965–80, the availability of fast computers enabled numerical methods for an exact solution of the Boltzmann equation, which were soon developed to a high degree of refinement, "in primis" the Monte Carlo method [1.8,9] and the iterative procedure [1.10]. These, in turn, led to a more rigorous interpretation of experiments [1.11]. In the same period new experimental techniques were introduced, and the materials which were more interesting and promising for application purposes (e.g., Si, Ge, GaAs and related III–V and II–VI compounds) were systematically characterized. In particular, new techniques were developed which, for the first time, enables reliable measurements of the diffusion coefficient to be performed.

Thus, these years witness, on the one hand, systematic analysis of results associated with bulk properties and, on the other, the opening up of a new area of research centered upon super-lattice (multi-layered) structures [1.12,13]. With respect to hot electron transport in bulk materials, we can retrospectively identify three "grouping arguments" which have catalyzed the researchers' efforts.

The first one is the generalization to arbitrary field strength of the basic kinetic coefficients: mobility μ, diffusion coefficient D, and spectral density of velocity fluctuations, S_v. At equilibrium these coefficients are related to each other by the fluctuation-dissipation theorem [1.14]. In its macroscopic formulation this theorem can be expressed by the Einstein relation (fluctuation-dissipation theorem of first kind) and by the Nyquist relation (fluctuation-dissipation theorem of second kind) [1.15]. For carrier concentrations far from degeneracy and neglecting the quantum correction factor (i.e., $\hbar\omega \ll K_B T_0$ will be assumed), these are, respectively, given by

$$D = \frac{1}{e}\,\mu K_B T_0 \tag{1.1}$$

$$S_v(\omega) \equiv 2\pi\,\frac{\overline{(\delta v)^2}}{\delta\omega} = \frac{4}{e}\,K_B T_0\,\mathrm{Re}\,\{\mu^*(\omega)\}. \tag{1.2}$$

Here e is the electron charge, K_B the Boltzmann constant, T_0 the absolute temperature of the thermal bath, v a component of the carrier velocity, ω the angular frequency; the bar signifies a time average, and the frequency dependent differential mobility $\mu^*(\omega)$ (in this case field independent) is expressed as a complex number in accordance with the usual notation. (It has to be noted that, owing to the assumed linearity with respect to external fields, the zero frequency chord mobility $\mu = v_d/E$ and $\mathrm{Re}\,\{\mu^*(\omega)\} = dv_d/dE$ coincide for $\omega\tau \ll 1$, τ being of the order of the momentum relaxation time). Thus, for thermal equilibrium, an independent determination of the noise spectral density or of the diffusion coefficient does not add information not otherwise available from the mobility.

A generalization to a high electric field of the Einstein relation was proved under the two auxiliary conditions [1.14]: (i) the system, seen as a two-terminal device, is electrically stable, that is $\mathrm{Re}\,\{\mu^*(E, \omega)\} > 0$, (in this case μ^* is field dependent); (ii) two-particle interaction is neglected.

Under these conditions the Price relationship [1.16] can be written as

$$D(E, \omega) = \frac{1}{e} \operatorname{Re} \left\{ \mu^*(E, \omega) \right\} K_B T_n(E, \omega). \tag{1.3}$$

Here the noise temperature T_n is a convenient way to express the spectral density of velocity fluctuations when the system is displaced from equilibrium due to the application of an external electric field. Its macroscopic meaning is related to the measurable quantity $K_B T_n \Delta f$, Δf being the frequency bandwidth, which is the maximum noise power at frequency f, which can be displayed by the network in an output circuit [1.17]. T_n represents a property of the ensemble of charge carriers which is, in general, different from both its "energy temperature" T_e (conveniently defined as $T_e = \langle \mathscr{E} \rangle 2/(3 K_B)$, $\langle \mathscr{E} \rangle$ being the carrier average-energy) and from the thermal bath temperature T_0. The hot-electron condition is therefore responsible for the introduction of kinetic coefficients which depend upon the electric field strength and are related to each other through (1.3).

The latter represents a generalization of the fluctuation-dissipation theorem under conditions far from equilibrium [1.18, 19]. The determination of these coefficients for different materials in a wide range of field strengths (up to 200,000 V/cm) and temperatures (from 6 up to 430 K) has been carried out successfully in these last years.

Furthermore, the satisfactory agreement with the macroscopic interpretation [1.20] has enabled improved knowledge of the different scattering mechanisms which charge carriers undergo in their motion in the crystal.

A second grouping argument is the analysis of instabilities related to the condition of negative differential mobility. Under this condition it is well known that a random fluctuation of carrier density produces a space charge that grows exponentially in time [1.21]. As a result direct conversion of energy from a dc to a microwave frequency (ac) is made possible. This self-organized phenomenon can give rise to interesting examples of broken symmetry [1.22, 23] when the band structure of the material is of many-valley type.

A third grouping argument is concerned with the analysis of transport properties in the streaming-motion limit. This physical condition, which corresponds to a carrier distribution function needle-shaped along the field direction, is made possible when the dominant scattering process is through optical-phonon emission. Thus the ensemble of carriers is characterized by the time a carrier takes to reach the optical phonon energy starting from rest, and the net effect of the external applied field is to order the carrier motion which, at equilibrium, is randomly spread. Under such a condition a lot of peculiar transport phenomena become possible, for instance, the practical vanishing of diffusion processes and the saturation of the drift velocity of charge carriers [1.24].

With regard to hot-electron phenomena in superlattices and related submicrometer structures, this argument represents a new area, of considerable practical interest, and presently in rapid development. As such, we are not in a

position to set out this argument systematically, as in the case of bulk phenomena. Aside from this, two grouping arguments are attracting researcher's
attention and these we shall propose for the attention of the reader. They are
usually referred to as real-space transfer [1.25] and ballistic transport [1.26].

Real-space transfer can be obtained in semiconductors heterostructures
which are modulation-doped; the $Al_x Ga_{1-x}$ As-GaAs heterostructure can be
taken as prototype.

If the $Al_x Ga_{1-x}$ As component is doped, then electrons move towards the
GaAs which is the material with the lower band gap, provided that the band edge
discontinuity between the materials is sufficiently large. The electrons are then
separated from their parent donors and experience a much reduced impurity
scattering. Under hot-electron conditions, when a high field is applied parallel to
the heterolayer, the electrons are accelerated until they attain enough energy to
propagate perpendicular to the layers and reunite with their parent donors.
Therefore, the electrons experience strong impurity scattering and negative
differential resistance can occur. This is the real-space analogy (real-space
transfer) to the Gunn effect.

Ballistic transport can occur in submicrometer structures, a simple prototype
being of the type n^+-n-n^+, with the active region n of submicrometer length.
When the active-region length becomes comparable to or less than the carrier
mean free path, transport may occur without collision (ballistic motion) in a
perfect analogy with the case of vacuum diodes.

This phenomenon, which seems to have the inherent possibility of improving
device performances in terms of low power dissipation and high speed logic, is
becoming highly attractive from an applied point of view.

1.2 Outline of the Book

The book is ideally divided into three parts which are intended to develop the
"grouping arguments" briefly reviewed above.

Part 1 (Chap. 2) presents the general theory underlying hot-electron transport and serves as the foundation of the subsequent chapters of the book. The
first principles of such a theory rely on the Boltzmann equation, the band
structure and the scattering mechanisms. The fundamental quantities, drift
velocity, diffusion coefficient and white noise temperature associated with
velocity fluctuations are rigorously defined. Then, their dependence upon
electric field strength and temperature is investigated for simple but valuable
models and for some real cases of interest.

Part 2 (Chaps. 3–6) reviews the most interesting results which have been
obtained to date for bulk properties and whose microscopic interpretation seems
to be well established. More emphasis has been given to the quantities measured
and their physical interpretation than to the materials under investigation,
through contributions in which the experiments and theory are appropriately
balanced.

Accordingly, Chap. 3 describes the time-of-flight technique, an experimental set-up which has provided quite reliable measurements of both drift velocity and diffusion coefficient. Furthermore, several results obtained on different materials are reported together with the available theoretical interpretation. Chapter 4 analyzes hot carrier transport and fluctuation phenomena as obtained through microwave carrier heating. Together with the determination of microwave conductivity and related relaxation effects, nonhomogeneous carrier heating is also investigated. In particular, this technique is shown to be complementary with the time-of-flight technique for the determination of the diffusion coefficient, thus providing an experimental proof of the Einstein and Price relationships, see (1.1, 3).

Chapter 5 is devoted to the study of the instabilities related to negative differential conductivity in the case of covalent many-valley semiconductors. The conditions for the existence of a multivalued electron distribution are discussed in the light of theoretical models which well agree with experiments. Chapter 6 deals with the properties of hot-electron transport under streaming motion conditions.

Starting from the simplest situation, when only an applied electric field is present, the inclusion of a transverse magnetic field is then considered. In this way a situation of population inversion occurring in the continuum of the energy band is shown to be possible. Furthermore, the case of intense microwave electric field coupled with a magnetic field is shown to predict a carrier bunching in momentum space.

Part 3 (Chaps. 7 and 8) deals with phenomena related to superlattices and submicrometer structures and gives some insight into their potential applicability to devices. Accordingly, Chap. 7 presents the general features of band structure and scattering mechanisms in multilayer structures, then treats of real space transfer. Furthermore, the possible microelectronic applications of multidimensional superlattices and heterostructures are discussed. Chapter 8 analyzes carrier transport under non-steady state conditions as it occurs in the presence of a very short time or very small space configuration of the electric field. New features are then found which characterize the carrier dynamics, and some interesting phenomena, such as ballistic motion, overshoot and undershoot of drift velocity, and negative diffusivity, are predicted. All these new aspects, which still await full experimental evidence, are expected to introduce a significant improvement in the performances of future devices.

References

1.1 S.Chapman, T.G.Cowling: *The Mathematical Theory of Non-Uniform Gases* (Cambridge U. Press, London 1970)
1.2 L.G.H. Huxley, R.W.Crompton: *The Diffusion and Drift of Electrons in Gases* (Wiley, New York 1974)
1.3 H. Fröhlich: Proc. R. Soc. **A188**, 521 (1947)

1.4 H. Fröhlich, F. Seitz: Phys. Rev. **79**, 526 (1950)

1.5 W. Shockley: Bell System. Tech. J. **30**, 990 (1951)

1.6 J.B.Gunn: Solid State Commun. **1**, 88 (1963)

1.7 E.M.Conwell: High Field Transport in Semiconductors in *Solid State Phys.*, Suppl., 9 (Academic, New York 1967)

1.8 T.Kurosawa: Proc. 8th Intern. Conf. Phys. Semicond., Kyoto, J. Phys. Soc. Jpn., Suppl., **A49**, 345 (1966)

1.9 K.Binder (ed.): *Monte Carlo Methods in Statistical Physics*, Topics Current Phys., Vol. 7 (Springer, Berlin, Heidelberg 1979)

1.10 H.Budd: Proc. 8th Intern. Conf. Phys. Semicond., Kyoto, J. Phys. Soc. Jpn., Suppl., **21**, 420 (1966)

1.11 C.Jacoboni, L.Reggiani: Rev. Mod. Phys. **55**, 645 (1983)

1.12 L.Esaki, L.L.Chang: Phys. Rev. Lett. **33**, 495 (1974)

1.13 G.Bauer, F.Kuchar, H.Heinrich (eds.): *Two-Dimensional Systems, Heterostructures and Superlattices*, Springer Ser. Solid-State Sci., Vol. 53 (Springer, Berlin, Heidelberg 1984)

1.14 H.B.Callen, T.A.Welton: Phys. Rev. **83**, 34 (1951)

1.15 R.Kubo: Response, Relaxation and Fluctuations in *Lecture Notes Phys. 31*, ed. by J.Ehlers, K.Hepp, H.A.Weidenmuller (Springer, Berlin, Heidelberg 1974) p. 74

1.16 P.J.Price: In *Fluctuation Phenomena in Solids*, ed. by R.E.Burgess (Academic, New York 1965) p. 325

1.17 J.P.Nougier: In *Physics of Nonlinear Transport in Semiconductors*, ed. by D.K.Ferry, J.R.Barker, C.Jacoboni (Plenum, New York 1980) p. 415

1.18 H.Grabert: *Projection Operator Techniques in Nonequilibrium Statistical Mechanics*, Springer Tracts Mod. Phys., Vol. 95 (Springer, Berlin, Heidelberg 1982)

1.19 J.J.Niez, D.K.Ferry: Phys. Rev. **28B**, 889 (1983)

1.20 L.Reggiani: Proc. 15th Intern. Conf. Phys. Semicond., Kyoto, J. Phys. Soc. Jpn. Suppl. **A49**, 317 (1980)

1.21 See for example, S.M.Sze: *Physics of Semiconductor Devices* (Wiley, New York 1981)

1.22 D.Forster: *Hydrodynamic, Fluctuations, Broken Symmtery, and Correlation Functions* (Benjamin, Reading, MA 1975)

1.23 H.Haken: *Synergetics*, An Introduction, 3rd ed., Springer Ser. Syn. 1 (Springer, Berlin, Heidelberg 1983)

1.24 S.Komiyama: Adv. Phys. **31**, 255 (1982)

1.25 K.Hess: Proc. 16th Intern. Conf. Phys. Semicond., ed. by M.Averous (North-Holland, Amsterdam 1983) p. 723

1.26 M.S.Shur, L.F.Eastman: IEEE Trans. **ED-26**, 1667 (1979)

2. General Theory

Lino Reggiani

With 21 Figures

This chapter aims at presenting a general theory of hot electrons as applied to a bulk semiconductor. It consists of two main parts. In the first (Sects. 2.2–6) the transport theory is developed in terms of the semiclassical Boltzmann equation with the objective of giving a rigorous definition of the macroscopic kinetic coefficients such as drift velocity, differential conductivity, diffusion, and noise temperature. In the second (Sects. 2.7–13) a microscopic description is given in terms of band structure and scattering mechanisms, and typical results are illustrated. As a valuable test of the theory, the recovery of well-established results for the ohmic case, obtained at decreasing electric field strengths, will be emphasized.

In Sect. 2.2 the Boltzmann equation, linearized in the collision operator, is presented, and the phenomenological current equation is derived from a particular solution under the condition that the gradient concentration is a small and slowly varying function. Then in Sects. 2.3, 4 the Monte Carlo and iterative methods are reported as techniques for obtaining an exact solution of the Boltzmann equation. In Sect. 2.5 relations between noise associated to velocity fluctuations and diffusion are calculated following the general method of the transfer of impedance. In Sect. 2.6 the Boltzmann equation including the two-particle interaction is considered, and rigorous definitions of the kinetic coefficients which include carrier-carrier scattering are given. Section 2.7 presents the band-structure model suited to most cubic semiconductors. Sections 2.8–12 summarize the different scattering mechanisms by reporting the analytical form of their squared matrix elements which in the Born approximation describes the interactions. Section 2.13 illustrates the calculations for the drift velocity, diffusion coefficient, and noise temperature obtained for a simple model semiconductor, and reports the results obtained for the most significant semiconductors: Si, Ge, and GaAs.

2.1 Hot-Electron Concept and Related Experimental Evidence

The name "hot electrons" (we shall often write "electrons" meaning in fact "charge carriers", that is electrons or holes, indifferently) given to the problem of transport at high electric fields in semiconductors, as pointed out in some fundamental works by *Fröhlich* [2.1], is associated with a rise of the electron mean energy above its thermal equilibrium value $(3/2) K_B T_0$ (K_B being the

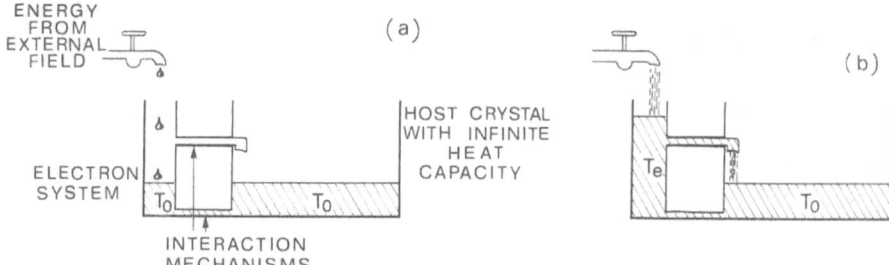

Fig. 2.1 a, b. Hydraulic analogue of the energy transfer from the external electric field to the electron system and from the electron system to the host crystal. In the low-field case (**a**) some interaction mechanisms are capable of maintaining the temperature of the electron system equal to that of the thermal bath. At high fields (**b**) the energy supplied to the electrons is higher, and a new stationary state is attained by an increased efficiency of the scattering mechanisms already active at low fields and, sometimes, by the onset of new mechanisms

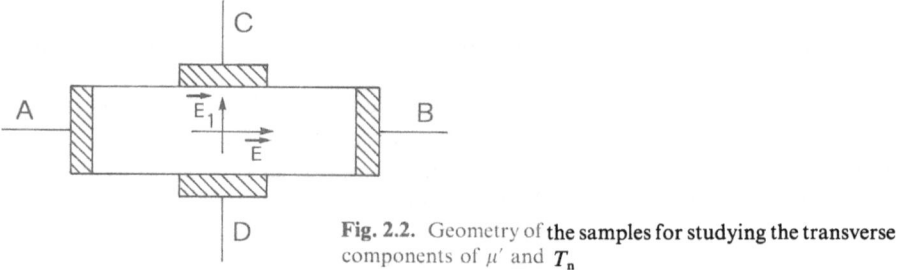

Fig. 2.2. Geometry of the samples for studying the transverse components of μ' and T_n

Boltzmann constant and T_0 the thermodynamic temperature). For many years after that, physicists spoke of an electron temperature T_e higher than the lattice temperature of the host crystal. It is now known that this idea does not correctly describe reality since the electronic distribution function very often prevents an unambiguous definition of an electron temperature; that is, it is not of a Maxwellian type. It must be recognized, however, that the concept of an electron temperature (hydraulic analogy is illustrated in Fig. 2.1) has greatly helped the understanding of high-field transport and still gives a useful terminology for a heuristic investigation of this problem.

When hot-electron conditions are determined by application of a static high electric field E, the three basic macroscopic quantities of interest are the drift velocity v_d (and the associated differential mobility $\mu' = dv_d/dE$), the diffusion coefficient D, and the white-noise temperature T_n (here understood as a measure of the noise associated to carrier velocity fluctuations). Depending upon the direction of the applied electric field, we shall speak of one longitudinal and two transverse components for μ', D, and T_n. As illustrated in Fig. 2.2, superimposed on the longitudinal field E between terminals A–B, the application of a transverse electric field E_1 (weak enough to justify a linear-response analysis) between terminals C–D allows one to measure the transverse components of μ' and T_n.

In the linear-response regime (i.e., $E \to 0$) the drift velocity depends linearly upon the electric field through the mobility μ (i.e., Ohm's law holds) as

$$v_d = \mu E. \tag{2.1}$$

Einstein's relation expresses diffusion in terms of mobility and thermal equilibrium temperature as

$$D = \frac{\mu K_B T_0}{e}, \tag{2.2}$$

e being the electron charge taken with its sign. Nyquist's relation expresses the available noise power P_{av} of a two-terminal network for unit bandwidth of frequency Δf in terms of the thermal equilibrium temperature as

$$\frac{P_{av}}{\Delta f} = K_B T_0. \tag{2.3}$$

At thermal equilibrium, and thus for continuity at $E \to 0$, it is

$$\frac{P_{av}}{\Delta f} = K_B T_0 = \frac{eD}{\mu}. \tag{2.4}$$

Equation (2.4) is nothing other than the macroscopic expression of the fluctuation-dissipation theorem [2.2, 3] (here the quantum correction [2.4] has been neglected, that is, the $hf \ll K_B T_0$ condition is satisfied, h being the Planck constant).

From the above we see that in the linear-response regime an independent determination of diffusion and white-noise temperature, in addition to drift velocity, does not add any particular information about the transport properties of the material with respect to the knowledge obtained from a determination of the ohmic mobility. Under hot-electron conditions, on the contrary, Einstein's and Nyquist's relations no longer hold in general; therefore an independent determination of drift, diffusion, and white noise provides new information. In particular, at increasing field strengths deviations from linearity of drift velocity and departures from thermal equilibrium values of diffusion and white-noise temperature have been experimentally measured. Figure 2.3 illustrates these effects which are to be considered the basis for the experimental evidence of hot-electron conditions.

A direct consequence of the dependence of transport parameters upon field strength is that, even for cubic semiconductors, the longitudinal component with respect to field direction of these parameters differs in general from the

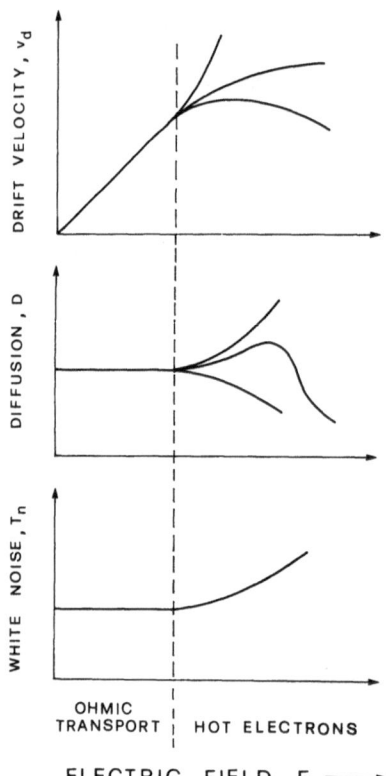

Fig. 2.3. Schematic illustration of the dependence upon electric field strength of drift velocity, diffusion coefficient, and noise temperature as taken along the field direction (scales are assumed to be logarithmic)

transverse components. Furthermore, depending upon the band structure of the given semiconductor, anisotropy of the transport parameters with the direction which the field makes with respect to crystallographic axes can arise. Table 2.1 reports a survey on the main effects related to hot-electron conditions, effects which have been experimentally demonstrated.

Besides the above-mentioned effects, other experimental evidence for hot-electron conditions can be found by the direct generation of hot carriers through injection from contacts [2.19] or by interaction with light pulses [2.20] and high energetic particles [2.21]. Furthermore, by coupling the electric field with other external fields the phenomenology of hot electrons spreads over electro-optical, galvagnomagnetic [2.22], piezoresistance [2.23], and thermoelectric phenomena [2.24].

2.2 The Boltzmann Equation and Its Solution

Here we shall limit ourselves to a semiclassical approach, thus the kinetic equation which will be used to describe hot-electron transport is the Boltzmann

Table 2.1. Historical survey on the experimental evidence of the main effects related to hot-electron conditions in cubic semiconductors

1 – Deviation from Ohm's law through a nonlinear dependence of the drift velocity upon electric field strength [2.5].

2 – Deviation from Einstein's relation of the diffusion coefficient through a diffusivity which depends on electric field strength [2.6,7].

3 – Deviation from Nyquist's relation of the noise-power per unit bandwidth (measured in equivalent noise temperature) through an increase of the white-noise temperature at increasing electric field [2.8].

4 – Anisotropy with respect to the direction of the applied electric field of the kinetic coefficients: differential mobility μ' [2.9], diffusion coefficient D [2.10,11], white-noise temperature T_n [2.12].

5 – Anisotropy with respect to the direction which the applied electric field makes with the crystallographic axes of the drift velocity (Sasaki effect [2.13]) and of the kinetic coefficients [2.12,14,15].

6 – Negative differential mobility for electric field strengths above a threshold value (Gunn effect) [2.16].

7 – Avalanche phenomena with an increase of the free-carrier concentration at increasing electric field strengths [2.17].

8 – Hot-electron emission from a p–n junction [2.18].

equation. Let us recall the main assumptions which justify such an approach in the following [2.25, 26]:

i) Carrier density should be sufficiently low so that only binary collisions occur.
ii) Time between successive collisions τ should be long enough when compared with the duration of a collision τ_{coll}, i.e., $\tau_{coll}/\tau \ll 1$.
iii) Density gradients should be small over the range of the interparticle potential.

2.2.1 The Time-Dependent Boltzmann Equation

In the operator form, the Boltzmann equation can be written as

$$\left(\frac{\partial}{\partial t}+\dot{k}\,\frac{\partial}{\partial k}+v\,\frac{\partial}{\partial r}+I_k^{th}\right)f(k,r,t)=0, \tag{2.5}$$

where $f(k,r,t)$ is the single particle distribution function. In all the problems considered here, $f(k,r,t)$ is the same for both the possible spin orientations, therefore $(4\pi^3)^{-1}f(k,r,t)dk\,dr$ give the expected number of particles in dr about position r and dk about wavevector k at time t. The introduction of the concept of a quantum state of wavevector k indicates that we are using the one-electron approximation. Closely connected with this approximation is the band structure which describes the energy $\mathscr{E}(k)$ of state k as a function of k. The equation of motion under the influence of an applied electric field $E(r, t)$ is

$$\dot{k} = \frac{e}{\hbar} E(r, t), \tag{2.6}$$

the dot indicating time derivative and $\hbar$ being the Planck constant divided by 2π. In (1.5) v is the group velocity of the carrier given by

$$v(k) = \frac{1}{\hbar} \frac{\partial}{\partial k} \mathscr{E}(k). \tag{2.7}$$

The collision operator I_k^{th} describes the rate of change of $f(k, r, t)$ due to scattering with thermal bath via phonons and imperfections. None of the interactions with which we are concerned produces spin flips, and we shall only deal with nondegenerate semiconductors, in which $f \ll 1$. Furthermore, the effect of long-range Coulomb interaction between electrons (space charge) as well as carrier-carrier scattering are neglected (Sect. 2.6 will consider these effects). Consequently, I_k^{th} is linear in the distribution function and is given by

$$I_k^{th} f(k, r, t) = \lambda(k) f(k, r, t) - \frac{V_0}{(2\pi)^3} \int dk' \, W(k, k') f(k', r, t), \tag{2.8}$$

where V_0 is the volume of the crystal, $W(k, k')$ is the total transition rate, that is, the probability per unit time of scattering of an electron from state k to state k', and

$$\lambda(k) = \frac{V_0}{(2\pi)^3} \int W(k, k') dk' \tag{2.9}$$

is the scattering rate.

The total transition rate $W(k, k')$ is the sum of the transition rates due to different independent processes:

$$W(k, k') = \sum_i W_i(k, k'). \tag{2.10}$$

For any ith process the transition rate is assumed to be independent from field strength and is calculated by means of the time-dependent perturbation theory (Sect. 2.8).

It should be noted that, consistent with our previous assumptions, applying the perturbation theory restricts the validity of the present treatment to cases in which scattering events are rare and the mean free path of an electron is several atomic distances. In other words, we are assuming that the order of magnitude of operators in different terms of the Boltzmann equation (2.5) (($\partial/\partial t$), ($eE/\hbar$) ($\partial/\partial k$), $v(\partial/\partial r)$, I_k^{th}) is small when compared with $1/\tau_{coll}$.

One further approximation is to neglect the deviation from equilibrium of phonon population. Owing to the very short times of phonon-phonon interactions, at normal temperatures (> 30 K) this approximation is generally well justified, and for a survey of the effects related to nonequilibrium phonon population under particular circumstances we refer to the review paper of *Kocevar* [2.27].

Once the distribution function has been obtained, one can readily find all relevant macroscopic variables, that is to say, the carrier concentration:

$$n(r, t) = \frac{1}{4\pi^3} \int f(k, r, t) dk, \tag{2.11}$$

the current density

$$j(r, t) = \frac{e}{4\pi^3} \int v(k) f(k, r, t) dk, \tag{2.12}$$

the drift velocity

$$v_d(r, t) = \frac{j(r, t)}{en(r, t)}, \tag{2.13}$$

and the mean energy

$$\langle \mathscr{E}(r, t) \rangle = \frac{1}{n(r, t)} \int \mathscr{E}(k) f(k, r, t) dk. \tag{2.14}$$

In all the above equations the integration in k space is performed over the volume of the first Brillouin zone.

2.2.2 The Phenomenological Current Equation

To study the electrical behaviour of materials when high electric fields are applied, a phenomenological current equation is usually introduced in place of (2.12) and the current density is given by

$$j(r, t) = en(r, t) v_{do}(E) - eD(E) \frac{\partial}{\partial r} n(r, t), \tag{2.15}$$

in which $E = E(r, t)$. The drift velocity v_{do} is that obtained in absence of concentration gradients, and the diffusion term accounts for concentration gradient effects throguh the diffusion coefficient $D(E)$. Equation (2.15), usually called Fick's law in presence of an electric field, is an obvious generalization of the low-field case for which v_{do} depends on the field through the field-independent mobility, see (2.1), and also D is field-independent.

By substituting (2.15) into the continuity equation

$$\frac{\partial n(r.t)}{\partial t} = \frac{1}{e} \, \text{div} j(r,t), \tag{2.16}$$

the well-known diffusion equation is obtained:

$$\frac{\partial n(r,t)}{\partial t} = -v_{do,i} \frac{\partial n(r,t)}{\partial x_i} + D_{ij} \frac{\partial^2 n(r,t)}{\partial x_i \partial x_j}, \tag{2.17}$$

where D_{ij} is the diffusion tensor, and the sum over repeated indices is implied. With the initial distribution

$$n(r,0) = \delta(x)\delta(y)\delta(z), \tag{2.18}$$

the solution of (2.17) is standard, and assuming isotropic crystals with the field along the x direction, one obtains

$$n(x,y,z;t) = \frac{A_0}{t^{3/2}} \exp \left[-(x-v_{do}t)^2/(4D_{lo}t)\right] \exp \left[-(y^2+z^2)/(4D_{tr}t)\right], \tag{2.19}$$

where A_0 is an appropriate normalizing constant, D_{lo} is the longitudinal and D_{tr} the transverse component of the diffusion coefficient, respectively.

The phenomenological current equation (2.15) is only an approximation of the more fundamental equation (2.12). In the following, its derivation from the Boltzmann equation will be carried out in order to illustrate which specific approximations are to be made and, when possible, to discuss their importance. The derivation closely follows the one given by *Vinter* [2.28].

Without loss of generality, we write the distribution function as a product

$$f(k,r,t) = n(r,t)g(k,r,t), \quad \text{so that} \tag{2.20}$$

$$\frac{1}{4\pi^3} \int g(k,r,t)dk = 1. \tag{2.21}$$

By substituting (2.20) into (2.5), noting that I_k^{th} operates on k only, when dividing by $n(r,t)$ we obtain

$$\left\{ \frac{\partial}{\partial t} + \frac{eE}{\hbar} \frac{\partial}{\partial k} + I_k^{th} - \frac{1}{n(r,t)} \left[\frac{\partial n(r,t)}{\partial t} + v(k) \frac{\partial n(r,t)}{\partial r} \right] \right.$$

$$\left. + v(k) \frac{\partial}{\partial r} \right\} g(k,r,t) = 0. \tag{2.22}$$

By integrating over k and using the definition of v_d as

$$v_d(r,t) = \frac{1}{4\pi^3} \int v(k) g(k,r,t) dk, \qquad (2.23)$$

the well-known continuity equation is obtained as

$$\frac{\partial n(r,t)}{\partial t} + \mathrm{div}\,[n(r,t)v_d(r,t)] = 0. \qquad (2.24)$$

By substituting in (2.22) $\partial n(r,t)/\partial t$ given by (2.24), we obtain

$$\left\{ \frac{\partial}{\partial t} + \frac{eE}{\hbar}(r,t)\frac{\partial}{\partial k} + I_k^{th} + \frac{1}{n(r,t)}\frac{\partial n(r,t)}{\partial r}[v(k) - v_d(r,t)] \right.$$

$$\left. + v(k)\frac{\partial}{\partial r} - \mathrm{div}\, v_d(r,t) \right\} g(k,r,t) = 0. \qquad (2.25)$$

In order to obtain a local equation, that is, an equation in which $g(k,r,t)$ depends on r only through $E(r,t)$ and the relative concentration gradient γ, which is defined as

$$\gamma = \frac{1}{n(r,t)}\frac{\partial n(r,t)}{\partial r} \qquad (2.26)$$

in (2.25), the last two terms $v(\partial/\partial r)g$ and $g \cdot \mathrm{div}\, v_d$ should be disregarded. This corresponds to neglecting the effect of electric field gradients and changes of γ in space. A second approximation which must be made is the assumption that changes with time in field and concentration gradients are sufficiently slow to allow $g(k,r,t)$ to attain a stationary state. Thus in (2.25) the first term $(\partial/\partial t)g$ is also neglected. An empirical current equation which should account for these three terms has been recently proposed by *Thornber* [2.29].

An equivalent way to state the above approximations is to consider the concentration gradient as a perturbation characterized by the wavevector q and angular frequency ω, and to limit the analysis of its effect to cases for which the condition $\omega\tau \ll 1$ and $ql \ll 1$ holds, τ and l representing average values of the time between successive collisions and of the mean free path of carriers, respectively. This point of view will be considered in Sect. 2.6.

Under the above two approximations, (2.25) reduces to

$$\left\{ \frac{e}{\hbar}E(r,t)\frac{\partial}{\partial k} + I_k^{th} + \gamma[v(k) - v_d(\gamma,E)] \right\} g(k,\gamma,E) = 0. \qquad (2.27)$$

Equation (2.27) is suitable for being studied perturbatively. To do so we define a linear operator $\hat{L}$ as

$$\hat{L} = \left(\frac{e}{\hbar} \, \boldsymbol{E} \, \frac{\partial}{\partial \boldsymbol{k}} + I_{\boldsymbol{k}}^{\text{th}} \right). \tag{2.28}$$

With $\gamma = 0$ and a homogeneous field, the stationary form of (2.27) is

$$\hat{L}g_0(\boldsymbol{k}, E) = 0, \tag{2.29}$$

with the drift velocity corresponding to the homogeneous case given by

$$v_{d_0}(E) = \frac{1}{8\pi^3} \int v(\boldsymbol{k}) g_0(\boldsymbol{k}, E) d\boldsymbol{k}. \tag{2.30}$$

We now treat the remaining term of (2.27) perturbatively, that is, we look for the small correction to be applied to drift velocity due to diffusion effect. To this end we shall introduce the numerical parameter $\gamma(K_B T_0/eE)$, which under $\gamma(K_B T_0/eE) \ll 1$ will justify our perturbative approach. In fact, when $\gamma = (eE/K_B T_0)$, it can be seen that the solution of (2.27) is the equilibrium Maxwell-Boltzmann energy distribution, and the situation described becomes that of a p–n junction without an applied field in which the diffusion current exactly balances the drift current. Taking for the sake of simplicity the field gradient along the x direction, for the change in distribution $g_1(\boldsymbol{k}, \gamma, E)$ we have

$$\hat{L}g_1(\boldsymbol{k}, \gamma, E) = [v_{d_0}(E) - v(\boldsymbol{k})]\gamma g_0(\boldsymbol{k}, E). \tag{2.31}$$

In terms of the Green function $G(\boldsymbol{k}', \boldsymbol{k}, E)$ of the operator $\hat{L}$ defined by

$$\hat{L}G(\boldsymbol{k}', \boldsymbol{k}, E) = \delta(\boldsymbol{k} - \boldsymbol{k}'), \tag{2.32}$$

the formal solution of (2.31) is

$$g_1(\boldsymbol{k}, \gamma, E) = \gamma \int G(\boldsymbol{k}', \boldsymbol{k}, E) \, [v_{d_0}(E) - v(\boldsymbol{k}')]g_0(\boldsymbol{k}', E) d\boldsymbol{k}', \tag{2.33}$$

and the corresponding change in drift velocity will be

$$v_{d_1}(\gamma, E) = \frac{\int v(\boldsymbol{k})g_1(\boldsymbol{k}, \gamma, E)d\boldsymbol{k}}{\int g_0(\boldsymbol{k}, E)d\boldsymbol{k}} = -\gamma D_{lo}(E), \tag{2.34}$$

where through (2.33) the longitudinal diffusion coefficient is given by

$$D_{lo}(E) = \frac{\int d\boldsymbol{k} \int d\boldsymbol{k}' v(\boldsymbol{k})G(\boldsymbol{k}', \boldsymbol{k}, E) \, [v(\boldsymbol{k}') - v_{d_0}(E)]g_0(\boldsymbol{k}', E)}{\int g_0(\boldsymbol{k}, E)d\boldsymbol{k}}. \tag{2.35}$$

Generalization of (2.35) to the transverse case, for which $v_{do} = 0$, is straightforward. It is instructive to note that in the absence of an external field and in the relaxation time approximation, $G(k', k) = \tau(k)\delta(k - k')$ and g_0 is the Maxwell-Boltzmann distribution function; thus (2.35) reproduces the well-known definition of diffusion under the low-field condition [2.30].

Through (2.35) we have thus found an expression for the diffusion coefficient at high fields from a special solution of the Boltzmann equation (2.5) which may be written as

$$f(k, r, t) = n_0 \exp \{\gamma[r - v_d(\gamma, E)t]\} g(k, \gamma, E), \qquad (2.36)$$

where n_0 is a constant average concentration.

By considering a small gradient of concentration (i.e., $\gamma(K_B T_0/eE) \ll 1$) the distribution function in (2.36) gives, in agreement with the phenomenological current equation (2.15),

$$v_d(\gamma, E) = v_{do}(E) - \gamma D(E). \qquad (2.37)$$

2.2.3 Equivalent Definitions of Fick's Diffusion Coefficient

Because of the prohibitive difficulty in calculating diffusion through (2.35), in the following, three equivalent ways of defining such a quantity at a hydrodynamic level are reported. For simplicity, let us deal with the components of x, v_{do}, and v in some specific but arbitrary direction.

On defining the mth moment of $n(x, t)$ as

$$M^{(m)} = \frac{1}{N} \int_{-\infty}^{+\infty} x^m n(x, t) dx \equiv \langle x^m \rangle, \qquad (2.38)$$

N being the total number of particles, by using the diffusion equation (2.17), the following recursion relations are found:

$$M^0 = 1,$$

$$\frac{d}{dt} M^{(1)} = v_{do},$$

$$\frac{d}{dt} M^{(2)} = 2(v_{do} M^{(1)} - D), \qquad (2.39)$$

$$\frac{d}{dt} M^{(m \geq 2)} = m v_{do} M^{(m-1)} + m(m-1) D M^{(m-2)}.$$

For the second central moment $\langle(x-\langle x\rangle)^2\rangle$, we obtain

$$D_{ij} = \lim_{t\to\infty} \frac{1}{2}\frac{d}{dt}\left[\langle x_i(t)x_j(t)\rangle - \langle x_i(t)\rangle\,\langle x_j(t)\rangle\right], \tag{2.40}$$

where the long-time limit ensures that Fick's law is satisfied [2.31].

Equation (2.40) is consistent with the interpretation of diffusion as a spreading in space of some initial particle distribution; diffusion is thus related to a second central moment which increases linearly with time.

By carrying out the time derivative of (2.40), and using the fact that an ensemble average commutes with the time derivative, an equivalent expression of diffusion is obtained as [2.32, 33]

$$D_{ij} = \lim_{t\to\infty}\left[\langle x_i(t)v_j(t)\rangle - \langle x_i(t)\rangle\,\langle v_j(t)\rangle\right]. \tag{2.41}$$

In this way diffusion is correlated to the achievement of a stationary value for the space velocity covariance of the initial particle distribution.

By writing the trajectory in terms of velocity, a third equivalent expression of diffusion is obtained as [2.33, 34]

$$D_{ij} = \lim_{t\to\infty}\int_0^t \langle\delta v_i(t)\delta v_j(t-t')\rangle\,dt' + \langle\delta x_i(0)\delta v_j(t)\rangle. \tag{2.42}$$

Equation (2.42) relates diffusion to the autocorrelation function of velocity fluctuations; it has to be noted that under the long-time limiting condition the integrand becomes a function of t' only, and the last term vanishes.

At very short times, comparable or less than an average scattering time, the time dependence of the quantities inside the limit in (2.40–42) can be used to describe a diffusion coefficient which will depend upon time and initial conditions of motion. Figure 2.4 illustrates the dependence upon time of the quantities defined by (2.40–42) when there is no applied electric field and each particle is assumed to start at $t=0$ at the position $x=0$ with an equilibrium Maxwell-Boltzmann momentum distribution.

In the figure, τ indicates the time necessary for the correct space-velocity correlation to be achieved so that Fick's law holds. It is worth noting that since the space-velocity covariance and the velocity-fluctuations autocorrelation function achieve a constant value in the long-time limit, they are better suited to check the validity of Fick's law than the second central moment which exhibits a long-time trend.

Figure 2.5 illustrates the transient behavior of the spreading in space and of the related diffusion coefficient, which depends upon time, when at $t=0$ an electric field is switched on; Curve 1 refers to a small applied electric field and Curves 2 and 3 to a high applied electric field so that hot-electron conditions are established. In particular, under transient conditions hot-electron diffusion is

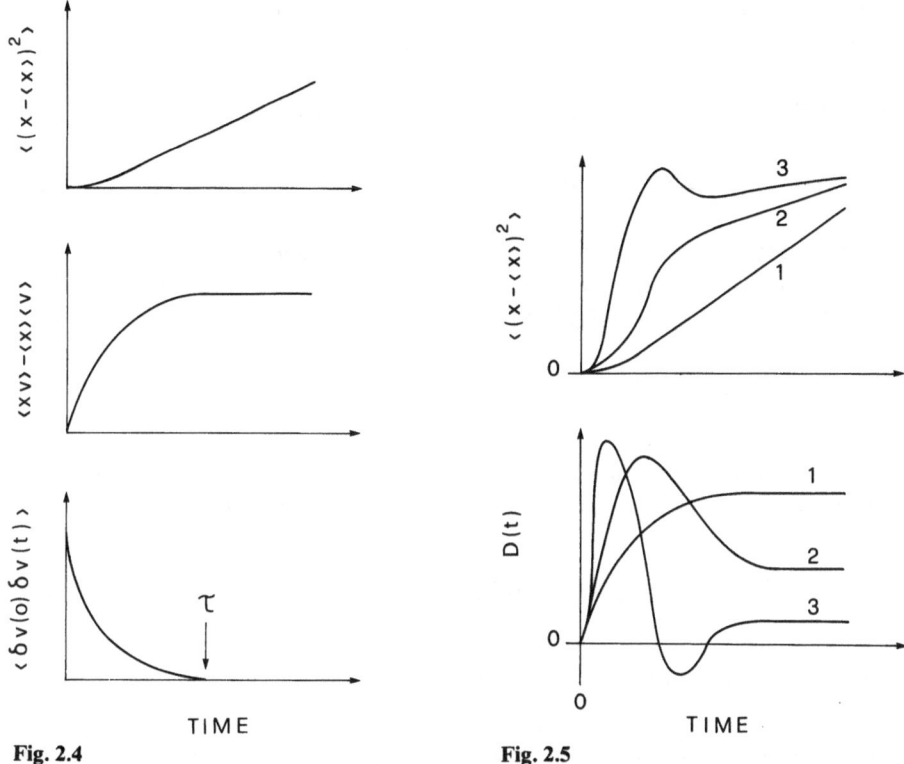

Fig. 2.4

Fig. 2.5

Fig. 2.4. Schematic representation of the time dependence of spatial spreading, space-velocity covariance, and velocity-fluctuations autocorrelation function

Fig. 2.5. Spatial spreading and related diffusion coefficient as a function of time when transient conditions are analyzed: (*1*) refers to an ohmic case; (*2*) and (*3*) refer to possible hot-electron conditions (see text)

shown to exhibit values greater than those under long-time limit conditions (overshoot effect) [2.35] as well as negative values [2.35, 36]. For more details, see Chaps. 3 and 7.

2.2.4 The Generalized Diffusion Equation

In real physical systems the particle density $n(x, t)$ is never exactly of the Gaussian form given by (2.19) [2.31]. If $n(x, t)$ satisfies the diffusion equation it does so only in an asymptotic sense, that is, only on a macroscopic time length scale. In the following we present a generalization of the diffusion equation (2.17) to arbitrary space time variations of the density gradient. This allows us to define a frequency and wavevector-dependent diffusion coefficient $D(q, \omega)$

[2.37] which is in general a complex number. The limit $D(0,0)$ will be real, being the ordinary diffusivity as described in the two previous subsections.

Let us consider a system of particles subject to an external uniform static electric field E along x. A particle source of sinusoidal type,

$$S(x,t) = S_0[1 + \cos (qx - \omega t)], \tag{2.43}$$

is then introduced together with a trapping mechanism with a constant rate τ_d^{-1}. The source amplitude S_0 and τ_d are connected through the condition of particle conservation over a wavelength λ:

$$\int_\lambda S_0[1 + \cos (qx - \omega t)]dx = \int_\lambda \tau_d^{-1} n(x,t)dx. \tag{2.44}$$

The particles are thus trapped and redistributed in space. By making use of the more convenient complex formalism, the diffusion equation (2.17) becomes

$$\frac{\partial n}{\partial t} = D\frac{\partial^2 n}{\partial x^2} - v_{do}\frac{\partial n}{\partial x} + S_0\{1 + \exp [i(qx - \omega t)]\} - \tau_d^{-1} n. \tag{2.45}$$

We look for a solution of (2.45) of the type

$$n(x,t) = n_0 + n_1 \exp [i(qx - \omega t + \varphi)], \tag{2.46}$$

where n_0 is a constant average term, n_1 and φ are the amplitude and the phase shift of the harmonic disturbance, respectively. By substituting (2.46) into (2.44) we obtain

$$S_0 = n_0\tau_d^{-1}, \tag{2.47}$$

and (2.45) becomes

$$-i\omega + q^2 D + iqv_{do} + \tau_d^{-1} = \frac{n_0}{n_1}\tau_d^{-1} \exp (-i\varphi). \tag{2.48}$$

As long as problems where n is linear are considered, this equation can also be used at frequencies and wavelengths so high that Fick's law does not hold; in this case D must be taken as a function of q and ω. Thus from (2.48) we obtain an expression for $D(q, \omega, \tau_d^{-1})$ in terms of the relative amplitude n_1/n_0 and of the phase shift φ of the disturbance:

$$D = \frac{1}{q^2}\left[\frac{n_0}{n_1}\tau_d^{-1} \exp (-i\varphi) + i(\omega - qv_{do}) - \tau_d^{-1}\right]. \tag{2.49}$$

By the path integral method [2.38] we can obtain the following integral expression for $n(x,t)$:

$$n(x,t)=n_0 \int_{-\infty}^{t} dt'\tau_d^{-1} \int_{-\infty}^{+\infty} dx'\{1+\exp[i(qx'-\omega t')]$$

$$\cdot P(x',t';x,t)\exp[-\tau_d^{-1}(t-t')]\}, \tag{2.50}$$

where $P(x',t';x,t)dx$ is the probability that without trapping, a particle in x' at t' will be in dx around x at t.

The integral of the first term in brackets in (2.50) is equal to n_0, since from the time $-\infty$ to t all the particles have been trapped and redistributed by the source. Here $P(x',t';x,t)$ is a function of x', t' and x, t only through the differences $x''=x-x'$ and $t''=t-t'$ so that (2.46, 50) yield

$$n(x,t)-n_0=n_0\exp[i(qx-\omega t)]\int_0^{\infty}\tau_d^{-1}dt\int_{-\infty}^{+\infty}dx\exp[i(qx-\omega t)-t\tau_d^{-1}]$$

$$\cdot P(x,t). \tag{2.51}$$

From (2.46, 50) we then obtain

$$R\equiv\frac{n_1}{n_0}\exp(i\varphi)=\int_0^{\infty}\tau_d^{-1}dt\int_{-\infty}^{+\infty}dx\exp[-i(qx-\omega t)-t\tau_d^{-1}]P(x,t), \tag{2.52}$$

and (2.49) finally yields

$$D(q,\omega,\tau_d^{-1})=\frac{1}{q^2\tau_d R}[1-R+i\tau_d R(\omega-qv_{do})]. \tag{2.53}$$

Equations (2.52, 53) provide an expression for $D(q,\omega,\tau_d^{-1})$ valid for any wavevector, frequency, and particle lifetime τ_d. When τ_d is much greater than both ω^{-1} and any microscopic time, $D(q,\omega)=D(q,\omega,0)$ is obtained for steady-state conditions. Then $P(x,t)$ must be derived from knowledge of the particle dynamics.

Figures 2.6 and 7 report the results for the real part of $D(q,\omega)$ as a function of q and ω, respectively, as obtained from a Monte Carlo simulation for the case of electrons in Si [2.39]. As a general trend, $D(q,\omega)$ decreases at asymptotically high values of both q and ω because the average over several wavelengths as well as over several periods of carrier concentrations in the past is not affected by contributions from electrons coming from far away [2.39, 40]. The peaks of $D(q,\omega)$ at intermediate frequency (Fig. 2.7) are related to the presence of high-energy intervalley phonons whose emission at high fields leads to some streaming character of the carrier motion. The peaks in $D(q,\omega)$ of Fig. 2.6 have the same interpretation.

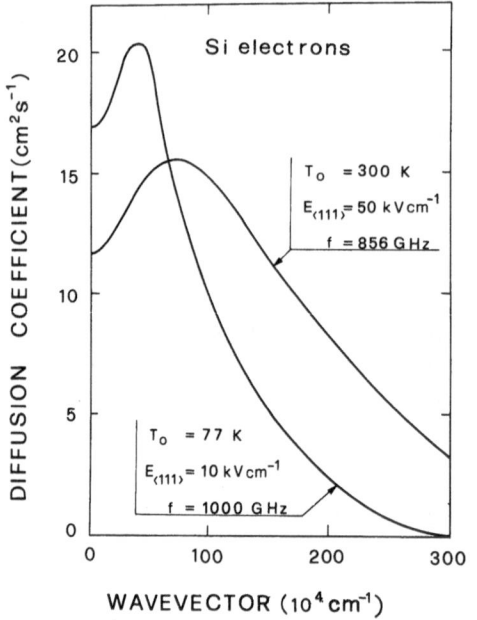

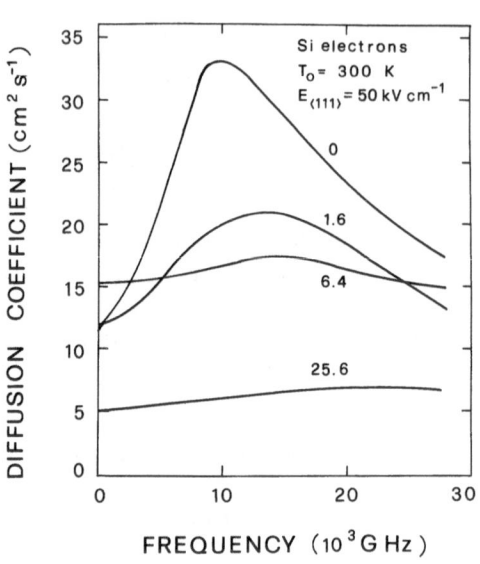

Fig. 2.6. High-field, low-frequency diffusion coefficient as a function of wavevector at the indicated temperatures for the case of electrons in Si [2.39]

Fig. 2.7. High-field diffusion coefficient as a function of angular frequency ω at the different values of wavevector q indicated by the numbers on the curves in units of 10^5 cm^{-1} for the case of electrons in Si [2.39]

2.2.5 Comparison with Previous Definitions for $D(q)$ and $D(\omega)$

The problem of defining a q-dependent diffusion coefficient $D(q) \equiv D(q,0)$ and a ω-dependent diffusion coefficient $D(\omega) \equiv D(0,\omega)$ has been analyzed in [2.39, 41]. As to the limit $\omega \to 0$, (2.53) gives

$$D(q) = \lim_{\tau_d \to \infty} \frac{1}{q^2 \tau_d R} (1 - R - iq\tau_d v_{do} R), \tag{2.54}$$

which by using (2.52) and carrying out the limit gives

$$D(q) = \frac{1}{q^2} (P_q^{-1} - iq v_{do}), \tag{2.55}$$

with

$$P_q = \int_0^\infty dt \int_{-\infty}^{+\infty} \exp(-iqx) P(x,t) dx, \tag{2.56}$$

which is in agreement with [2.40].

As regards the expression of $D(\omega)$, let us consider the limit for q approaching zero of (2.53). We will expand in series of qR given by (2.52) as

$$R = R_0 + qR_1 + q^2 R_2 + \dots . \tag{2.57}$$

Expressions for R_0, R_1, and R_2 can be obtained from (2.52) as

$$R_0 = \int_0^\infty \exp\left[-(1-i\omega\tau_d)t\tau_d^{-1}\right]\tau_d^{-1}dt = \frac{1}{1-i\omega\tau_d}, \tag{2.58}$$

$$R_1 = -i\int_0^\infty \exp\left[-(1-i\omega\tau_d)t\tau_d^{-1}\right]\langle x\rangle_t \tau_d^{-1}dt = -\frac{iv_{do}\tau_d}{(1-i\omega\tau_d)^2}, \tag{2.59}$$

$$R_2 = -\frac{1}{2}\int_0^\infty \exp\left[-(1-i\omega\tau_d)t\tau_d^{-1}\right]\langle x^2\rangle_t \tau_d^{-1}dt$$

$$= \frac{(v_{do}\tau_d)^2}{(1-i\omega\tau_d)^3} - \frac{\tau_d}{(1-i\omega\tau_d)^2}\int_0^\infty C(t)\exp\left(i\omega t - t\tau_d^{-1}\right)dt, \tag{2.60}$$

where $C(t)$ is the autocorrelation function of velocity fluctuations

$$C(t) = \langle \delta v(0)\delta v(t)\rangle. \tag{2.61}$$

By substituting (2.58–60) into (2.53), for vanishing small q we obtain

$$D(\omega) = \lim_{\tau_d \to \infty} \lim_{q \to 0} D(q, \omega, \tau_d^{-1}) = \int_0^\infty C(t)\exp\left(i\omega t\right)dt. \tag{2.62}$$

Thus we find that for a density gradient which depends upon time, a frequency-dependent diffusion coefficient is obtained from the Fourier transform of the autocorrelation function of velocity fluctuations, in agreement with that proposed by *Price* [2.41] and *Schlup* [2.42].

2.3 The Monte Carlo Method

The Monte Carlo method is at present widely used in hot-electron studies [2.32, 43], therefore the basic principles underlying this method will be presented in detail in the following. The method consists of a simulation of the motion of one or more electrons inside the crystal, subject to the action of an external applied electric field and of given scattering mechanisms. The duration of a carrier's free flight (i.e., the time between two successive collisions) and the scattering processes involved in the simulation are selected stochastically according to a given probability describing the microscopic process.

As a consequence, the foundation of any Monte Carlo method relies on the generation of a sequence of random numbers with given distribution probabilities.

When the purpose of the analysis is the investigation of a steady-state, homogeneous phenomenon, it is sufficient in general to simulate the motion of a single electron then, from ergodicity, we may assume that a sufficiently long path of this sample electron will give information on the behavior of the entire electron gas. When, on the other hand, the transport process under investigation is not homogeneous and/or is not stationary, it is necessary to simulate a large number of electrons and follow them in their dynamic histories in order to obtain the desired information on the process of interest.

2.3.1 A Typical Monte Carlo Program

Let us summarize here the major structure of a typical Monte Carlo program suited to the simulation of a stationary, homogeneous, transport process. The details of each step of the procedure will be given in the following sections. For the sake of simplicity we shall refer to the case of a cubic semiconductor with an externally applied electric field *E*. The simulation starts with one electron in

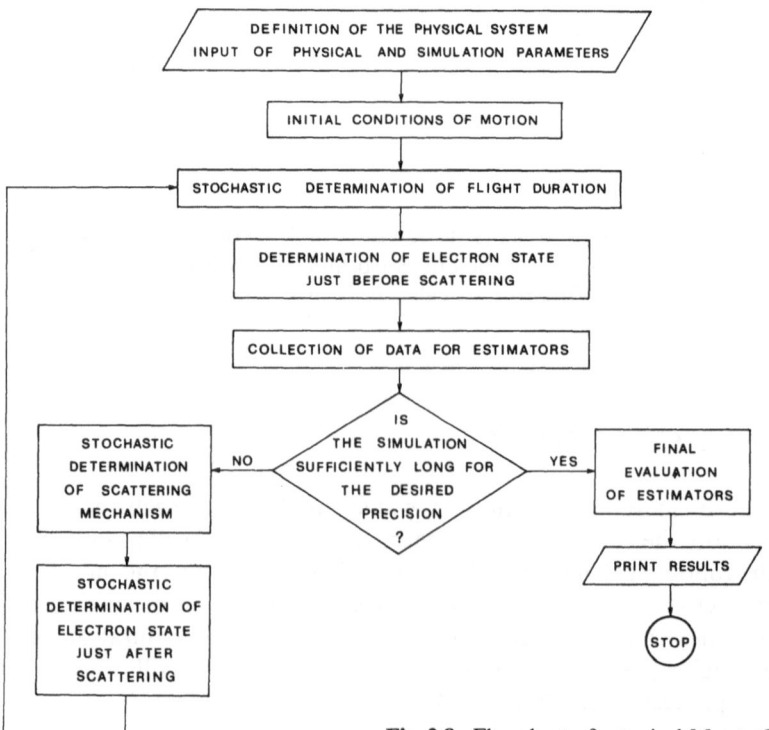

Fig. 2.8. Flowchart of a typical Monte Carlo program

given initial conditions with wavevector k_0 ; then, the duration of the first free flight is determined stochastically from a probability distribution determined by the scattering probabilities. During the free flight the external force is made to act according to (2.6).

In this part of the simulation all quantities of interest, velocity, energy, etc. are recorded. Then a scattering mechanism is chosen as responsible for the end of the free flight, according to the relative probabilities of all possible scattering mechanisms. From the transition rate of this scattering mechanism, a new k state after scattering is stochastically determined as the initial state of the new free flight, and the entire process is iteratively repeated. The results of the calculation become more and more precise as the simulation goes on, and the simulation ends when the quantities of interest are known with the desired precision.

A simple way to determine the precision, that is, the statistical uncertainty of transport quantities, may consist of splitting the entire history into a number of successive sub-histories of equal time duration, and determining a quantity of interest for each of them. We then determine the average value of each quantity and take its standard deviation as an estimate of its statistical uncertainty.

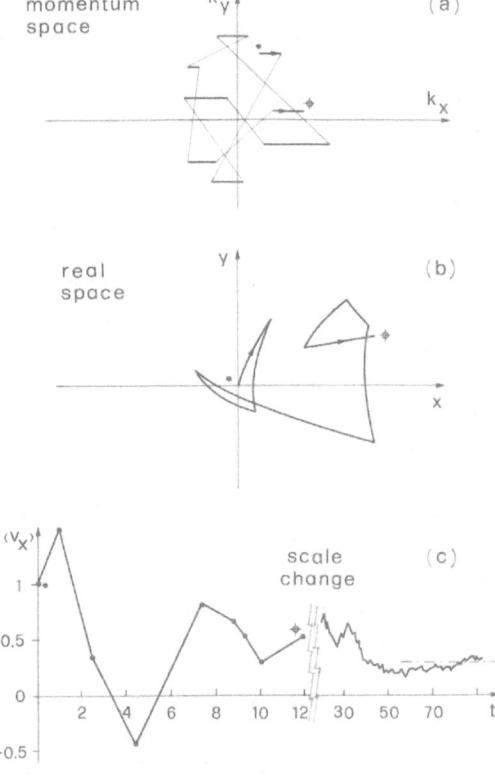

Fig. 2.9a–c. The principles of the Monte Carlo method. For simplicity a two-dimensional model is considered here. (**a**) Simulation of the sampling particle, in the wavevector space, subject to an accelerating force (field) oriented along the positive x direction. The heavy segments are due to the effect of the field during free flights; curves represent discontinuous variations of k due to scattering processes. (**b**) The path of the particle in real space. It is composed of eight fragments of parabolas corresponding to the eight free flights in part (**a**) of the figure. (**c**) The average velocity of the particle obtained as a function of simulation time. The left section of the curve ($t < 12$) is obtained by the simulation illustrated in parts (**a**) and (**b**) of the figure. ($-\cdot-$) represents the "exact" drift velocity obtained with a very long simulation time. Special symbols indicate corresponding points in the three parts of the figure; ($*$) is the starting point. All units are arbitrary

Figure 2.8 shows a flowchart of a simple Monte Carlo program suitable for the simulation of a stationary, homogeneous transport process.

Figure 2.9 illustrates the principles of the method by showing the simulation in k space and real space and the effect of collecting statistics in the determination of the drift velocity.

2.3.2 Definition of the Physical System

The starting point of the program consists of defining the physical system of interest as regards the parameters of the material and the values of the physical quantities such as lattice temperature and electric field. It is worth noting that among the parameters to be entered in order to characterize the material the least known ones, usually taken as adjustable parameters, are the coupling strengths describing the interactions of the electron with the lattice and/or extrinsic defects inside the crystal. At this level we also define the parameters which control the simulation, such as the duration of each sub-history, the desired precision of the results, and so on.

The next step in the program is a preliminary calculation of each scattering rate as a function of electron energy. This step will provide information on the maximum value of these functions which will be useful for optimizing the efficiency of the simulation. Finally, all cumulative quantities must be set at zero.

2.3.3 The Initial Conditions of Motion

In the case under consideration, in which a steady-state situation is simulated, the time of simulation must be sufficiently long so that the initial conditions of the electron motion do not influence the final results.

The choice of an appropriate time of simulation is thus a compromise between the need for ergodicity ($t \rightarrow \infty$) and the need to save computer time. The longer the simulation time, the less the initial conditions will influence the average results; however, in order to avoid the undesired effects of an inappropriate initial choice and to obtain a better convergence, the elimination of the first part of the simulation from the collection of the statistics may be convenient.

When the simulation is split into many sub-histories, better convergence to steady state can be achieved by taking the initial state of each new sub-history as equal to the final state of the previous one. In this way, only the initial condition of the first sub-history will influence the final results in a biased way.

On the other hand, when a simulation is used to study a transient phenomenon and/or a transport process in a nonhomogeneous system (for example, when electron transport in a small device is analyzed), then it is necessary to simulate many electrons separately. In this case the distribution of the initial electron states must be taken into account according to the particular physical situation under investigation, and the initial transient becomes an essential part of the desired results.

2.3.4 Flight Duration, Self-Scattering

The electron wavevector k changes continuously during a free flight because of the applied field according to (2.6). Thus, if $P[k(t)] \, dt$ is the probability that an electron in the state k suffers a collision during the time dt, the probability that an electron which suffered a collision at time $t = 0$ has not yet suffered another collision after a time t is

$$\exp\left\{ -\int_0^t P[k(t')] dt' \right\},$$

which, generally, gives the probability that the interval $(0, t)$ does not contain a scattering event. Consequently, the probability $\mathscr{P}(t) dt$ that the electron will suffer its next collision during dt around t is given by

$$\mathscr{P}(t) dt = P[k(t)] \exp\left\{ -\int_0^t P[k(t')] dt' \right\} dt. \tag{2.63}$$

Because of the complexity of the integral at the exponent, it is very impractical to generate stochastic free flights with the distribution of (2.63) starting from evenly distributed random numbers r. *Rees* [2.44,45] has devised a very simple method to overcome this difficulty. If $\Gamma \equiv 1/\tau_0$ is the maximum value of $P(k)$ in the region of interest in k space, a new fictitious "self-scattering" is introduced so that the total scattering probability, including the self-scattering, is constant and equal to Γ. If the carrier undergoes such a self-scattering, its state k' after the collision is taken to be equal to its state k before the collision, so that in practice the electron path continues unperturbed as if no scattering occurred at all. It is worth noticing that, more generally, it is sufficient that Γ is not less than the maximum value of $P(k)$; furthermore, as we shall see below, Γ can be a convenient function of energy.

Now, with a constant $P(k) = \tau_0^{-1}$, (2.63) reduces to

$$\mathscr{P}(t) = \frac{1}{\tau_0} \exp(-t/\tau_0), \tag{2.64}$$

and random numbers r can be used very simply to generate stochastic free flights t_r through [2.43]

$$t_r = -\tau_0 \ln(1 - r) \equiv -\tau_0 \ln(r), \tag{2.65}$$

where we have made use of the fact that as r is evenly distributed between 0 and 1, the same applies to $(1 - r)$.

The computer time wasted in taking care of self-scattering events is more than compensated for by the simplification of the calculations necessary to evaluate the free-flight durations by means of (2.65).

As regards the choice of the constant Γ, we may note that, in general, $P(k)$ is simply a function of the electron energy $P(\mathscr{E})$; a suitable choice for Γ is then the maximum value of $P(\mathscr{E})$ in the region of energies which are expected to be sampled during the simulation. When $P(\mathscr{E})$ is not a monotonic function of $\mathscr{E}$, its maximum value must be evaluated in some way, for example, with a tabulation at the beginning of the computer program. When $P(\mathscr{E})$ is an increasing function of $\mathscr{E}$, as is often the case, this is accomplished by taking $\Gamma = P(\mathscr{E}_M)$, where $\mathscr{E}_M$ is a maximum electron energy with negligible probability of being achieved by the carrier during the simulation. It must be observed, however, that the range of energy visited by the electron during the simulation is not known at the beginning, when Γ is chosen. Therefore, an estimate must be made for $\mathscr{E}_M$ keeping in mind that $\mathscr{E}_M$ cannot be taken as too large, otherwise a correspondingly unnecessary large value of Γ would result in a waste of computer time for self-scattering events. It is therefore necessary to decide what action to take when, during the simulation, the electron energy $\mathscr{E}$ happens to exceed the maximum value $\mathscr{E}_M$ set up at the beginning of the computer run. A safe method is to increase the value of $\mathscr{E}_M$ and correspondingly of Γ, if required, for the simulation to come, without intervening on the electron energy. Since Γ must be independent of the simulated flight, it is always necessary to check that Γ has not been changed too many times. This is guaranteed because when $\mathscr{E}_M$ has been underestimated at the beginning of the program run, it will quickly increase to a value above which the electron energy will rarely go.

2.3.5 The Choice of the Scattering Mechanism

During the free flight the electron dynamics is governed by (2.6) so that at the end the electron wavevector and energy are known, and each scattering rate λ_i (i indicates the ith mechanism) can be evaluated.

The probability of a self-scattering event will be the complement to Γ of the sum of the λ_i's. A mechanism must then be chosen from all possible ones; to this end, given a random number r, the product $r\Gamma$ is compared with the successive sums of the λ_i's, and the jth mechanism is chosen if j is such that the first of these successive sums which is greater than $r\Gamma$ is $\lambda_1 + \lambda_2 + \lambda_3 + \ldots + \lambda_j$.

If all real scattering events have been tried and none of them has been selected, it means that $r\Gamma > \lambda(\mathscr{E})$, and self-scattering occurs. Thus, in such a procedure, a self-scattering event is the most time consuming, since all λ_i's must be explicitly calculated. However, an expedient can be used to overcome this shortcoming, which can be called fast self-scattering. It consists of setting up a mesh of the energy range consideration at the beginning of the simulation and then recording in a vector the maximum total scattering rate $\bar{\lambda}^{(l)}$ in each energy interval $\Delta\mathscr{E}^{(l)}$ (energy interval equally large in a logarithmic scale may be useful).

At the end of the flight, if the electron energy falls in the nth interval, before trying all λ_i's separately, $r\Gamma$ is compared with $\bar{\lambda}^{(n)}$; at this stage if $r\Gamma > \bar{\lambda}^{(n)}$ then a self-scattering event certainly occurs, otherwise all λ_i's will be successively evaluated. Consequently, the improvement is obtained that only when $\lambda(\mathscr{E}) < r\Gamma < \bar{\lambda}^{(n)}$ does a self-scattering event occur which requires the evaluation of all λ_i's.

As regards the scattering probabilities of the various mechanisms, we shall survey them in Sects. 2.8–12.

2.3.6 The Choice of the State After Scattering

Once the scattering mechanism which causes the end of the electron's free flight is determined, the new state after scattering of the electron, k', must be chosen as the final state of the scattering event. When the free flight ends with a self-scattering event, k' must be taken as equal to k, the state before scattering. When, on the contrary, a true scattering occurs, then k' must be determined stochastically, according to the transition rate of the particular mechanism.

2.3.7 Time Average for the Collection of Results Under Steady-State Conditions

Generally speaking, we may obtain the time-average value of a quantity $\mathscr{A}[k(t)]$ (e.g., the drift velocity, the mean energy, etc) during a single history of duration T as

$$\bar{\mathscr{A}} = \frac{1}{T} \int_0^T \mathscr{A}[k(t)]dt = \frac{1}{T} \sum_i \int_0^{t_i} \mathscr{A}[k(t')]dt', \tag{2.66}$$

where the bar indicates time average, and the integral over the whole simulation time T is separated into the sum of integrals over all free flights of duration t_i. When a steady state is investigated, T should be taken as sufficiently long so that $\bar{\mathscr{A}}$ in (2.66) represents an unbiased estimator of the average of the quantity $\mathscr{A}$ over the electron gas.

In a similar way we may obtain the electron distribution function: a mesh of k space (or of energy) is set up at the beginning of the computer run; during the simulation the time spent by the sample electron in each cell of the mesh is recorded and, for large T, this time conveniently normalized will represent the electron distribution function, that is, the solution of the Boltzmann equation [2.46]. This evaluation of the distribution function can be considered a special case of (2.66) in which we choose for $\mathscr{A}$ the functions $n_j(k)$ with a value of 1 if k lies inside the jth cell of the mesh, and zero if otherwise.

This method is more generally used to obtain transport quantities from Monte Carlo simulations.

2.3.8 Synchronous-Ensemble Method for the Collection
of Results Under Steady-State Conditions

Another possible method of obtaining an average quantity $\mathscr{A}$ from the simulation of a steady-state phenomenon has been introduced by *Price* [2.47], and it is called the synchronous-ensemble method. Let us consider Fig. 2.10 in which for each of the N electrons of the physical system, a time axis is considered. Circles in these axes indicate scattering events. In principle, each of the axes can represent a simulated electron and self-scattering can be included, if desired. In general, the average value of a quantity $\mathscr{A}$ is defined as the ensemble average at time t over the N electrons of the system:

$$\langle \mathscr{A} \rangle = \frac{1}{N} \sum_i \mathscr{A}_i(t_i = t). \tag{2.67}$$

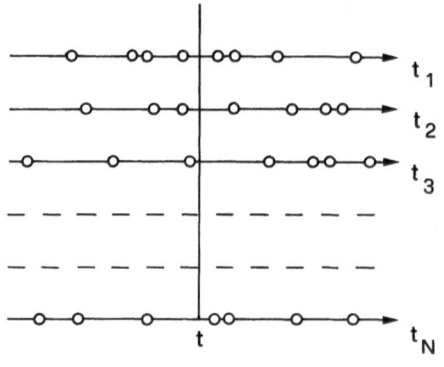

Fig. 2.10. Sketch illustrating the synchronous-ensemble method. Subscripts i $(1, 2, 3, \ldots, N)$ label the time axis of the ith particle; (O) indicate scattering events; t is the generic time of a steady-state condition

TIME (arb. units)

In particular, the steady-state distribution function is proportional to the number of electrons $n(\mathbf{k})\Delta\mathbf{k}$ that at time t are found to be in a cell of fixed volume $\Delta\mathbf{k}$ around $\mathbf{k}$. Then (2.67) can also be evaluated as

$$\langle \mathscr{A} \rangle = C_0 \sum_{\mathbf{k}} n(\mathbf{k}) \mathscr{A}(\mathbf{k}), \tag{2.68}$$

where C_0 is an appropriate normalization constant, and $n(\mathbf{k})$ can be considered as proportional to the probability of finding any given electron in state $\mathbf{k}$.

Now in the spirit of the path integral method [2.38], an electron will be found in $\mathbf{k}$ at time t, if it has been scattered in the suitable $\mathbf{k}'$ an appropriate time interval t' before t, and has not been scattered again between $(t - t')$ and t.

Let $n_a(\mathbf{k})$ be the so-called after-scattering distribution function, proportional to the probability that an electron is found in $\mathbf{k}$ immediately after a scattering event; then $n(k)$ is proportional to

$$n(k) \propto \int_0^\infty n_a[k'(t')] \mathscr{P}[k'(t'), t'] dt',$$

where $\mathscr{P}(k', t')$ is the probability that an electron in k' at a given time will not be scattered before an interval of time t'.

If we now consider the before-scattering distribution function $n_b(k)$, proportional to the probability that an electron is found in k immediately before a scattering event, by an argument similar to that given above we shall find that

$$n_b(k) \propto \frac{1}{\tau(k)} \int_0^\infty n_a[k'(t')] \mathscr{P}[k'(t'), t'] dt',$$

where $\tau^{-1}(k)$ is the scattering rate for an electron in k. Thus the steady-state distribution function is given by

$$n(k) = \frac{1}{\tau_0} n_b(k) \tau(k), \tag{2.69}$$

where τ_0 is an appropriate normalization constant. If, by including self-scattering events, we use a constant $\tau(k)$, then the steady-state distribution function becomes proportional to the before-scattering distribution $n(k) \propto n_b(k)$. [It may seem strange that with a constant $\tau(k)$, the before-scattering distribution $n_b(k)$ is equal to the steady-state distribution since the state just before the scattering events seems to be influenced "at the maximum" by the applied fields. However, one should consider that, while $n_b(k)$ weighs equally all free flights (short and long ones) with average duration τ, when an instantaneous picture of the electron gas is taken at any time t, longer free flights are more probably caught. In other words, in the latter case the vertical line in Fig. 2.10 crosses free flights whose mean duration is longer than the average over all free flights; in fact, the distribution of the hemi-flights on the right and left of the line t reproduces the distribution of flight durations so that the average length of the flights crossed by t is 2τ.] Moreover (2.68) can be used in a Monte Carlo simulation in the following form:

$$\langle \mathscr{A} \rangle = \frac{1}{N} \sum_i \mathscr{A}_{b_i}, \tag{2.70}$$

where the sum covers all N electron free flights, and $\mathscr{A}_{b_i}$ indicates the value of the quantity $\mathscr{A}$ evaluated at the end of the free flight immediately before the ith scattering event.

2.3.9 Statistical Uncertainty

In order to estimate the statistical uncertainty on a time-average result $\bar{\mathscr{A}}$ due to the finite value of the simulation time T, as anticipated in Sect. 2.3.1, the whole history can be split (this splitting requires the interruption of a free flight at the end of each sub-history simulation; the remaining part of the flight is thus used as the initial part of the simulation of the successive sub-history) into N sub-histories of duration T/N; for each of them a value $\langle\mathscr{A}\rangle_l$ is obtained. We may then take as the most probable value of $\langle\mathscr{A}\rangle$ the average of the $\langle\mathscr{A}\rangle_l$'s (which will be equal to $\bar{\mathscr{A}}$) and its standard deviation as the statistical uncertainty on $\bar{\mathscr{A}}$.

The sub-histories must be sufficiently long to be considered independent of each other, but sufficiently short to allow us to simulate a large number of them. Furthermore, it is possible to carry out, on the series of partial results $\langle\mathscr{A}\rangle_l$, those tests [2.48] which verify the statistical nature of their fluctuations.

2.3.10 Time- and Space-Dependent Phenomena

For time and/or space-dependent problems we cannot rely on the ergodic property of the system, and an ensemble of particles must be explicitly simulated [2.49, 50]. It must be stressed that the estimators given by (2.66, 70) are based on the hypothesis of steady-state conditions and cannot be used when a time-dependent phenomenon is analyzed: the ensemble average of a quantity $\mathscr{A}$ must actually be estimated according to its definition given by (2.67).

One exception to this rule, which will be discussed in Sect. 2.3.13, is that of phenomena which are periodic in space and/or time. In this case the different situations which the electron will experience in equivalent positions or times will take into account the many possibilities of various particles, and again the simulation of a single electron may yield the necessary information about the entire electron gas.

2.3.11 Transients

We shall consider here a situation of homogeneous electron gas with a time-dependent behavior. In particular, it is of interest to study the case of a sudden change in the value of an applied electric field and to investigate the transient dynamic response to such a change.

In this situation many particles must be independently simulated with appropriate distribution of initial conditions. Provided that the number of simulated particles is sufficiently large, the average value of a quantity of interest, obtained on this sample ensemble as a function of time, will be representative for the average of the entire gas.

The duration of the transient response is not known a priori and will be of the order of the largest of the characteristic times of the electron system. This time may be called the "transient transport time" and, in general, depends upon the values of the applied field and temperature; in our case of high-field transport in semiconductors, it may roughly correspond to the energy relaxation time or to the time for repopulation of different valleys [2.51].

To determine the precision of the results obtained, the entire ensemble can be separated into a certain number of sub-ensembles; for each of them an estimate of the quantity of interest $\mathscr{A}$ will be performed. Then their average and standard deviation can be taken as the most probable value and the statistical uncertainty of $\mathscr{A}$, respectively.

The transient dynamic response, which will be obtained by means of the simulation, will of course depend upon the initial conditions of the carriers, and these must be assumed according to the situation which is to be explored. Initial velocity distributions which may be of interest are, for example, a Maxwellian distribution at the lattice temperature and no drift, or a Maxwellian distribution with an electron temperature higher than that of the lattice with or without drift.

2.3.12 Space-Dependent Phenomena

The simulation of a steady-state phenomenon in a physical system where the electron transport depends upon the position in space is of particular interest for the analysis and the modeling of devices. Also in this case, an ensemble of independent particles must be used, and averages must be performed over particles at given positions. Statistical uncertainties of the results can be obtained with sub-ensembles as indicated in the previous section.

In a steady-state situation particles will enter the region of interest continuously, and in the simulation some initial wavevectors must be assumed when a new particle is considered, according to the momentum distribution of the particles in the physical system under investigation. For example, a cold hemi-Maxwellian distribution, that is, with only positive components of the velocities along a given direction, may be convenient in order to simulate a metallic contact of a device. If the simulated electron entering the device goes back to the contact, a new electron must be generated, but the information relative to the "lost" electron must be accounted for in the final results in a way which depends on the particular analysis being carried out.

Space- and time-dependent phenomena may present similar characteristics, and they have sometimes been confused in simulation problems. For instance, if a field is suddenly switched on from zero to a large value, during the initial transient, the drift velocity may reach values larger than in steady-state conditions (overshoot effect). This is sometimes used to discuss the behavior of an electron stream, coming from a low-field region and entering a region with a high applied field.

Even if from a qualitative point of view this mode of reasoning is correct, in rigorous calculations the kind of average that must be done is different whether time- or space-dependent phenomena are considered. In a steady-state phenomenon with a space dependence of the applied fields it is, in general, necessary to consider the electron properties at given positions [2.52, 53], so that the simulation must perform records and averages for given points at any time. On the other hand, when the object of the investigation is the evolution in time of a homogeneous system, average quantities must be evaluated at given times independently of the particle positions.

2.3.13 Periodic Fields

The Monte Carlo simulation can be extended to permit the calculation of the response of charge carriers to periodic external fields of any strength [2.54, 55]. When a field

$$E = E_0 + E_1 \sin (\omega t) \tag{2.71}$$

is applied, and the ac term is small enough for a linear response analysis to be carried out, the average electron velocity will be of the form

$$\langle v(t) \rangle = v_0 + v_1 \sin (\omega t) + v_2 \cos (\omega t). \tag{2.72}$$

The coefficients v_1 and v_2 of the fundamental response in (2.72) can be obtained as sine and cosine Fourier transforms, respectively, of the velocity of the simulated electron over its history. Since the equation of motion of the

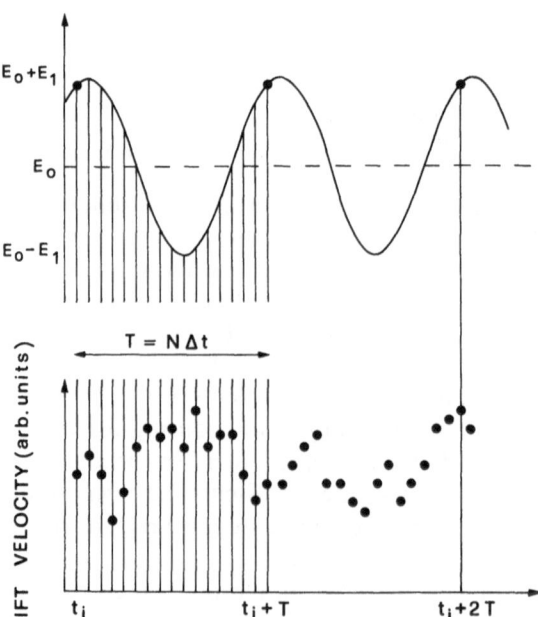

Fig. 2.11. Simulation of transport with periodic fields. A random variable (for example, instantaneous velocity at time t) is recorded in response to the periodic input signal $E(t)$. Period T is divided into N intervals of equal duration Δt, and the average response is obtained by averaging the values at t_i, $t_i + T$, $t_i + 2T, \ldots$ [2.55]

particle subject to a field given by (2.71) is known in explicit terms, the Fourier coefficients v_1 and v_2 can easily be obtained by the simulation [2.54].

For large periodic fields, the periodic part of the current will contain higher harmonics, besides the fundamental frequency. These components can also be obtained from the Fourier analysis of the simulated velocity, but in this case statistical-noise problems become very severe.

It is also possible to obtain the total response of the electron gas without Fourier analysis, by sampling the electron velocity at fixed times, corresponding to definite phases in the period of the external force. To be more explicit, as illustrated in Fig. 2.11, let us suppose that we pick out, from the simulation, the electron velocity at times given by

$$0, \Delta t, 2\Delta t, \ldots, l\Delta t, \ldots,$$

where Δt (with typical values within $0.5 - 1 \times 10^{-14}$ s) is a fraction $1/N$ of the period $2\pi/\omega$ of the ac field:

$$\Delta t = \frac{1}{N} \frac{2\pi}{\omega}. \tag{2.73}$$

Then we average the values of v obtained at times

$$l\Delta t, (l+N)\Delta t, (l+2N)\Delta t, \ldots.$$

The result is an estimation of the average electron velocity $\langle v \rangle$, which is a periodic function of t with the same period $2\pi/\omega$, at the times given above.

The synchronous-ensemble method described in Sect. 2.3.8 for static fields has been extended by *Lebwohl* [2.54] to include the case of periodic fields. The result is a simple parametric dependence of (2.69) upon time:

$$n(\mathbf{k}, t) = \frac{1}{\tau_0} n_b(\mathbf{k}, t) \tau(\mathbf{k}), \tag{2.74}$$

and in the applications it is possible to operate as for static fields with the additional care of assigning each analyzed event to the proper phase (or time within the period).

2.3.14 Proof that the Monte Carlo Method Leads to a Distribution Function Which Satisfies the Space-Homogeneous, Time-Dependent Boltzmann Equation

Here we shall report a proof originally given by *Reklaitis* [2.56]. Let us define $P_n(\mathbf{k}_0; \mathbf{k}, t)$ to be the probability density that the electron will pass through the state $\mathbf{k}$ at time t during the course of the nth free flight in the electric field, having started at the state $\mathbf{k}_0$ at time zero. The explicit time dependence in the definition must be retained since the electron can pass through $\mathbf{k}$ during the nth free flight after any time between 0 and ∞.

Then

$$P(k_0; k, t) = \sum_{n=0}^{\infty} P_n(k_0; k, t) \tag{2.75}$$

is the probability density that the electron which started at time zero from the state k_0 will appear in the state k at time t after an arbitrary number of scattering events. The distribution function calculated by the Monte Carlo technique can be expressed as

$$f(k, t) = \int f_0(k_0) P(k_0; k, t) dk_0, \tag{2.76}$$

where $f_0(k_0)$ is the initial distribution function at time zero. The probability function P_n must satisfy the equation

$$P_n(k_0; k, t) = \frac{V_0}{(2\pi)^3} \int dk' \int_0^t dt' P_{n-1}(k_0; k', t') \int dk'' W^+(k', k'')$$

$$\cdot \exp\left[-\int_0^{t-t'} \Gamma\left(k'' - \frac{eE}{\hbar}\tau\right) d\tau \right] \delta\left(k - k'' - \frac{eE}{\hbar}(t-t')\right). \tag{2.77}$$

The right-hand side of (2.77) is the product of three probabilities: (i) the probability $P_{n-1}(k_0; k', t')$ that the electron will pass through some state k' at time $t' < t$ during the $(n-1)$th free flight. (ii) the probability per unit time $W^+(k', k'')$ that the electron will be scattered from this state to some other state k'' (which is the sum of all transition rates from k' to k'', including self-scattering). (iii) the probability $\exp\left[-\int_0^{t-t'} \Gamma\left(k'' + \frac{eE}{\hbar}\tau\right) d\tau \right]$ that the electron will not be scattered while drifting from k'' to k during the nth free flight (Γ is the total scattering rate including self-scattering). The delta function ensures that the vector $(k - k'')$ is parallel to the field and has the correct magnitude for the electron to arrive at k from k'' in the time $(t - t')$.

Using the delta function to perform the integration over k'', we obtain

$$P_n(k_0; k, t) = \frac{V_0}{(2\pi)^3} \int dk' \int_0^t dt' P_{n-1}(k_0; k', t') W^+\left[k', k - \frac{eE}{\hbar}(t-t')\right]$$

$$\cdot \exp\left[-\int_0^{t-t'} \Gamma\left(k - \frac{eE}{\hbar}\tau\right) d\tau \right]. \tag{2.78}$$

Differentiating (2.78) with respect to time, by using Leibnitz's theorem [2.57] we have

$$\frac{\partial P_n}{\partial t} = \frac{V_0}{(2\pi)^3} \left[\int dk' P_{n-1} W^+ + \int dk' \int_0^t dt' P_{n-1} \exp\left(-\int_0^{t-t'} \Gamma d\tau \right) \frac{\partial W^+}{\partial t} \right.$$

$$\left. + \int dk' \int_0^t dt' P_{n-1} W^+ \frac{\partial}{\partial t} \exp\left(-\int_0^{t-t'} \Gamma d\tau \right) \right]. \tag{2.79}$$

Now, noting that

$$\frac{\partial W^+}{\partial t} = \frac{\partial W^+}{\partial k} \frac{\partial k}{\partial t} = \frac{eE}{\hbar} \frac{\partial W^+}{\partial k}, \tag{2.80}$$

and that

$$\frac{\partial}{\partial t} \left\{ \exp\left[-\int_0^{t-t'} \Gamma\left(k - \frac{eE}{\hbar}\tau\right) d\tau \right] \right\} = -\Gamma(k) \exp\left[-\int_0^{t-t'} \Gamma\left(k - \frac{eE}{\hbar}\tau\right) d\tau \right]$$

$$-\frac{eE}{\hbar} \frac{\partial}{\partial k} \left\{ \exp\left[-\int_0^{t-t'} \Gamma\left(k - \frac{eE}{\hbar}\tau\right) d\tau \right] \right\}, \tag{2.81}$$

we obtain

$$\frac{\partial P_n}{\partial t} = \frac{V_0}{(2\pi)^3} \int dk' P_{n-1}(k_0; k', t) W^+(k', k) - \Gamma(k) P_n(k_0; k, t)$$

$$-\frac{eE}{\hbar} \frac{\partial}{\partial k} P_n(k_0; k, t). \tag{2.82}$$

Summing (2.82) over n from $n=1$ to ∞ and taking into account (2.75) we find

$$\frac{\partial P}{\partial t} = -\frac{eE}{\hbar} \frac{\partial}{\partial k} P(k_0; k, t) - \Gamma(k) P(k_0; k, t)$$

$$+\frac{V_0}{(2\pi)^3} \int dk' P(k_0; k, t) W^+(k, k') + \frac{\partial P_0}{\partial t} (k_0; k, t)$$

$$+\frac{eE}{\hbar} \frac{\partial P_0}{\partial k} (k_0; k, t) + \Gamma(k) P_0(k_0; k, t), \tag{2.83}$$

where $P_0(k_0; k, t)$ is the probability density that the electron which started at time zero from the state k_0 will appear at time t in state k without scattering:

$$P_0(k_0; k, t) = \delta\left(k - k_0 - \frac{eE}{\hbar}t\right) \exp\left[-\int_0^t \Gamma\left(k_0 + \frac{eE}{\hbar}\tau\right) d\tau \right]. \tag{2.84}$$

The total scattering rate $\Gamma(k)$ and the total transition rate $W^+(k', k)$ are expressed in the following way:

$$\Gamma(k) = \lambda_0(k) + \lambda(k), \tag{2.85}$$

$$W^+(k', k) = \frac{(2\pi)^3}{V_0} \lambda_0 \delta(k' - k) + W(k', k), \tag{2.86}$$

where $(2\pi)^3 V_0^{-1} \lambda_0(\boldsymbol{k})\delta(\boldsymbol{k}'-\boldsymbol{k})$ is the transition rate for the self-scattering process.

Substituting (2.84–86) into (2.83) we obtain the equation for $P(\boldsymbol{k}_0;\boldsymbol{k},t)$:

$$\frac{\partial P}{\partial t} = -\frac{eE}{\hbar}\frac{\partial P}{\partial k} - \lambda P + \frac{V_0}{(2\pi)^3} \int P(\boldsymbol{k}_0;\boldsymbol{k}',t)W(\boldsymbol{k}',\boldsymbol{k})d\boldsymbol{k}'. \tag{2.87}$$

Multiplying both sides of (2.87) by $f_0(\boldsymbol{k}_0)$, integrating over $\boldsymbol{k}_0$ and taking into account (2.76), we find that the distribution function calculated by Monte Carlo satisfies the space-homogeneous and time-dependent Boltzmann equation:

$$\frac{\partial f(\boldsymbol{k},t)}{\partial t} = -\frac{eE}{\hbar}\frac{\partial f(\boldsymbol{k},t)}{\partial k} - \lambda f(\boldsymbol{k},t) + \frac{V_0}{(2\pi)^3}\int f(\boldsymbol{k}',t)W(\boldsymbol{k}',\boldsymbol{k})d\boldsymbol{k}'$$

$$= -\left(\frac{eE}{\hbar}\frac{\partial}{\partial k} + I_k^{\text{th}}\right)f(\boldsymbol{k},t). \tag{2.88}$$

It is worth noting that this solution is independent of the self-scattering rate. Analogous proof for steady-state conditions can be found in [2.46, 52].

2.4 The Iterative Technique

Let us consider the integro-differential Boltzmann equation for a spatially homogeneous system as given in (2.88). Following [2.58], we introduce the path variables

$$\begin{cases} \tilde{\boldsymbol{k}} = \boldsymbol{k} - \dfrac{eE}{\hbar}t \\[2mm] \tilde{t} = t, \end{cases} \tag{2.89}$$

which represent the collisionless trajectory in $\boldsymbol{k}$ space of the electron.

Equation (2.88) becomes

$$\frac{df}{d\tilde{t}}\left(\tilde{\boldsymbol{k}} + \frac{eE}{\hbar}\tilde{t},\tilde{t}\right) + \lambda\left(\tilde{\boldsymbol{k}} + \frac{eE}{\hbar}\tilde{t}\right)f\left(\tilde{\boldsymbol{k}} + \frac{eE}{\hbar}\tilde{t}\right)$$

$$= \frac{V_0}{(2\pi)^3}\int d\boldsymbol{k}'f(\boldsymbol{k}',t)W\left(\tilde{\boldsymbol{k}} + \frac{eE}{\hbar}\tilde{t},\boldsymbol{k}'\right). \tag{2.90}$$

Multiplying through by the factor

$$\exp\left[\int_0^{\tilde{t}} \lambda\left(\tilde{\boldsymbol{k}} + \frac{eE}{\hbar}\theta\right)d\theta\right],$$

and integrating between $\tilde{t}_1$ and $\tilde{t}_2$, we obtain

$$f\left(\tilde{k}+\frac{eE}{\hbar}\tilde{t}_2,\tilde{t}_2\right)\exp\left[\int_0^{\tilde{t}_2}\lambda\left(\tilde{k}+\frac{eE}{\hbar}\theta\right)d\theta\right]$$

$$=f\left(\tilde{k}+\frac{eE}{\hbar}\tilde{t}_1,\tilde{t}_1\right)\exp\left[\int_0^{\tilde{t}_1}\lambda\left(\tilde{k}+\frac{eE}{\hbar}\theta\right)d\theta\right]+\frac{V_0}{(2\pi)^3}\int_{\tilde{t}_1}^{\tilde{t}_2}d\tilde{t}$$

$$\cdot\exp\left[\int_0^{\tilde{t}}\lambda\left(\tilde{k}+\frac{eE}{\hbar}\theta\right)d\theta\right]\int dk'f(k',\tilde{t})\,W\left(\tilde{k}+\frac{eE}{\hbar}\tilde{t},k'\right). \tag{2.91}$$

Coming back to the variables (k,t) by setting

$$\begin{cases} k=\tilde{k}+\dfrac{eE}{\hbar}\tilde{t}_2 \\[2mm] t=\tilde{t}_2, \end{cases} \tag{2.92}$$

and writing t_1 instead of $\tilde{t}'$, for simplicity, we have from (2.91):

$$f(k,t)=f\left[k-\frac{eE}{\hbar}(t-t'),\,t'\right]\exp\left\{-\int_{t'}^t\lambda\left[k-\frac{eE}{\hbar}(t-\theta)\right]d\theta\right\}+\frac{V_0}{(2\pi)^3}\int_{t'}^t d\tilde{t}$$

$$\cdot\exp\left\{-\int_{\tilde{t}}^t\lambda\left[k-\frac{eE}{\hbar}(t-\theta)\right]d\theta\right\}\int dk'f(k',t)\,W\left[k-\frac{eE}{\hbar}(t-\tilde{t}),k'\right]. \tag{2.93}$$

Equation (2.93), which is an integral form of the Boltzmann equation, can be interpreted in simple physical terms as follows: two different contributions concur to determine the carrier distribution function $f(k,t)$:

i) the contribution of the electrons which were in the state $k-eE(t-t')/\hbar$ at some previous time t', and have drifted to the state k in the interval $(t-t')$ under the influence of the field E, without being scattered [first term on the right-hand side of (2.93)];

ii) the contribution of the electrons which were scattered from any state k' to the new state $k-eE/\hbar(t-\theta)$ at any time θ between t' and t, and which arrive at the state k at the time t without being scattered again [second term on the right-hand side of (2.93)].

Since, provided that $t'<t$, the time t' is arbitrary, we can consider the limiting form of (2.93) when $t'\to-\infty$,

$$f(k,t)=\frac{V_0}{(2\pi)^3}\int_0^\infty dt''\exp\left[-\int_0^{t''}\lambda\left(k-\frac{eE}{\hbar}t'''\right)dt'''\right]$$

$$\cdot\int dk'f(k',t-t'')\,W\left(k-\frac{eE}{\hbar}t'',k'\right), \tag{2.94}$$

where we have used the rearrangement of variables

$$\begin{cases} t''=t-\tilde{t} \\ t'''=t-\theta. \end{cases} \tag{2.95}$$

The iterative technique for the solution of (2.94) consists of substituting an arbitrary function $f_0(k,t)$ into the right-hand side of (2.94), and calculating $f(k,t)$ as the right-hand side of (2.94) itself. This function is then resubstituted on the right-hand side, and the procedure is repeated until convergence is achieved.

Numerical iterative solutions of (2.94) have been discussed for the stationary conditions, in which the dependence of $f(k,t)$ upon t vanishes [2.45, 59, 60]. The stability of the iterative procedure and refinements for quicker computations have been discussed by *Rees* [2.45] and *Vassel* [2.61].

For the steady-state conditions, taking advantage of the self-scattering process with a constant Γ, (2.85, 86), the iterative scheme of (2.94) can be written as

$$f^{(i+1)}(k)=\frac{V_0}{(2\pi)^3} \int\limits_0^\infty \exp{(-\Gamma t)}dt \int W^+\left(k-\frac{eE}{\hbar} t,k'\right)f^{(i)}(k')dk'. \tag{2.96}$$

It is noteworthy that (2.96) can still be of use for the time-dependent case when Γ is assumed to be large enough compared with the scattering rate λ. To see this, the integral over k' in (2.96) is first carried out to give

$$f^{(i+1)}(k)=\int\limits_0^\infty \exp{(-\Gamma t)} (\Gamma-I_k^{th})f^{(i)}\left(k-\frac{eE}{\hbar} t\right)dt. \tag{2.97}$$

Now for large values of Γ, the contributions to the integral over t only come from very small t, and the right-hand side of (2.97) can be approximated through a series expansion such as

$$(\Gamma-I_k^{th})f^{(i)}\left(k-\frac{eE}{\hbar} t\right)\simeq(\Gamma-I_k^{th})f^{(i)}(k)-\frac{eE}{\hbar} t \frac{\partial}{\partial k} (\Gamma-I_k^{th})f^{(i)}(k). \tag{2.98}$$

Substituting (2.98) into (2.97), carrying out the integration, and keeping only terms up to first order in $1/\Gamma$, we obtain

$$f^{(i+1)}(k)=f^{(i)}(k)-\frac{1}{\Gamma}\left(I_k^{th}+\frac{eE}{\hbar}\frac{\partial}{\partial k}\right)f^{(i)}(k). \tag{2.99}$$

The comparison between (2.99) and the space-homogeneous, time-dependent Boltzmann equation (2.88) shows that $f^{(i+1)}(k)$ is related to $f^{(i)}(k)$ through the same equation, provided ∂t is set equal to Γ^{-1}. Thus we can write

$$f^{(i+1)}(k)=f^{(i)}(k)+\frac{1}{\Gamma}\frac{\partial f^{(i)}(k,t)}{\partial t}. \tag{2.100}$$

The approximation made on assuming that Γ was large in comparison with the scattering rate λ can be arbitrarily improved by increasing Γ, but only at the expense of needing more iterations to obtain the steady state. Practical calculations [2.62] show, however, that very large values of Γ are not needed since Γ only has to be much greater than the values of the scattering rates at energies where a substantial amount of electrons is present.

2.5 Noise Associated with Velocity Fluctuations

2.5.1 General Definitions and the Transfer Impedance Method

The system under investigation is a two-terminal device of length L and cross-sectional area A polarized at dc bias. We are interested in evaluating the noise current spectral density $S_I(\omega)$. We recall that, through the Wiener-Khintchine theorem, $S_I(\omega)$ is related to the autocorrelation function of current fluctuations by the relation

$$S_I(\omega) = 2 \int\limits_{-\infty}^{+\infty} \exp\,(\mathrm{i}\omega\tau)\,\overline{\delta I(t)\delta I(t+\tau)}\,d\tau, \qquad (2.101)$$

where $\delta I(t) = I(t) - I_0$ is the current fluctuation around the average value I_0, and the bar indicates time average (which from ergodicity is equivalent to ensemble average [2.63]). Steady-state conditions, that is invariance with respect to a time translation, ensure that $S_I(\omega)$ is only a function of ω. Furthermore, $S_I(\omega)$ is a macroscopic quantity which is conveniently represented by an equivalent noise temperature T_n, defined through the following relation [2.30]:

$$S_I(\omega) = \frac{4\,K_B T_n}{\mathrm{Re}[Z(\omega)]}, \qquad (2.102)$$

where $Z(\omega)$ is the small signal impedance at the bias point relating the voltage and current fluctuations, δV and δI, measured between the terminals of the device, through a generalized Ohm's law [2.64, 65]:

$$\delta V = Z(\omega)\delta I. \qquad (2.103)$$

In analogy with Nyquist's relation (2.3), $T_n(\omega)$ can be determined experimentally through the measurement of the maximum power dissipated by the noise source on the matched output impedance at frequency $\omega = 2\pi f$ per unit bandwidth Δf [2.12, 66] (Chap. 3):

$$P_{av}(\omega) = K_B T_n(\omega)\Delta f. \qquad (2.104)$$

By inverting (2.101) and setting $\tau = 0$, the power spectrum relation is found:

$$\int_0^\infty S_I(\omega)d\omega = \overline{(\delta I)^2}. \tag{2.105}$$

By simply exchanging current I with voltage V, voltage fluctuations can be defined through (2.101, 105). Furthermore, in a theory which accounts only for the first-order terms in the fluctuating quantities, the following relation between S_I and S_V can be proven to hold [2.67]:

$$S_V(\omega) = |Z(\omega)|^2 S_I(\omega). \tag{2.106}$$

Here we are assuming Re $\{Z(\omega)\} > 0$, that is, the electrical system is considered stable with respect to fluctuations in charge density.

A general theory for the calculation of the small signal impedance and the noise makes use of the transfer impedance method [2.68, 69]. This method allows us to relate a small perturbation in local current density at frequency ω, $\delta j(r, \omega) \exp(i\omega t)$, superimposed to the dc value $j(r)$, with the instantaneous small ac local field, $\delta E(r, \omega) \exp(i\omega t)$, superimposed to the dc value $E(r)$. To first-order terms in the local equation, we obtain a relation of the form

$$\hat{L}\delta E(r, \omega) = \delta j(r, \omega), \tag{2.107}$$

where $\hat{L}$ is a linear operator.

The Green matrix $z(r, r', \omega)$ of the operator $\hat{L}$ is defined by

$$\hat{L}z(r, r', \omega) = \mathbb{1}\,\delta(r - r'), \tag{2.108}$$

where $\delta(r - r')$ is the Dirac function and $\mathbb{1}$ the unitary matrix in r-space. Applying (2.108) to the vector $\delta j(r)$, integrating over r', and taking into account (2.107), we obtain

$$\delta E(r, \omega) = \int dr' z(r, r', \omega)\delta j(r', \omega). \tag{2.109}$$

The kernel $z(r, r', \omega)$ is called the transfer impedance matrix; it is a propagator which describes how the source perturbation (current, in this case) is related to the proper response field (electric, in this case). A further line integral from one electrode to the other of the two terms in (2.109) gives the induced voltage fluctuation as a function of the current fluctuation, see (2.103). Therefore we find

$$Z(\omega) = \int_0^L \int_0^L z(r, r', \omega)dr\, dr'. \tag{2.110}$$

When $\mathrm{Re}\{Z(\omega)\}>0$ and carrier-carrier interaction are neglected, exact relationships between noise associated with velocity fluctuations and diffusion can be found. In this case we are interested in the region of the spectrum for which ω is high enough to neglect Flicker as well as generation-recombination noise [2.30] but sufficiently low to satisfy $\omega\tau\ll1$, τ being the collision time, so that $\exp(i\omega\tau)$ is always well approximated by unity. In the following sections the two limiting cases relating to homogeneous-field and space-charge-limited-current conditions, respectively, will be considered to illustrate the use of the transfer impedance method.

2.5.2 The Homogeneous-Field Case and the Modified Einstein Relation

Using a one-dimensional treatment, justified by the simple geometry considered, the current equation is

$$j=en\mu(E)E, \tag{2.111}$$

which by linearization (n is constant) gives

$$\delta j=en\mu'(E)\delta E, \tag{2.112}$$

μ' being the differential mobility $dv_d(E)/dE$.

Comparing (2.112) with (2.107), through (2.108) the transfer impedance results as

$$z(x,x')=\frac{\delta(x-x')}{eAn\mu'}. \tag{2.113}$$

From (2.110) we find

$$Z=\frac{L}{eAn\mu'}. \tag{2.114}$$

Let us evaluate the spectral density of voltage fluctuations under the condition $\omega\to0$:

$$S_V(0)=2\int_{-\infty}^{+\infty}\overline{\delta V(t)\delta V(t+\tau)}\,d\tau$$

$$=4\int_0^\infty d\tau\int_0^L dx_1\int_0^L \overline{dx'z(x_1,x')\delta I(x',t)\int_0^L dx_2\int_0^L dx''z(x_2,x'')\delta I(x'',t+\tau)}.$$
$$\tag{2.115}$$

Since, owing to uniform conditions, δI is independent from x, we have

$$S_V(0) = 4Z^2 \int_0^\infty \overline{\delta I(t) \delta I(t+\tau)} \, d\tau. \tag{2.116}$$

From the Ramo-Shockley theorem of electrostatics [2.70, 71], we obtain

$$\delta I = \frac{e}{L} \sum_{i=1}^N \delta v_i, \tag{2.117}$$

where $\delta v_i(t)$ is the fluctuation of the one-carrier instantaneous velocity in the field direction, and the sum is evaluated over all carriers N. By substituting (2.117) into (2.116), we find

$$S_V(0) = 4Z^2 \left(\frac{e}{L}\right)^2 \int_0^\infty \overline{\sum_i \sum_j [\delta v_i(t) \delta v_j(t+\tau)]} \, d\tau, \tag{2.118}$$

which, by neglecting the carrier-carrier correlation, reduces to

$$S_V(0) = 4Z^2 \left(\frac{e}{L}\right)^2 N \int_0^\infty \overline{\delta v(t) \delta v(t+\tau)} \, d\tau = 4 \frac{e^2 N Z^2}{L^2} D, \tag{2.119}$$

where for the last equality use has been made of (2.42) in the long-time limit. By using (2.106), it is easy to recover the well-known formula [2.68]:

$$S_I(0) = \frac{4e^2 N}{L^2} D. \tag{2.120}$$

From (2.102) we find

$$D = \frac{K_B T_n}{e} \mu', \tag{2.121}$$

which we call the generalized Einstein relation. Eq. (2.121) can be generalized to the three-dimensional case for which we shall introduce one parallel and two transverse components of D, T_n, and μ' with respect to the field direction. It can be immediately recognized that at equilibrium, that is, when $T_n = T_0$ and $\mu' = \mu$, (2.121) recovers the Einstein relation.

The generalized Einstein relation, originally proposed by *Price* [2.41], has provided a useful method for the experimental determination of the diffusion coefficient through noise-conductivity measurements [2.12, 72, 73].

2.5.3 The Space-Charge-Limited-Current Case and the Gisolf-Zijlstra Relation

Using a one-dimensional treatment as in the previous section, the current equation, with the diffusion term neglected, and the Poisson equation are, respectively,

$$j = e n(x) \mu(x) E(x), \tag{2.122}$$

$$\frac{dE(x)}{dx} = \frac{e}{\varepsilon \varepsilon_0} n(x), \tag{2.123}$$

where ε is the permittivity of free space and ε_0 the relative static dielectric constant of the material. Let us note that owing to stationarity, j is x independent.

Linearization of (2.122, 123) gives

$$\delta j = \varepsilon \varepsilon_0 \left(\frac{dE}{dx} \mu' + \mu E \frac{d}{dx} \right) \delta E. \tag{2.124}$$

It can be easily verified that the Green's function giving the transfer impedance is

$$z(x, x') = \frac{u(x - x')}{A \varepsilon \varepsilon_0 \mu E} = \frac{u(x - x')}{I} \frac{dE}{dx}, \tag{2.125}$$

where $u(x - x')$ is the unit step function. The fact that x must be larger than x' for a nonzero result corresponds to field propagation to the right.

Let us evaluate the spectral density of voltage fluctuations given by (2.115). Now by considering slices of material of volume $A \Delta x$, providing that Δx remains larger than the carrier mean free path, but small enough to justify a local field analysis [2.62], we argue that carrier velocity fluctuations in different slices at x' around Δx, and x'' around Δx, are uncorrelated unless the slices overlap. Under this assumption, in analogy with (2.120), we obtain

$$\int_0^\infty \overline{\delta I(x', t) \delta I(x'', t + \tau)} \, d\tau = e^2 A n(x') D(x') \delta(x' - x''), \tag{2.126}$$

where $n(x')$ is the carrier concentration within a slice of material at position x'.

Substituting (2.126) into (2.125), performing the line integration, and using the Poisson equation, the Gisolf-Zijlstra formula [2.74] is obtained as

$$S_V(0) = \frac{4 e \varepsilon \varepsilon_0 A}{I^2} \int_0^{E(L)} [E(L) - E]^2 D(E) \, dE. \tag{2.127}$$

Equation (2.127) relates noise and diffusion under space-charge-limited-current conditions. Its validity requires that the carrier mean free path remains small with respect to the length of the sample and that the diffusion current is negligible.

This relation, generalized to the case of weak injection conditions [2.68], has provided a useful method for the determination of the diffusion coefficient through noise-conductivity measurements under nonuniform field conditions [2.75–77].

2.6 The Effect of Carrier-Carrier Interaction

This section reports on a general definition of the kinetic coefficients which includes carrier-carrier interaction. The treatment closely follows the review paper of *Gantsevich* et al. [2.25].

2.6.1 The Boltzmann Equation with the Two-Particle Collision Term

The Boltzmann equation for steady-state and spatially homogeneous systems including two-particle collisions is written

$$\left(\frac{eE}{\hbar}\frac{\partial}{\partial k}+I_k^{\text{th}}\right)\bar{f}(k)+I_k^{\text{ee}}[\bar{f}(k),\bar{f}(k)]=0. \tag{2.128}$$

Here the bar denotes ensemble average (in general, the interpretation of the single-particle distribution function as the ensemble average of a function $f(k,r,t)$, defined as the physical density in a six-dimensional position-wavevector space, enables fluctuations about the average to be treated [2.78]). The two-particle collision integral is

$$I_k^{\text{ee}}(\bar{f},\bar{f})=\left(\frac{V_0}{(2\pi)^3}\right)^3\int dk'\int dk_1'\int dk_1\,[W(k,k_1;k',k_1')\bar{f}(k)\bar{f}(k_1)$$
$$-W(k',k_1';k,k_1)\bar{f}(k')\bar{f}(k_1')], \tag{2.129}$$

where $W(k,k_1;k',k_1')$ is the transition rate for a particle-particle collision which carries two particles from the occupied states k and k_1 to the empty states k' and k_1'. Let us note that, owing to the lack of distinguishability, it is

$$W(k,k_1;k',k_1')=W(k_1,k;k',k_1')=W(k,k_1;k_1',k')=W(k_1,k;k_1',k'). \tag{2.130}$$

Let us point out that (2.128) is nonlinear due to the interparticle collision integral. Furthermore, the two-particle collision integral of (2.129) vanishes after substitution of the displaced Maxwellian distribution function with arbitrary drift momentum $\hbar k_d$ and electron temperature T_e:

$$f(k)=D_0\exp\left\{-[\hbar^2(k-k_d)^2/(2mK_BT_e)]\right\}, \tag{2.131}$$

where D_0 is an appropriate normalization constant and m the carrier effective mass.

Indeed the transition rates in (2.130) contain δ functions ensuring energy and wavevector conservation in a single collision ('Umklapp' processes are neglected), $\mathscr{E}(k) + \mathscr{E}(k_1) = \mathscr{E}(k') + \mathscr{E}(k_1')$; $k + k_1 = k' + k_1'$, and this leads to the expression

$$I_k^{ee}(\bar{f}, \bar{f}) = \left(\frac{V_0}{(2\pi)^3}\right)^3 \int dk_1 D_0^2 \exp\left\{-\hbar^2[(k-k_d)^2 + (k_1-k_d)^2]/(2mK_BT_e)\right\}$$

$$\cdot \int dk' \int dk_1' [W(k,k_1;k',k_1') - W(k',k_1';k,k_1)], \qquad (2.132)$$

which is equal to zero owing to the Stueckelberg's property [2.79]. This, in turn, ensures the zero value of the last integral in the right-hand side of (2.132).

2.6.2 Fluctuations Near a Non-Equilibrium Steady-State

Let us consider a semiconductor in which, along with a strong and constant electric field E, there also exists a weak variable field $\delta E(r,t)$ of the form

$$\delta E(r,t) = \delta E_{q\omega} \exp(-i\omega t + iq \cdot r). \qquad (2.133)$$

We are interested in evaluating the corresponding correction $\delta f(k,r,t)$ to the homogeneous steady-state distribution function $\bar{f}(k)$, which satisfies (2.133) so that

$$\delta f(k,r,t) = \delta f_{q\omega} \exp(-i\omega t + iq \cdot r). \qquad (2.134)$$

The kinetic equation for $\delta f(k,r,t)$ is obtained by linearizing the time- and space-dependent Boltzmann equation (2.5) which includes the two-particle collision integral given by (2.129), and which results in [2.25]

$$\left(\frac{\partial}{\partial t} + v \cdot \frac{\partial}{\partial r} + I_k\right)\delta f(k,r,t) + \frac{e}{\hbar} \delta E(r,t) \cdot \frac{\partial}{\partial k} \bar{f}(k) = 0, \qquad (2.135)$$

where I_k is the linearized collision operator given by

$$I_k = \frac{eE}{\hbar} \cdot \frac{\partial}{\partial k} + I_k^{th} + I_k^{ee}(\bar{f}), \qquad (2.136)$$

with the linearized particle-particle (e−e) collision operator $I_k^{ee}(\bar{f})$ given by [2.80]

$$I_k^{ee}(\bar{f})\varphi = I_k^{ee}(\bar{f},\varphi) + I_k^{ee}(\varphi,\bar{f})$$

$$= \left(\frac{V_0}{(2\pi)^3}\right)^3 \int dk_1 \int dk' \int dk_1' \{W(k,k_1;k',k_1')$$

$$\cdot [\bar{f}(k)\varphi(k_1) + \bar{f}(k_1)\varphi(k)]$$

$$- W(k',k_1';k,k_1)[\bar{f}(k')\varphi(k_1') + \bar{f}(k_1')\varphi(k')]\}. \tag{2.137}$$

Thus the operator I_k^{ee} (and hence I_k) is a functional of the steady-state distribution $\bar{f}(k)$.

In general, by differentiating (2.128) with repsect to the total number of electrons N and accounting for the property of the linearized two-particle scattering operator given by (2.137), we obtain

$$I_k\left[\frac{\partial}{\partial N}\bar{f}(k)\right] = 0, \tag{2.138}$$

where we use the distribution function normalized to the number of particles N; therefore we obtain

$$\frac{V_0}{4\pi^3}\int \bar{f}(k)dk = N; \qquad \frac{V_0}{4\pi^3}\int \frac{\partial}{\partial N}\bar{f}(k)dk = 1. \tag{2.139}$$

Let us point out that the function $\partial(\bar{f}(k))/\partial N$ coincides with $(1/N)\bar{f}(k)$ only in the equilibrium state or when interelectronic collisions can be neglected.

Poisson's equation can be incorporated by representing the perturbing field $\delta E(r,t)$ in the form of a sum of two terms: the external field δE^{ext} and the self-consistent field δE^{scf}, which arises due to the redistribution of the charges in the semiconductor. Thus we obtain

$$\text{div } \delta E^{scf}(r,t) \equiv iq\delta E^{scf}(r,t) = \frac{e}{\varepsilon\varepsilon_0}\delta n(r,t), \tag{2.140}$$

where $\delta n(r,t)$ is the fluctuation of concentration given by

$$\delta n(r,t) = \delta n_{q\omega}\exp(-i\omega t + iq\cdot r) = \frac{1}{4\pi^3}\int \delta f(k,r,t)dk. \tag{2.141}$$

By substituting (2.133, 134, 140) in (2.135), we find

$$(-i\omega + iq\cdot v + I_k)\delta f_{q\omega}(k) - i\frac{e^2}{\hbar\varepsilon\varepsilon_0 q^2}\delta n_{q\omega}q\cdot\frac{\partial\bar{f}(k)}{\partial k}$$

$$+ \frac{e}{\hbar}\delta E_{q\omega}^{ext}\cdot\frac{\partial\bar{f}(k)}{\partial k} = 0. \tag{2.142}$$

Equation (2.142) is the kinetic equation for the space-time Fourier transform of the perturbed distribution function. By setting $\delta E_{q\omega}^{ext} = 0$, (2.142) will describe the spontaneous fluctuations of the system around the steady state. This case can be treated by using the Langevin method; to this end we shall add on the right-hand side of (2.142) the random term $y_{q\omega}(k)$, which satisfies the two following properties:

i) it has an average value equal to zero:

$$\langle y_{q\omega}(k) \rangle = 0, \tag{2.143}$$

ii) its spectral density according to [2.25] is given by

$$[y(k)y(k_1)]_{q\omega} = (I_k + I_{k_1})\bar{f}(k)\delta_{k,k_1} - I_{kk_1}^{ee}(\bar{f},\bar{f}), \tag{2.144}$$

where $[\ ]_{q\omega}$ stands for the space-time Fourier transform of the two-particle (with wavevector k and k_1, respectively) correlator $\langle y(k,r,t)y(k_1,r,t+\tau) \rangle$, and $I_{kk_1}^{ee}(\bar{f},\bar{f})$ is the two-particle collision term without including one summation given by

$$I_{kk_1}^{ee}(\bar{f},\bar{f}) = \left(\frac{V_0}{(2\pi)^3}\right)^2 \int dk' \int dk'_1 [W(k,k_1;k',k'_1)\bar{f}(k)\bar{f}(k_1)$$

$$- W(k',k'_1;k,k_1)\bar{f}(k')\bar{f}(k'_1)]. \tag{2.145}$$

When electron-electron interaction is neglected, only the first term in the right-hand side of (2.145) survives.

The Boltzmann-Langevin equation in $q-\omega$ space is thus obtained as

$$(-i\omega + iq \cdot v + I_k)\delta f_{q\omega}(k) - i\frac{e^2}{\hbar\varepsilon\varepsilon_0 q^2}\delta n_{q\omega}q \cdot \frac{\partial \bar{f}(k)}{\partial k} = y_{q\omega}(k). \tag{2.146}$$

By integrating (2.146) over k and recalling the zero-sum property, see (2.152), of the random term, we obtain the continuity equation in $q-\omega$ space:

$$i\omega\delta n_{q\omega} + \frac{i}{e}q \cdot \delta j_{q\omega} = 0, \tag{2.147}$$

where the space-time Fourier transform of the density-current fluctuation $\delta j_{q\omega}$ is given by

$$\delta j_{q\omega} = \frac{e}{4\pi^3}\int v\delta f_{q\omega}(k)dk. \tag{2.148}$$

By defining the external random-current density related to the random term $y_{q\omega}(k)$ as

$$g_{q\omega} = \frac{e}{4\pi^3} \int v I_k^{-1} y_{q\omega}(k) dk, \qquad (2.149)$$

Gantsevitch et al. [2.25] demonstrated the following theorem:

$$(\delta j_\alpha \delta j_\beta)_{\omega \to 0} = (g_\alpha g_\beta)_{q\omega}. \qquad (2.150)$$

Equation (2.150) reveals the identity of the spectral density of local random-current density g, which provokes the spatially inhomogeneous fluctuations, with the spectral density of the spatially homogeneous density-current fluctuations δj in the limit $\omega \to 0$.

2.6.3 Definitions of the Kinetic Coefficients from the Low-Frequency and Long-Range Fluctuations

In order to obtain $\delta f_{q\omega}(k)$ and $\delta j_{q\omega}$ from (2.146, 148), we use the Chapman-Enskog procedure [2.81], taking advantage of the inequalities $\omega\tau \ll 1$, $ql \ll 1$. Here we recall that $1/\tau$ gives the order of magnitude of the linearized collision operator I_k.

The idea of the Chapman-Enskog method is to reduce the solution of the space- and time-dependent Boltzmann equation to the solution of a number of differential equations with the coefficients expressed through the solutions of some Boltzmann equations of the type of space- and time-independent response. These last equations have the general form

$$I_k z(k) = x(k), \qquad (2.151)$$

I_k being a linear operator.

To manage (2.146) we need now a proper definition of an inverse operator I_k^{-1}. The collisions and the electric field conserve the number of electrons, redistributing them in k space; thus, while acting on a distribution function $z(k)$, the operator I_k generates a function with a zero-sum property

$$\int I_k z(k) dk = 0. \qquad (2.152)$$

Therefore, from (2.151) we should require

$$\int x(k) dk = 0. \qquad (2.153)$$

Since the function $\partial/\partial N(\bar{f}(k))$ is the solution of the homogeneous equation (2.138), the solution of (2.151) is not unique; in fact, if $z^0(k)$ is a particular solution of (2.151), then $z^1(k) = z^0(k) + \text{const} \times \partial/\partial N(\bar{f}(k))$ is also a solution. Thus, for the present, an inverse operator I_k^{-1} is not defined uniquely. To define it uniquely, the following is required:

$$\int I_k^{-1} x(k) dk = 0,\qquad (2.154)$$

that is, when the particular solution $z^1(k)$ obeying the zero-sum property is named as $I_k^{-1} x(k)$. Such a solution can always be constructed from any particular solution $z^0(k)$ by choosing $\text{const} = -(V_0/4\pi^3) \int z^0(k) dk$. Thus the operators I_k and I_k^{-1} generate the functions with the zero-sum property, and the operator I_k^{-1} is allowed to act only on the functions with the same property. It is convenient to require the operator I_k to act only on functions having the zero-sum property. This requirement can always be fulfilled because of the identity

$$I_k z(k) = I_k\left[z(k) - \frac{V_0}{4\pi^3} \int z(k') dk' \frac{\partial}{\partial N} \bar{f}(k) \right].\qquad (2.155)$$

According to this convention, we should rewrite (2.151) as follows:

$$I_k\left[z(k) - \frac{V_0}{4\pi^3} \int z(k') dk' \frac{\partial}{\partial N} \bar{f}(k) \right] = x(k).\qquad (2.156)$$

Now the general solution of (2.151) can be obtained from (2.156), after multiplying by I_k^{-1}, as

$$z(k) = \frac{V_0}{4\pi^3} \int z(k') dk' \frac{\partial}{\partial N} \bar{f}(k) + I_k^{-1} x(k).\qquad (2.157)$$

At this stage we are ready to treat the Boltzmann-Langevin equation (2.146). Multiplying the continuity equation (2.147) by $V_0(\partial/\partial N)\bar{f}(k)$ and subtracting the result from (2.146), we obtain

$$-i\omega\left[\delta f_{q\omega}(k) - V_0 \delta n_{q\omega} \frac{\partial \bar{f}(k)}{\partial N} \right] + iq\left[v\delta f_{q\omega}(k) - \frac{V_0}{e} \delta j_{q\omega} \frac{\partial}{\partial N} \bar{f}(k) \right]$$

$$-i\frac{e^2}{\hbar\varepsilon\varepsilon_0 q^2}\, \delta n_{q\omega} q \cdot \frac{\partial f(k)}{\partial k} + I_k \delta f_{q\omega}(k) = y_{q\omega}(k).\qquad (2.158)$$

By rewriting the last term on the left-hand side of the above equation in accordance with the convention of (2.155) and multiplying by I_k^{-1}, we obtain

$$\delta f_{q\omega}(\boldsymbol{k}) = V_0 \delta n_{q\omega} \frac{\partial}{\partial N} \bar{f}(\boldsymbol{k}) + \mathrm{i}\, \frac{e^2}{\hbar \varepsilon \varepsilon_0 q^2}\, \delta n_{q\omega} q I_k^{-1} \frac{\partial \bar{f}}{\partial \boldsymbol{k}}(\boldsymbol{k})$$

$$- \mathrm{i} \boldsymbol{q} I_k^{-1} \left[v \delta f_{q\omega}(\boldsymbol{k}) - \frac{V_0}{e}\, \delta j_{q\omega} \frac{\partial \bar{f}(\boldsymbol{k})}{\partial N} \right]$$

$$+ \mathrm{i} \omega I_k^{-1} \left[\delta f_{q\omega}(\boldsymbol{k}) - V_0 \delta n_{q\omega} \frac{\partial \bar{f}}{\partial N}(\boldsymbol{k}) \right] + I_k^{-1} y_{q\omega}(\boldsymbol{k}). \qquad (2.159)$$

All expressions under the sign I_k^{-1} in (2.159) are functions with the zero-sum property. This allows us to treat them separately, and is the reason for our subtracting procedure. Multiplying (2.159) by $e v_\alpha/(4\pi^3)$, we obtain, after integration over $\boldsymbol{k}$,

$$\delta j_{q\omega,\alpha} = e v'_{d,\alpha} \delta n_{q\omega} + \sigma'_{\alpha\beta} \xi_\beta - \mathrm{i}\, \frac{e}{4\pi^3}\, q_\beta \int d\boldsymbol{k}\, v_\alpha I_k^{-1} \left[v_\beta \delta f_{q\omega}(\boldsymbol{k}) \right.$$

$$\left. - \frac{V_0}{e}\, \delta j_{q\omega,\beta} \frac{\partial}{\partial N} \bar{f}(\boldsymbol{k}) \right] + \mathrm{i}\, \frac{e}{4\pi^3}\, \omega \int d\boldsymbol{k}\, v_\alpha I_k^{-1} \left[\delta f_{q\omega}(\boldsymbol{k}) \right.$$

$$\left. - V_0 \delta n_{q\omega} \frac{\partial \bar{f}(\boldsymbol{k})}{\partial N} \right] + g_{q\omega,\alpha}. \qquad (2.160)$$

where $v'_{d,\alpha}$ is the differential (with respect to the number of carriers) drift velocity given by

$$v'_{d,\alpha} = \frac{V_0}{4\pi^3} \int v_\alpha \frac{\partial \bar{f}(\boldsymbol{k})}{\partial N}\, d\boldsymbol{k}. \qquad (2.161)$$

Here $\sigma'_{\alpha\beta} \xi_\beta$ is the current density caused by a fluctuational field ξ_β given by

$$\xi_\beta = -\mathrm{i} q_\beta \frac{e}{\varepsilon \varepsilon_0 q^2}\, \delta n_{q\omega}, \qquad (2.162)$$

$\sigma'_{\alpha\beta}$ being the differential conductivity given by

$$\sigma'_{\alpha\beta} = \frac{e^2}{4\pi^3 \hbar} \int v_\alpha I_k^{-1} \frac{\partial}{\partial k_\beta} \bar{f}(\boldsymbol{k})\, d\boldsymbol{k}. \qquad (2.163)$$

Here $g_{q\omega,\alpha}$ is the external random density current given by (2.149).

Equations (2.159, 160) are suitable for the Chapman-Enskog – like iteration. The repeated substitution of (2.159) into (2.160) generates the power series in ω and q for the current density $\delta j_{q\omega}$. After the first iteration [i.e., by substituting $\delta f_{q\omega}(\boldsymbol{k}) \simeq V_0 \delta n_{q\omega} \partial f(\boldsymbol{k})/\partial N$ and $\delta j_{q\omega} = e v'_d \delta n_{q\omega}$], the third term in (2.158) gives the diffusion density current

$$ie\,\frac{V_0}{4\pi^3}\,q_\beta \int dk\,v_\alpha I_k^{-1}(v_\beta - v_{\mathrm{d},\beta}')\,\frac{\partial}{\partial N}\,\bar{f}(k)\delta n_{q\omega}=ieq_\beta D_{\alpha\beta}\delta n_{q\omega}, \tag{2.164}$$

with the diffusion constant $D_{\alpha\beta}$, which accounts for carrier-carrier scattering, defined by

$$D_{\alpha\beta}=\frac{V_0}{4\pi^3}\int dk\,v_\alpha I_k^{-1}(v_\beta - v_{\mathrm{d},\beta}')\,\frac{\partial}{\partial N}\,\bar{f}(k), \tag{2.165}$$

while the fourth term vanishes.

Thus, as first approximation in the parameters $\omega\tau$ and ql for the fluctuating density current, we have

$$\delta j_{q\omega,\alpha}=ev_{\mathrm{d},\alpha}'\delta n_{q\omega}+\sigma_{\alpha\beta}\xi_\beta-ieq_\beta D_{\alpha\beta}\delta n_{q\omega}+g_{q\omega,\alpha}. \tag{2.166}$$

According to (2.144, 150), the spectral density for the random density current $g_{q\omega}$ is given by

$$(g_\alpha g_\beta)_{q\omega}=\left(\frac{e}{4\pi^3}\right)^2\int dk \int dk_1\,v_\alpha v_{1\beta} I_k^{-1} I_{k_1}^{-1}[(I_k$$
$$+I_{k_1})\bar{f}(k)\delta_{k,k_1}-I_{kk_1}^{ee}(\bar{f},\bar{f})]. \tag{2.167}$$

To make use of the identity $I_k^{-1}I_k=1$ by accounting for (2.138), we rewrite (2.167) as follows:

$$(g_\alpha g_\beta)_{q\omega}=\left(\frac{e}{4\pi^3}\right)^2\int dk \int dk_1\,v_\alpha v_{1\beta} I_k^{-1} I_{k_1}^{-1}\Bigg\{(I_k+I_{k_1})N\,\frac{\partial}{\partial N}\,\bar{f}(k_1)$$
$$\cdot\left[\delta_{k,k_1}-\frac{\partial}{\partial N}\,\bar{f}(k_1)\right]+\left[(I_k+I_{k_1})\left(\bar{f}(k)-N\,\frac{\partial}{\partial N}\,\bar{f}(k_1)\right)\delta_{k,k_1}\right.$$
$$-I_{kk_1}^{ee}(\bar{f},\bar{f})\Bigg]\Bigg\}. \tag{2.168}$$

Now the first term in curly brackets can be written as

$$\int dk \int dk_1\,v_\alpha v_{1\beta} I_k^{-1} I_{k_1}^{-1}(I_k+I_{k_1})N\,\frac{\partial}{\partial N}\,\bar{f}(k)\left[\delta_{k,k_1}-\frac{\partial}{\partial N}\,\bar{f}(k_1)\right]$$
$$=\frac{4\pi^3}{V_0}\left\{N\int dk\,v_\alpha I_k^{-1}(v_\beta-v_{\mathrm{d},\beta}')\,\frac{\partial}{\partial N}\,\bar{f}(k)+N\int dk_1\,v_{1\beta}I_{k_1}^{-1}(v_{1\alpha}-v_{\mathrm{d},\alpha}')\,\frac{\partial}{\partial N}\,\bar{f}(k_1)\right\}$$
$$=\left(\frac{4\pi^3}{V_0}\right)^2 N(D_{\alpha\beta}+D_{\beta\alpha}); \tag{2.169}$$

and the second term in curly brackets of (2.168) can be written as

$$\int dk \int dk_1 v_\alpha v_{1\beta} I_k^{-1} I_{k_1}^{-1} \left\{ I_{kk_1}^{ee}(\bar{f},\bar{f}) - (I_k + I_{k_1}) \left[\bar{f}(k) - N \frac{\partial}{\partial N} \bar{f}(k_1) \right] \delta_{k,k_1} \right\}$$

$$= \left(\frac{4\pi^3}{V_0} \right)^2 N \Theta_{\alpha\beta}.$$

(2.170)

Making use of the relation (2.148), (2.167–170) finally give

$$(g_\alpha g_\beta)_{q\omega} \equiv (\delta j_\alpha \delta j_\beta)_{\omega \to 0} = \frac{e^2}{V_0^2} N(D_{\alpha\beta} + D_{\beta\alpha} - \Theta_{\alpha\beta}).$$

(2.171)

Thus from (2.171) it is seen that no simple relation exists between diffusion and noise spectral density when carrier-carrier collisions are effective; the term $\Theta_{\alpha\beta}$ given by (2.170) explicitly accounts for these collisions. Calculations of $\Theta_{\alpha\beta}$ were performed under an electron temperature model [in which case the difference between $\bar{f}(k)$ and $N(\partial/\partial N)\bar{f}(k)$ can be neglected] [2.82,83]. Results show that positive as well as negative values of $\Theta_{\alpha\beta}$ are possible according to the different dependence on electron energy of the energy and momentum relaxation times.

When electron-electron collisions are negligible [i.e., $I_{kk_1}^{ee}(\bar{f},\bar{f}) = 0$, $N(\partial\bar{f}/\partial N) = \bar{f}$], and/or they are so determinant that the distribution function is a displaced Maxwellian one [1.84], then $\Theta_{\alpha\beta} = 0$ and we obtain

$$(\delta j_\alpha \delta j_\beta)_{\omega \to 0} = \frac{e^2}{V_0^2} N(D_{\alpha\beta} + D_{\beta\alpha}).$$

(2.172)

Equation (2.172) is equivalent to Einstein's generalized relation (2.121), since for the simple geometry considered, the current spectral density $S_I(0) = 2A^2(\delta j \delta j)_{\omega \to 0}$, A being the cross-sectional area of the device. In this case, according to (2.35), the diffusion tensor is given by

$$D_{\alpha\beta} = \frac{V_0}{4\pi^3} \frac{1}{N} \int dk \, v_\alpha I_k^{-1} (v_\beta - v_{d,\beta}) \bar{f}(k),$$

(2.173)

where

$$v_d = \frac{V_0}{4\pi^3} \frac{1}{N} \int dk \, v \bar{f}(k).$$

(2.174)

In thermal equilibrium we always have $\Theta_{\alpha\beta} = 0$, since $f^{eq}(k) = N(\partial/\partial N)\bar{f}^{eq}(k)$ and $I_{kk_1}^{ee}(f^{eq}, f^{eq}) = 0$. Moreover, in equilibrium $v_d = 0$ and $v f^{eq}(k) = -(K_B T_0/\hbar)$

$(\partial/\partial \boldsymbol{k}) f^{eq}(\boldsymbol{k})$; thus recalling (2.163), we get

$$(\delta j_\alpha \delta j_\beta)^{eq}_{\omega \to 0} = \frac{2e^2}{V_0^2} \, ND^{eq}_{\alpha\beta} = 2 \, \frac{K_B T_0}{V_0} \, \sigma^{eq}_{\alpha\beta}, \tag{2.175}$$

in accordance with the Nyquist and Einstein relations.

2.7 The Model Semiconductor

2.7.1 Band Structure

The energy region of the band structure of a semiconductor which is of interest in high-field transport is centered around the energy gap and extends some $\mathscr{E}_g$ (the width of the energy gap) above the conduction-band minimum and below the valence-band maximum. In the following we shall present a general model for the band structure which enables us to interpret the macroscopic properties for the whole class of cubic semiconductors belonging to both diamond and zinc-blende structures.

The model consists of one conduction band, with three sets of minima, and three valence bands. The minima of the conduction band are located at the Γ point $(\boldsymbol{k}=0)$; at the L points $(\boldsymbol{k}=(\pi/a_0, \ \pi/a_0, \ \pi/a_0)$, a_0 being the lattice parameter); and along Δ lines $(\boldsymbol{k}=k, 0, 0)$. The top of the valence band is located at Γ (for the case of zinc-blende structures, owing to their smallness linear $\boldsymbol{k}$ terms arising from the lack of inversion symmetry are neglected [1.85]). Two of these bands are degenerate at this point, while the third one is split off by spin-orbit interaction.

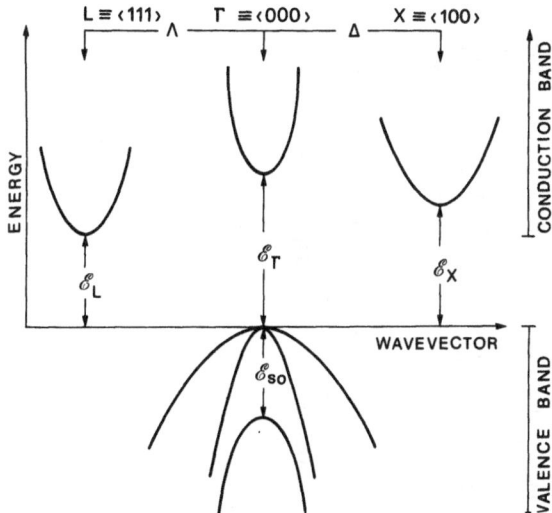

Fig. 2.12. Band structure of the cubic-model semiconductor

Table 2.2. Band gaps and mass parameters of some cubic semiconductors. Temperature dependences are neglected. Unless otherwise noted all values are taken from [2.86]

	$\mathscr{E}_\Gamma$ [eV]	$\mathscr{E}_L$ [eV]	$\mathscr{E}_\Delta$ [eV]	$\mathscr{E}_{so}$ [eV]	m_{lo} [m_0]	m [m_0]	m_{tr} [m_0]	α [eV^{-1}]	$|A|$	$|B|$	$|C|$	$m_h(\mathscr{E}/\mathscr{E}_{so}\ll1)$ [m_0]	$m_h(\mathscr{E}/\mathscr{E}_{so}\gg1)$ [m_0]	$m_{li}(\mathscr{E}/\mathscr{E}_{so}\ll1)$ [m_0]	$m_{li}(\mathscr{E}/\mathscr{E}_{so}\gg1)$ [m_0]
C	11.67[a]	12.67[a]	5.45[a]	0.006[b]	1.4[b]		0.36[b]		3.61[b]	0.18[b]	3.76[b]	0.53[b]	1.08[b]	0.19[b]	0.36[b]
Si	4.08	1.87	1.13	0.044	0.98		0.19	0.5	4.22	0.78	4.80	0.53[b]	1.26[b]	0.16[b]	0.36[b]
Ge	0.89	0.76	0.96	0.29	1.64		0.082	0.65	13.35	8.50	13.11	0.35[b]	0.73[b]	0.04[b]	0.25[b]
AlP	3.3	3.0	2.1	0.05					3.47	0.12	3.98	0.62[c]	2.28[c]	0.19[c]	0.38[c]
AlAs	2.95	2.67	2.16	0.28	2.0		0.23		4.04	1.56	4.71	0.75[c]	2.19[c]	0.15[c]	0.50[c]
AlSb	2.5	2.39	1.6	0.75	1.64				4.15	2.02	4.95	0.94[d]		0.14[d]	
GaP	2.7[c]	2.7[c]	2.2[c]	0.08	1.12		0.22		4.20	1.96	4.65	0.79[c]	1.78[c]	0.14[c]	0.55[c]
GaAs	1.42	1.71	1.90	0.34		0.067		0.64	7.65	4.82	7.71	0.62[c]	1.47[c]	0.07[c]	0.43[c]
GaSb	0.67	1.07	1.30	0.77		0.045		1.36	11.80	8.06	11.71	0.49[d]		0.046[d]	
InP	1.26	2.0	2.3	0.13		0.080		0.67	6.28	4.16	6.35	0.86[c]	2.56[c]	0.09[c]	0.59[c]
InAs	0.35	1.45	2.14	0.38		0.023		2.73	19.67	16.74	13.96	0.60[c]	1.57[c]	0.02[c]	0.42[c]
InSb	0.23[f]	0.98[f]	0.73[f]	0.81		0.014		5.72	35.08	31.28	22.27	0.47[d]		0.015[d]	
ZnS	3.8	5.3	5.2	0.07		0.28		0.14	2.54	1.50	2.75	1.80[c]	5.23[c]	0.22[c]	1.21[c]
ZnSe	2.9	4.5	4.5	0.43		0.14		0.26	3.77	2.48	3.87	1.47[c]	5.17[c]	0.15[c]	0.98[c]
ZnTe	2.56	3.64	4.26	0.92		0.18		0.26	3.74	2.14	4.30	0.63[g]		0.15[g]	
CdTe	1.80	3.40	4.32	0.91		0.096		0.45	5.29	3.78	5.46	0.7[g]		0.10[g]	

[a] [2.87]; [b] [2.88]; [c] [2.89]; [d] [2.90]; [e] [2.91]; [f] [2.92]; [g] [2.93]

The main features of the band structure in cubic semiconductors are summarized in Fig. 2.12. Table 2.2 reports the band parameters for most materials of interest.

2.7.2 Energy-Wavevector Relationship for Parabolic Bands

The particular form of the energy-wavevector relationship $\mathscr{E} = \mathscr{E}(k)$ of charge carriers determines their dynamical properties under the influence of an external force. In the following we shall explicitly refer to electrons or holes when we consider k states belonging to the conduction or valence bands, respectively.

In the region around the minima of the conduction band, usually called valley, or around the maximum of the valence band, the function $\mathscr{E}(k)$ is given by a quadratic expression of k (parabolic bands), which may assume one of these different forms:

$$\mathscr{E}(k) = \frac{\hbar^2 k^2}{2m}, \tag{2.176}$$

$$\mathscr{E}(k) = \frac{\hbar^2}{2} \left(\frac{k_{lo}^2}{m_{lo}} + \frac{k_{tr}^2}{m_{tr}} \right), \tag{2.177}$$

$$\mathscr{E}(k) = ak^2 [1 \mp g(\vartheta, \psi)]. \tag{2.178}$$

When (2.176–178) are used for electrons, k is measured from the center of the valleys.

Equation (2.176) (spherical case) represents a band with spherical equienergetic surfaces with a single scalar effective mass m, and it is appropriate for the minimum of the conduction band located at Γ and for the maximum of the split-off valence band. This case is the simplest one, and it is generally adopted as a "simple model" for any material when a rough estimate of transport properties is sought.

Equation (2.177) (ellipsoidal case) represents a band with ellipsoidal equienergetic surfaces with a tensor effective mass ($1/m_{lo}$ and $1/m_{tr}$ are the longitudinal and transverse components of the inverse effective-mass tensor). The ellipsoids have rotational symmetry around the crystallographic directions which contain the center of the valleys. This case is appropriate for the minima of the conduction band located at L and along Δ; for symmetry reasons several equivalent valleys are present (many-valley model – in this model it is assumed that electrons cannot move from one valley to another with continuous variation of their k values, because of the existence of intermediate regions of k space with energies which are too high).

Equation (2.178) (warped case) represents a band with a warped equienergetic surface, and it is appropriate for the two degenerate maxima of the valence band (here $\mp$ refers to heavy and light holes, respectively). Moreover, $g(\vartheta, \psi)$, ϑ and ψ are the polar and azimuthal angles of k with respect to the crystallographic axes, contains the angular dependence of the effective mass; and is given by [2.94]

$$g(\vartheta,\psi)=[b^2+c^2(\sin^4\vartheta\cos^2\psi\sin^2\psi+\sin^2\vartheta\cos^2\vartheta)]^{1/2}, \qquad (2.179)$$

with

$$a=\frac{\hbar^2|A|}{2m_0}, \qquad b=\frac{|B|}{|A|}, \qquad c=\frac{|C|}{|A|}, \qquad (2.180)$$

where A, B, and C are the inverse valence-band parameters [2.95], and m_0 the free-electron mass.

The different shapes of the surfaces of constant energy for the three cases considered above are shown in Fig. 2.13.

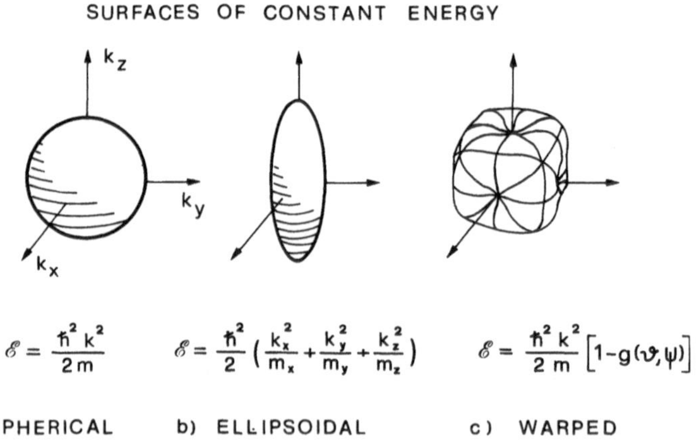

SURFACES OF CONSTANT ENERGY

$$\mathcal{E}=\frac{\hbar^2 k^2}{2m} \qquad \mathcal{E}=\frac{\hbar^2}{2}\left(\frac{k_x^2}{m_x}+\frac{k_y^2}{m_y}+\frac{k_z^2}{m_z}\right) \qquad \mathcal{E}=\frac{\hbar^2 k^2}{2m}\left[1-g(\vartheta,\psi)\right]$$

a) SPHERICAL   b) ELLIPSOIDAL   c) WARPED

Fig. 2.13a–c. Different shapes of surfaces of constant energy for electrons and holes in cubic semiconductors

2.7.3 Nonparabolicity

For values of k far from the minima of the conduction band and/or from the maxima of the valence band, the energy deviates from the simple quadratic expressions seen above, and nonparabolicity occurs.

For the conduction band, a simple analytical way of introducing nonparabolicity is to consider an energy-wavevector relationship of the type [2.96]

$$\mathcal{E}(1+\alpha\mathcal{E})=\frac{\hbar^2 k^2}{2m}, \qquad (2.181)$$

where the right-hand side of (2.181) can be substituted by one of the right-hand sides of (2.176, 177); α is a nonparabolic parameter, which can be related to other band quantities. In particular, the following approximate expressions have been given for minima at Γ [2.46], L [2.97], and Δ [2.43]:

$$\alpha(\Gamma) = \frac{1}{\mathscr{E}_\Gamma} \left(1 - \frac{m_\Gamma}{m_0}\right)^2, \tag{2.182}$$

$$\alpha(L) = \frac{1}{\mathscr{E}_{L_{3v}'} - \mathscr{E}_{L_{1c}}}, \tag{2.183}$$

$$\alpha(\Delta) = \frac{1}{2(\mathscr{E}_{\Delta_{2c}'} - \mathscr{E}_{\Delta_{1c}})} \left(1 - \frac{m_{lo}}{m_0}\right)^2, \tag{2.184}$$

where m_Γ is the value of the effective mass at the bottom of the Γ valley; $\mathscr{E}_\Gamma$ is the direct energy gap at Γ; L_{3v}' and L_{1c} are the states of valence and conduction bands with the given symmetry; and $\Delta_{2'c}$ and Δ_{1c} are states of the conduction band with the given symmetry for minima along Δ.

For the valence band, nonparabolicity cannot be parametrized in a form like that of (2.181). For this case, nonparabolicity exhibits two main features [2.98]:

i) it is more pronounced along the $\langle 110 \rangle$ and $\langle 111 \rangle$ directions for heavy and light holes, respectively;

ii) if $\mathscr{E}_{so}$ is the split-off energy of the lowest valence band, nonparabolicity is mostly effective at energies near $(1/3)\mathscr{E}_{so}$; in the limits of $\mathscr{E}/\mathscr{E}_{so} \ll 1$ and $\mathscr{E}/\mathscr{E}_{so} \gg 1$, the bands are parabolic.

For a nonparabolic band of the type given in (2.181) the velocity associated with a state k proves to be

$$v(k) = \frac{1}{\hbar} \frac{\partial \mathscr{E}}{\partial k} = \frac{\hbar k}{m(1 + 2\alpha \mathscr{E})}, \tag{2.185}$$

so that the conductivity effective mass m_c, defined by

$$v = \frac{\hbar k}{m_c} \tag{2.186}$$

is given by

$$m_c = m(1 + 2\alpha \mathscr{E}). \tag{2.187}$$

The effect of nonparabolicity on the density-of-states effectives mass m_d can be calculated following [2.99] as

$$m_{de}^{3/2} = (m_{tr}^2 m_{lo})^{1/2} (2\beta/\pi)^{1/2} \exp(\beta) K_2(\beta), \tag{2.188}$$

$$m_{di}^{3/2} = \pi^{-3/2} \int \exp \left[\mathcal{E}_i^v(x)/(K_B T_0) \right] dx, \tag{2.189}$$

$i \equiv h, li, so$, where suffixes e and i refer to conduction and valence bands, respectively, $\beta = (2\alpha K_B T_0)^{-1}$; K_2 is the modified Bessel function of order 2, $x = \hbar k/(2m_0 K_B T_0)^{1/2}$; and $\mathcal{E}_i^v$ taken with its sign is the energy wavevector relationship of the heavy $(i = h)$, light $(i = li)$ and split-off $(i = so)$ valence bands, respectively (according to *Kane* [2.98], to a good approximation, $\mathcal{E}_i^v$ can be evaluated from the $k \cdot p$ calculation; see also *Humphreys* [2.100]). The integration is carried out over the whole k space.

The temperature dependence of the density-of-states effective mass is sketched in Fig. 2.14 for the case of electrons and holes. From this figure it can be seen how nonparabolicity effects lead to an unlimited increase of the effective mass for the case of electrons, while for the case of all three types of holes the increase (or decrease for the split-off band) of the effective mass is confined within a well-defined region of energy.

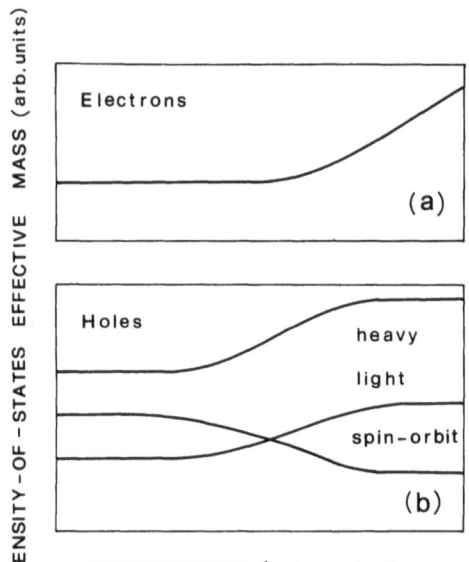

Fig. 2.14a, b. Schematic representation of the dependence upon temperature of the carrier density-of-states effective mass: (a) electrons, (b) holes

2.7.4 The Herring and Vogt Transformation

When considering the ellipsoidal case of (2.177), in order to simplify analytical calculations, it is useful to introduce the Herring and Vogt transformation [2.101], which reduces the ellipsoidal equienergetic surfaces to spheres, and is defined by

$$k_i^{*(m)} = T_{ij} k_j^{(m)}, \tag{2.190}$$

where $k^{*(m)}$ is the transformed wavevector. For an electron in the mth valley, the transformation matrix T_{ij} takes the form

$$T^{(m)} = \begin{vmatrix} \left(\dfrac{m_0}{m_{\text{tr}}}\right)^{1/2} & 0 & 0 \\ 0 & \left(\dfrac{m_0}{m_{\text{tr}}}\right)^{1/2} & 0 \\ 0 & 0 & \left(\dfrac{m_0}{m_{\text{lo}}}\right)^{1/2} \end{vmatrix} \qquad (2.191)$$

in the valley frame of reference, i.e., in the frame positioned at the center of the valley, with the z axis along its symmetry axis. Consequently, the energy-wavevector relationship in the starred space becomes of spherical type:

$$\mathscr{E}(k) = \frac{\hbar^2}{2m_0} T_{ij} T_{il} k_j k_l = \frac{\hbar^2 k^{*2}}{2m_0}, \qquad (2.192)$$

and the volume element dk is modified into $dk^* = (m_0/m_d)^{3/2} dk$, where $m_d = (m_{\text{lo}} m_{\text{tr}}^2)^{1/3}$ is the density-of-states effective mass.

To preserve vector equations, the transformation in (2.190) must be applied to other vector quantities such as driving forces and phonon wavevectors. Thus the equation of motion for an electron under the influence of an external force F becomes

$$\frac{d}{dt}(\hbar k^*) = F^*. \qquad (2.193)$$

The electron velocity as a function of k^* is given by

$$v_i = \frac{\hbar}{m_0} T_{ij} k_j^*, \qquad (2.194)$$

which is again generalized to the nonparabolic case by simply substituting m_0 with the expression $m_0 (1 + 2\alpha\mathscr{E})$.

2.8 Scattering Mechanisms

In this section we shall analyze the scattering mechanisms which carriers undergo during their motion in the host crystal. Since the dynamics of the electronic interaction is considered to be independent of the applied field (for interesting exceptions see, for example, [2.102]), what follows holds in general for the transport theory and it is not limited to hot-electron phenomena. All scattering calculations presented here will be carried out with a first-order and time-dependent perturbation approach (golden rule); consequently, only two-body interactions will be analyzed.

2.8.1 Classification Scheme

The electronic transitions of interest in charge transport in semiconductors can be classified as intravalley transitions, when the initial and final states of the electron lie in the same valley, or intervalley transitions, when the initial and final states lie in different valleys. For the case of holes, transitions are correspondingly called intraband or interband.

The most important scattering sources which determine these transitions are

- phonons;
- defects;
- impact-ionization;
- carrier-carrier.

The interaction of phonons with charge carriers can be due to the deformation of the otherwise perfect crystal produced by the phonons, or to the electrostatic forces produced by the polarization waves which accompany the phonons. The first kind of mechanism, typical of covalent semiconductors, is called deformation-potential interaction for both acoustic and optical phonons. The electrostatic interaction, typical of polar materials, is called piezoelectric interaction for the case of acoustic phonons, and polar interaction for optical phonons. Often the attribute of piezoelectric or polar is given directly to the phonon involved. Thus we speak of piezoelectric phonons or polar optical

Table 2.3. Scattering mechanisms in cubic semiconductors

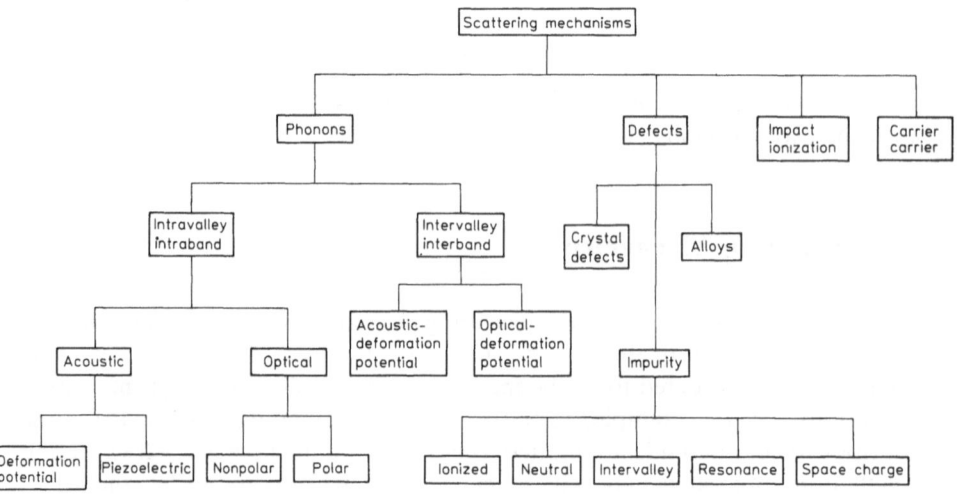

Table 2.4. Bulk constants of some cubic semiconductors

	ϱ_0 [gr/cm^3]	u_{lo} [10^5 cm/s]	u_{tr} [10^5 cm/s]	θ_{op} [K]	ε_0	ε_∞	a_0 [Å]	p [C/m^2]
C	3.51[a]	18.2[a]	12.3[a]	1940[a]	5.7[a]	—	3.57[a]	—
Si	2.33[b]	9.0[b]	5.3[b]	735[b]	11.7[b]	—	5.43[c]	—
Ge	5.32[b]	5.4[b]	3.2[b]	430[b]	16 [b]	—	5.66[c]	—
AlP	2.40[d]	8.0[e]	4.5[e]	725[d]	9.9[d]	7.6[d]	5.43[c]	—
AlAs	3.60[d]	6.3[e]	3.6[e]	550[d]	10.9[d]	8.5[d]	5.63[c]	—
AlSb	4.26[d]	5.0[e]	2.7[e]	493[d]	12.0[d]	10.2[d]	6.14[c]	—
GaP	4.13[d]	6.4[e]	3.7[e]	582[d]	11.1[d]	9.1[d]	5.45[c]	0.10[f]
GaAs	5.31[b]	5.2[b]	3.0[b]	418[b]	13.5[b]	11.6[d]	5.65[c]	0.160[g]
GaSb	5.61[d]	4.3[e]	2.5[e]	347[d]	15.7[d]	14.4[d]	6.10[c]	0.126[g]
InP	4.79[d]	5.1[e]	2.7[e]	498[d]	12.4[d]	9.5[d]	5.87[c]	0.035[g]
InAs	5.67[d]	4.2[e]	2.3[e]	350[d]	14.6[d]	11.8[d]	6.06[c]	0.045[g]
InSb	5.78[d]	3.7[e]	2.0[e]	284[d]	17.9[d]	15.7[d]	6.48[c]	0.071[g]
ZnS	4.08[c]	5.6[e]	2.9[e]	315[c]	8.32[h]	5.13[h]	5.41[c]	0.17[g]
ZnSe	5.42[c]	4.4[e]	2.4[e]	400[c]	8.33[h]	5.90[h]	5.67[c]	0.045[g]
ZnTe	5.72[c]	3.9[e]	2.0[e]	298[i]	9.7[i]	7.3[i]	6.10[c]	0.027[g]
CdTe	6.06[c]	3.3[e]	1.6[e]	250[i]	9.7[i]	7.2[i]	6.48[c]	0.034[g]

[a] [2.103]; [b] [2.86]; [c] [2.104]; [d] [2.90]; [e] Data are calculated from the elastic constants reported in [2.90, 105]; [f] [2.106]; [g] [2.107]; [h] [2.108]; [i] [2.93]

phonons as we speak of intervalley phonons to indicate phonons which induce intervalley electron transitions.

Defects can arise in the crystal owing to dislocations of different types and geometries, to impurities (neutral or ionized) due to extra atoms, and to the random distribution of component atoms among available lattice sites in alloys.

Impact ionization occurs when the carrier energy is approximately above the energy gap of the semiconductor under investigation. Related to this kind of scattering is the possibility of charge carrier multiplication, a phenomenon which leads to the electronic breakdown of the material.

Finally, carrier-carrier interaction is a very difficult type of scattering to include because, as seen in Sect. 2.6, it makes the Boltzmann equation nonlinear. As a general trend, the effect of this collision is, however, very limited under hot-electron conditions because of its electrostatic nature which results in a smaller efficiency at increasing carrier velocity. Table 2.3 summarizes the various scattering mechanisms treated here, and Table 2.4 reports the bulk constants of most of the cubic semiconductors under interest.

2.8.2 Transition Rates

By using the golden rule, the transition rate from the initial state k to a final state k' with total energies $\mathscr{E}$ and $\mathscr{E}'$, respectively, due to a perturbation with

Hamiltonian H', is given by

$$W(k, k') = \frac{2\pi}{\hbar} |H'|^2 \delta(\mathscr{E}' - \mathscr{E}). \tag{2.195}$$

The squared matrix element for nonumklapp processes can be factorized in the following way:

$$|H'|^2 = |V(q)|^2 \mathscr{G}(k, k'), \tag{2.196}$$

where $|V(q)|^2$ contains the dependence upon $q = k - k'$ of the squared Fourier transform of the interaction potential, and $\mathscr{G}$ is the overlap factor [2.109]

$$\mathscr{G}(k, k') = \left| \int_{\text{cell}} u_{k'}^*(r) u_k(r) dr \right|^2 \tag{2.197}$$

between the periodic part $u_k(r)$ of the Bloch wave functions of initial and final states.

2.8.3 The q-Dependence of the Transition Rates

The dependence of $V(q)$ upon the momentum transfer q is due to the nature of the scattering; Table 2.5 summarizes the expressions of $|V(q)|^2$ for the different scattering mechanisms of interest calculated for a simple spherical and parabolic band model without distinguishing electrons from holes. When the band model becomes more complex so that ellipsoidal or warped shapes of the equienergetic surfaces are considered and/or nonparabolicity occurs, to a good approximation $V(q)$ can still be taken from the simple model. The overlap factor and the use of the appropriate equation for $\mathscr{E} = \mathscr{E}(k)$ (Sect. 2.7) will take care of most of the band complexity.

In the following subsections we shall analyze the overlap factor for different types of bands and discuss some features of particular interest for each scattering mechanism.

2.8.4 Overlap Factor

The overlap factor in (2.196) is equal to one for exact plane waves or for wave functions formed with pure s states. When lower symmetries (for example, p states in our case of cubic semiconductors) are involved in the Bloch wave functions, an overlap factor less than one is obtained, which depends mainly upon the angle θ between the initial and final k and k' states.

In the many-valley case, owing to the large distance of the valley centers from the center of the Brillouin zone, the angle θ depends mostly on the valley involved in the transition, and $\mathscr{G}$ is thus almost constant for each type of (intra- or intervalley) scattering process [2.122]. The values for $\mathscr{G}$ in these cases may be included in the coupling constants.

In the case of a band extreme at Γ, analytical expressions for $\mathscr{G}$ can be given within a good approximation. For transitions of electrons in a central valley, $\mathscr{G}$ is strictly related to the nonparabolicity. Since both of them come from the mixture of p terms in the electron wave functions, it can be expressed as [2.46]

$$\mathscr{G}(k,k')=\frac{\{[1+\alpha\mathscr{E}(k)]^{1/2}[1+\alpha\mathscr{E}(k')]^{1/2}+\alpha[\mathscr{E}(k)\mathscr{E}(k')]^{1/2}\cos\theta\}^2}{[1+2\alpha\mathscr{E}(k)][1+2\alpha\mathscr{E}(k')]}. \qquad (2.198)$$

For transitions of holes within heavy or light bands, it can be expressed as [2.92]

$$\mathscr{G}(k,k')=\tfrac{1}{4}(1+3\cos^2\theta), \qquad (2.199)$$

while for interband transitions it can be expressed as

$$\mathscr{G}(k,k')=\tfrac{3}{4}\sin^2\theta. \qquad (2.200)$$

2.9 Phonon Scattering

2.9.1 Acoustic Scattering with Deformation-Potential Interaction

The squared matrix element used for acoustic scattering is given in Table 2.5. There E_1 is the acoustic-deformation-potential parameter, q the phonon wavevector, ϱ_0 the crystal density, and N_q the Bose-Einstein distribution function. The $\pm$ signs refer to absorption and emission processes, respectively. Also, u_{lo} is the longitudinal sound velocity since in the simple spherical case only longitudinal acoustic phonons interact with the electrons. If the same matrix element is to be used for ellipsoidal or warped shapes, u can be taken as an effective sound velocity $u=(1/3)u_{lo}+(2/3)u_{tr}$, which takes into account the possibility of the electrons interacting also with transverse acoustic phonons [2.90, 123].

For scattering of electrons from acoustic phonons via deformation potential, several approximations are usually made in the literature. These require some comments.

In a simple spherical and parabolic band, the energy and momentum conservations imply

$$k'=k\pm q, \qquad (2.201)$$

$$\mathscr{E}=\frac{\hbar^2 k_1^2}{2m}=\frac{\hbar^2 k^2}{2m}\pm\hbar q u_{lo}. \qquad (2.202)$$

Table 2.5. Squared matrix element $|V(q)|^2$ of different scattering mechanisms (for definition see text)

Acoustic-deformation potential [2.110]	$\dfrac{\hbar E_1^2 q}{2 V_0 \varrho_0 u_{lo}} \left(N_q + \dfrac{1}{2} \pm \dfrac{1}{2} \right)$
Acoustic piezoelectric [2.110]	$\dfrac{\hbar e^2 p^2}{2 V_0 (\varepsilon \varepsilon_0)^2 \varrho_0 u_{lo} q} \left(N_q + \dfrac{1}{2} \pm \dfrac{1}{2} \right)$
Nonpolar optical [2.110]	$\dfrac{\hbar^2 (D_t K)^2}{2 V_0 \varrho_0 K_B \theta_{op}} \left(N_q + \dfrac{1}{2} \pm \dfrac{1}{2} \right)$
Polar optical [2.110]	$\dfrac{2\pi e^2 K_B \theta_{op}}{V_0 \varepsilon q^2} \left(\dfrac{1}{\varepsilon_\infty} - \dfrac{1}{\varepsilon_0} \right) \left(N_q + \dfrac{1}{2} \pm \dfrac{1}{2} \right)$
Intervalley [2.110]	$\dfrac{\hbar^2 D_{jk}^2}{2 V_0 \varrho_0 K_B \theta_{jk}} \left(N_q + \dfrac{1}{2} \pm \dfrac{1}{2} \right)$
Edge dislocations [2.111, 112] (strain field)	$\dfrac{2\sigma_{li} a_s^2 (E_1^d)^2}{V_0} \left(\dfrac{1-2v}{1-v} \right)^2 \dfrac{\delta(q_z) \sin^2 \varphi}{q_\perp^2}$
Edge dislocations [2.113] (localized-line-charge)	$\dfrac{2\pi \sigma_{li}}{V_0} \left(\dfrac{eQ_{li}}{\varepsilon \varepsilon_0} \right)^2 \delta(q_z) \dfrac{1}{(q^2 + \lambda_D^{-2})^2}$
Ionized impurity [2.114]	$\dfrac{N_I Z_e^2 e^4}{V_0 (\varepsilon \varepsilon_0)^2 (q^2 + \lambda_D^{-2})^2}$
Neutral impurity [2.115]	$\dfrac{20\pi \hbar^4 N_N a_B^*}{V_0 m^2 k}$
Intervalley at impurities [2.116, 117]	$\dfrac{4\pi^2 N_I I^2 \mathscr{E}_I}{V_0 \mathscr{E}}$
Alloy scattering [2.118]	$\dfrac{3\pi^2 C_A (1 - C_A) (\Delta \mathscr{E}_A)^2}{16 N_A}$
Impact ionization [2.119]	$\dfrac{e^4}{V_0^2 (\varepsilon \varepsilon_0)^2 (|k_h - k_h'|^2 + \lambda_D^{-2})^2}$
Carrier-carrier [2.120, 121]	$\dfrac{e^4}{V_0^2 (\varepsilon \varepsilon_0)^2} \left(\dfrac{\delta_{k_1' + k_2'; k_1 + k_2}}{(k_1' - k_1)^2 + \lambda_D^{-2}} \right)^2 f(k_1) f(k_2)$

By substitution of (2.201) into (2.202) we can solve for q and obtain

$$q = \mp 2 (k \cos \theta - m u_{lo}/\hbar), \tag{2.203}$$

where θ is now the angle between k and q. The maximum value of q is obtained for absorption with backward scattering ($\cos \theta = -1$) and is given by

$$q_{max} = 2k + 2m u_{lo}/\hbar. \tag{2.204}$$

If the scattering mechanism were elastic, a backward collision would involve a $q_{max} = 2k$; thus the second term in (2.204) is the correction due to the energy of the phonon involved in the transition. The relative contribution of this term can be seen in (2.204) to be given by v/u_{lo}, where v is the actual velocity of the electron, which, in general, is much larger than u_{lo}. The maximum q involved in these transitions is therefore very close to $2k$; the corresponding maximum energy transfer is

$$\hbar q_{max} u_{lo} \approx 2\hbar k u_{lo} = 2mvu_{lo},$$ (2.205)

which, again, is in general much smaller than the electron energy $mv^2/2$.

Thus, very often, acoustic scattering is considered as an elastic process. However, we have to make some important observations in connection with this.

When ohmic transport is investigated by analytical means, the energy distribution function is assumed to be the equilibrium-Maxwellian distribution (for nondegenerate statistics), and no energy exchange of the electron with the heat bath is explicitly required. On the other hand, when hot-electron conditions are investigated, or a simulation analysis is undertaken, we need a mechanism which can exchange an infinitesimal amount of energy between electrons and the heat reservoir (the crystal): the role is physically played by the interaction with acoustic phonons. To consider this mechanism as elastic is therefore, in general, illegitimate.

When high temperatures or very high fields are considered, the average electron energy is larger than the optical-phonon energy, and this kind of phonon can assume the task of exchanging energy between the electrons and the crystal. In this case, the presence of the external field is essential for smearing out the energy of each single electron. In the absence of external fields, in fact, the electron energy would take on only its initial value or minus an integer number of optical-phonon energy quanta.

The above considerations are essential when the transport problem is solved with a Monte Carlo simulative technique (Sect. 2.3).

Another approximation usually made when dealing with acoustic-phonon scattering regards the phonon population:

$$N_q = \frac{1}{\exp\left[\hbar q u_{lo}/(K_B T_0)\right] - 1}.$$ (2.206)

This function is often approximated with the equipartition expression

$$N_q = \frac{K_B T_0}{\hbar q u_{lo}} - \frac{1}{2}.$$ (2.207)

[The term $-1/2$ in (2.207) represents the second term in a Laurent expansion of (2.206). For low values of the argument $x = \hbar q u_{lo}(K_B T_0)^{-1}$, the two-term formula gives a better approximation of the true N_q, but for large q it has the wrong limit. Neglecting the term $-1/2$, the strict one-term equipartition approximation tends to be zero for a large q.] The above expansion is valid when $\hbar q u_{lo} \ll K_B T_0$, i.e., when the thermal energy is much larger than the energy of the phonon involved in the transition. This approximation is not, therefore, independent of the elastic approximation discussed above. At very low temperatures and/or at very high fields this condition may break down, and again, when a precise balance of energy exchange of electrons with the heat bath via acoustic phonons is to be taken into account, the exact expression for N_q or a good approximation for it must be used [2.124].

From the condition $|\cos \theta| \leq 1$ and (2.203), the limits of variability of the phonon wavevector q in the transition are obtained. These limits are different whether the energy of the electron is below or above $m u_{lo}^2/2$, that is, whether the electron is faster or slower than sound. For slower electrons, there is no possibility of emitting acoustic phonons, and the possible values of absorbed q are included between

$$q_1 = 2(m u_{lo}/\hbar - k), \qquad q_2 = 2(m u_{lo}/\hbar + k). \tag{2.208}$$

For electrons faster than sound, there is no minimum q involved in the transitions, whereas its maximum value is

$$q_a = 2(k - m u_{lo}/\hbar), \qquad q_e = 2(k - m u_{lo}/\hbar) \tag{2.209}$$

for absorption and emission, respectively. However, unless very low temperatures ($\lesssim 1$ K) are considered, electrons slower than sound are so few that the above distinction is often neglected.

If the elastic and equipartition approximations are used, acoustic scattering becomes isotropic: any state k' in the energy-conserving sphere has the same probability of occurrence, independent of the angle formed with the initial state k. In this case at each scattering event the momentum of the electron is on average totally lost, and the momentum relaxation time coincides with the scattering time. We would like to emphasize, however, that for acoustic scattering this is true only in the elastic and equipartition approximations (and in spherical bands). If these approximations are relaxed, the electron keeps some memory of its previous momentum after the scattering process, and this effect becomes more and more pronounced as the temperature is lowered [2.125].

When from the simple spherical and parabolic case we move to more realistic band structures, several complications arise. The many-valley case, besides allowing the interaction with transverse as well as longitudinal sound waves, requires two deformation-potential parameters in treating acoustic-phonon scattering; they are usually called Ξ_d and Ξ_u. The full complexity of the theory was treated by *Herring* and *Vogt* [2.101], and *Gram* and *Joergensen* [2.126].

The valence band case also allows for the interaction with transverse as well as longitudinal sound waves. Usually a single deformation-potential parameter

E_1^0 is used [2.88] in place of the three parameters which are related to the warping of the equienergetic surfaces of holes [2.127, 128]. Furthermore, when the two degenerate bands at $k = 0$ are considered, much more complicated limits for the wavevectors of the phonons involved in the transitions are obtained [2.129].

Finally, corrections to the deformation-potential arising from quadrupole and octopole interactions have been considered [2.126, 130, 131].

2.9.2 Acoustic Scattering with Piezoelectric Interaction

The squared matrix element of acoustic piezoelectric scattering is given in Table 2.5. There p is the appropriate component of the piezoelectric tensor [2.110].

All the considerations in the previous section about the elastic and equipartition approximations in acoustic scattering and the variability of the phonon momentum involved in the transitions are based on energy and momentum conservations and are therefore independent of the type of interaction mechanism. Thus, they hold also for piezoelectric scattering.

It is worth noting that the electrostatic nature of this mechanism leads to a scattering efficency which decreases at increasing carrier energy, so that its importance under hot-carrier conditions is, in general, very limited, and this scattering should always be accompanied, in dealing with hot-electron problems, by scattering with acoustic phonons via deformation-potential interaction. Piezoelectric scattering has been considered in high-field transport, for example in [2.132, 133].

2.9.3 Optical-Phonon Scattering with Deformation-Potential Interaction

The perturbation Hamiltonian used for deformation-potential scattering from optical phonons is a generalization of that obtained (with a more sound theory) for acoustic phonons, and the squared matrix element calculated to zero order in the wavevector of the phonon is given in Table 2.5. There $D_t K$ (D_t being an interaction constant and K a reciprocal-lattice vector) is the optical deformation-potential and θ_{op} the equivalent optical phonon temperature which is assumed to be constant, since the dispersion curve of these phonons is usually quite flat for the q values involved in electronic intravalley transitions [2.110]. For the same reason N_q becomes q independent ($N_q = N_0$). No angular dependence is present, and the scattering is therefore isotropic.

When realistic materials are considered, one should account for the selection rules (Sect. 2.9.6). Furthermore, for the case of holes it has been shown that the scattering is still isotropic even when the symmetry properties of the hole wave functions are taken into account [2.128]. In this case the formalism makes use of an optical deformation potential d_0:

$$d_0 = a_0 (D_t K) \sqrt{2/3}, \tag{2.210a}$$

which can be microscopically calculated in terms of a band shift under strain [2.134–137].

Ferry [2.138] has calculated the matrix element to the first order in the phonon wavevector; this may become significant when the zero-order transition is forbidden by symmetry.

2.9.4 Optical-Phonon Scattering with Polar Interaction

The squared matrix element used for polar scattering from optical phonons is given in Table 2.5 [2.110], where ε_∞ is the relative high-frequency dielectric constant of the material.

To characterize the strength of the interaction one usually introduces the dimensionless Fröhlich coupling constant α_p defined by

$$\alpha_p = \frac{e^2}{\hbar\varepsilon}\left(\frac{m}{2K_B\theta_{op}}\right)^{1/2}\left(\frac{1}{\varepsilon_\infty}-\frac{1}{\varepsilon_0}\right). \tag{2.210b}$$

In general, if $\alpha_p < 1$ the electron-phonon coupling is considered weak, if $\alpha_p \simeq 5$ intermediate, and if $\alpha_p \geq 10$ strong.

The electrostatic nature of the interaction is such that forward scattering is favoured so that this mechanism is strongly anisotropic. The treatment of this scattering is again simplified by the constancy of the phonon energy in the transition. At high electron energy, the total scattering rate for polar optical scattering decreases with increasing energy due to the electrostatic nature of the interaction. This fact is responsible for the phenomenon called polar runaway [2.139].

In polar materials the lowest minimum in the conduction band is in general at the center of the zone. Thus polar scattering has not been treated in ellipsoidal valleys.

In the case of holes, the overlap factor plays an important role; in particular, for interband scattering it eliminates a divergency which would otherwise appear for vertical transitions [2.123].

2.9.5 Intervalley-Phonon Scattering

Electron transitions between states in two different equivalent valleys can be induced, as regards phonon scattering, by acoustic or optical modes. The possibility of intervalley scattering from impurities is discussed in Sect. 2.10.4. The phonon wavevector q involved in the transition is very close to the distance in the Brillouin zone between the minima of the initial and final valleys. Given these two valleys, q is almost constant, and therefore for a given branch of phonons, also the energy involved in the transition is approximately constant, as in the case of optical intravalley scattering. Thus intervalley scattering is usually treated, formally, in the same way as intravalley optical-phonon scattering with deformation-potential interaction [2.110]. The squared matrix element used for the scattering is reported in Table 2.5. There, D_{jk} is a suitable interaction constant for scattering from the jth valley to the kth induced by a phonon of energy $K_B \theta_{jk}$.

Polar or piezoelectric interaction is usually neglected for intervalley scattering since the large momentum transfer involved in these transitions makes its probability of occurrence very small.

2.9.6 Selection Rules

Intravalley optical and intervalley phonons are subject to selection rules when initial and final k states are along high symmetry directions (see, for example, the textbook of *Bir* and *Pikus* [2.85]). In dealing with cubic semiconductors, for electrons to zero order in the phonon wavevector, intravalley optical-phonon interaction is forbidden at Γ point and along $\langle 100 \rangle$ directions, while it is allowed along $\langle 111 \rangle$ directions [2.140, 97]. Similarly, for holes zero-order transitions are allowed at Γ [2.140, 127].

Concerning intervalley scattering, group-theoretical analysis including time-reversal symmetry [2.141, 142] shows that, in the case of valleys located along $\langle 100 \rangle$ directions, processes are allowed to zero order with phonons whose wave functions transform according to the representation Δ_2, and S_1 (for a review in the case of Si, see [2.143]). Accordingly, longitudinal optical modes assist g scattering (between parallel valleys), and longitudinal-acoustic as well as transverse modes assist f scattering (between perpendicular valleys). In the case of valleys located at L points, analogous arguments show that only phonons of symmetry X_1 can contribute [2.97]. Accordingly, longitudinal-acoustic and longitudinal-optical modes assist intervalley scattering.

It is worth noting that, in practice, initial and final states do not exactly coincide with high-symmetry points and consequently the selection rules need not be strictly fulfilled, as is confirmed in magnetophonon experiments [2.144]. However, for reasons of continuity it is reasonable to expect that the forbidden transitions will remain weak when compared with the allowed processes.

2.10 Scattering with Defects

This type of collision is elastic in nature and therefore it cannot control alone the transport process in the presence of an external field; it must be accompanied by some other dissipating scattering mechanism if the proper energy distribution of the electrons is to be derived from theory.

2.10.1 Scattering with Dislocations

Edge dislocations – a schematic picture is given in Fig. 2.15 – introduce a short-range field in the crystal which can be described by the deformation-potential

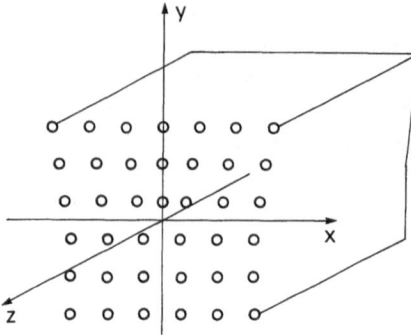

Fig. 2.15. Schematic arrangement of atoms in a crystal containing an edge dislocation: z is the dislocation line; x is the positive direction of the slip direction; xz is the slip plane

approach (strain type) [2.111, 112] and a long-range field due to the fact that dislocation lines are charged and surrounded by space-charge cylinders [2.145].

The square of the scattering matrix element for the strain field, following the results of *Dexter* [2.111], is reported in Table 2.5. Here a_s is the unit crystallographic slip distance, v the Poisson ratio, E_1^d a deformation-potential parameter, $q_\perp$ the component of $\boldsymbol{q}$ in the $x - y$ plane which is perpendicular to the dislocation line, φ the angle between $q_\perp$ and the x axis, and σ_{li} the density of dislocation lines per unit area.

The square of the scattering matrix element for localized line charge is calculated following the results of *Duster* and *Labusch* [2.113] and reported in Table 2.5. Here $Q_{li} = ef/a_i$ is the line charge of the dislocation (f being the occupation probability of the dislocation band and a_i the distance between imperfection centers along the dislocation line) and λ_D the Debye screening length given by

$$\lambda_D = \left(\frac{\varepsilon \varepsilon_0 K_B T_0}{4 \pi e^2 n} \right)^{1/2}. \tag{2.211}$$

We should note that the screening length given in (2.211) is evaluated with the assumption that the electrons have an equilibrium energy distribution. When at high fields the energy distribution deviates from equilibrium, (2.211) fails and the screened potential depends on the carrier distribution, so that the screening problem makes the kinetic equation nonlinear in the distribution function. As a general trend, when the average electron energy increases, the screening effect decreases, as can be seen from (2.211) in an electron temperature approximation.

2.10.2 Ionized Impurity Scattering

Two different formulations have initially been given at about the same time as the theory of ionized impurity scattering: the *Conwell* and *Weisskopf* theory [2.146] and the *Brooks* and *Herring* theory [2.114]. They differ for the model used to screen the Coulomb potential of the ion. Since then, several authors have presented refinements of the theory of impurity scattering and the interested reader is referred to [2.147] for a comprehensive review.

The square matrix element of the perturbation Hamiltonian of a scattering center for the *Brooks* and *Herring* case is given in Table 2.5. Here N_I is the net ionized impurity concentration, and Z_e is the number of charge units at the impurity.

Due to the electrostatic nature of the interaction, the efficiency of ionized impurity scattering decreases on increasing electron energy. This fact is accompanied by the increasing importance of phonon scattering. Thus, when at low fields the mobility is controlled by ionized impurity scattering, in a certain range of field strength a superohmic behavior of the drift velocity versus field is expected [2.110]. Experimentally, care must be taken in distinguishing this effect from a carrier multiplication, since both effects would result in a current increasing more than linearly with the applied field.

As for the case of polar and piezoelectric scattering, the Coulomb field of an ionized impurity is a negligible source of intervalley scattering owing to the large momentum transfer involved in these transitions.

2.10.3 Neutral Impurity Scattering

Neutral impurities have very small cross sections at normal concentrations (which can be quite high $\simeq 10^{16}\,\mathrm{cm}^{-3}$). Thus they can influence the transport process only at very low temperatures (see, for example, [2.107, 148–151]). The concomitance of other scattering mechanisms (such as ionized impurities and acoustic phonons), which at such low temperatures are not totally described by the present theoretical models, makes the analysis of impurity scattering from experimental data very difficult.

The square matrix element of the perturbation Hamiltonian, as given by *Erginsoy* [2.115], is reported in Table 2.5. Here N_N is the neutral impurity concentration and a_B^* the Bohr radius of the ground state of the impurity. This formulation leads to an energy-independent scattering rate.

Analyses of a possible energy dependence can be found in [2.107, 149, 152].

2.10.4 Intervalley Impurity Scattering

Impurities can act as a source of intervalley scattering in many-valley semiconductors [2.153]. The square matrix element as calculated by *Price* and *Hartman* [2.116] is given in Table 2.5. There N_I is the impurity concentration, $\mathscr{E}_I$ is the binding energy of the impurity center, and, following *Asche* and *Sarbei* [2.117], I can be expressed as

$$I = \frac{\pi \mathscr{E}_{cs} a^2 b}{u_0}, \tag{2.212}$$

where $\mathcal{E}_{cs}$ is the chemical-shift energy, a and b are length parameters for the orbit of the donor ground state in the effective-mass approximation, and u_0 is a dimensionless parameter denoting a correction to the effective-mass formalism and depending on the nature of the donor ground state.

2.10.5 Resonance Impurity Scattering

An impurity potential may give rise to levels, which are degenerate with states in the conduction band, in which the electron would not be truly localized and bound to the center, but only transiently so. In semiconductors such impurity states may arise, for example, because of the many-nonequivalent-valley structure of the conduction band. These states can be modeled in terms of resonant scattering states. For a neutral center, following the results of *El-Ghanem* and *Ridley* [2.154], the scattering rate τ_{RS}^{-1} can be written as

$$\frac{1}{\tau_{RS}} = \frac{2^{3/2}\,\pi\hbar^2\,N_N}{m^{3/2}}\,\mathcal{E}^{-1/2}\,\frac{\mathcal{E}_b^2}{(\mathcal{E}-\mathcal{E}_R)^2 + \mathcal{E}_b^2}\,, \tag{2.213}$$

where $(\mathcal{E}-\mathcal{E}_R)$ is the difference between the energy of the carrier and of the resonant level, which are both measured from the minimum of the conduction band, and $\mathcal{E}_b$ is the energy associated with the uncertainty broadening.

2.10.6 Space-Charge Scattering

In highly compensated semiconductors, for example, GaAs, GaSb, etc., an intrinsic region may occur in the material due to the local compensation of donors or acceptors. The intrinsic region will include not only the portion of material that is compensated, but a space-charge region that forms around it to provide the appropriate potential difference with respect to the uncompensated material. The potential difference thus created acts as an infinite wall for the carriers. *Conwell* and *Vassel* [2.155] estimated the diameter of the space charge as 200 Å for a free-carrier concentration of $10^{15}\,\text{cm}^{-3}$ in GaAs. Treating these scattering centers as large impenetrable-spheres of density N_{sc} and cross-sectional area Q_0, they estimated the scattering rate τ_{sc}^{-1} as

$$\frac{1}{\tau_{sc}} = N_{sc} Q_0\,\frac{\hbar k}{m}\,. \tag{2.214}$$

2.10.7 Alloy Scattering

Alloy scattering refers to the scattering present in alloys due to the random distribution of component atoms among the available sites. The typical model refers to III–V compounds with a composition of the type $A_xB_{1-x}C$ with $0 \leq x \leq 1$. The random arrangement of the constituent A and B atoms among the C atoms gives a random component to the crystal potential leading to the alloy scattering process. The squared matrix element for such a collision was evaluated in [2.118] by using the inner potential model, and is reported in Table 2.5. Here C_A is the fraction of atoms of type A, N_A is the number of atoms in the crystal, and $\Delta\mathscr{E}_A$ is taken as the difference in electron affinity of the two end-point binary compounds BC and AC.

2.11 Impact Ionization

For this interaction, which by involving an electron-hole pair generation is highly dissipative, we shall assume an electron-hole pair characterized by the wavevectors k_e, k_h before scattering and k_e', k_h' after scattering. Thus, the first electron is taken to make a transition from k_e to k_e' in the conduction band while the second makes an interband transition from k_h to k_h'. Accordingly, total energy conservation implies $\mathscr{E}(k_e) + \mathscr{E}(k_h) = \mathscr{E}(k_e') + \mathscr{E}(k_h')$. By taking the perturbing potential to be a screened Coulomb interaction between the conduction and the valence electron, the squared matrix element was calculated [2.119] and is reported in Table 2.5. Its similarity with the ionized impurity scattering can be easily recognized.

2.12 Carrier-Carrier Interaction

The matrix element for the carrier-carrier interaction between two identical particles belonging to the same band of initial wavevectors k_1, k_2 and final wavevectors k_1', k_2' is easily obtained by extension from the ionized impurity scattering [2.97, 121] and is reported in Table 2.5. Note that the k vector selection rule becomes a condition for conservation of k vector:

$$k_1' + k_2' = k_1 + k_2. \qquad (2.215)$$

Furthermore, $f(k_1)$ and $f(k_2)$ are the occupation probabilities of the two initial states where, according to the nondegenerate conditions used here, it has been assumed that the final states are always empty.

In such an interaction the total energy of the two colliding particles is conserved, thus no dissipation occurs. Momentum and energy are, however, redistributed among the particles so that the shape of the electron distribution function is influenced: a typical result is reported as an example in Fig. 2.16. It can be seen that electron-electron interaction tends to smooth the original shape of the distribution, due to phonon interaction only, towards a Maxwellian shape with an electron temperature higher than the equilibrium value.

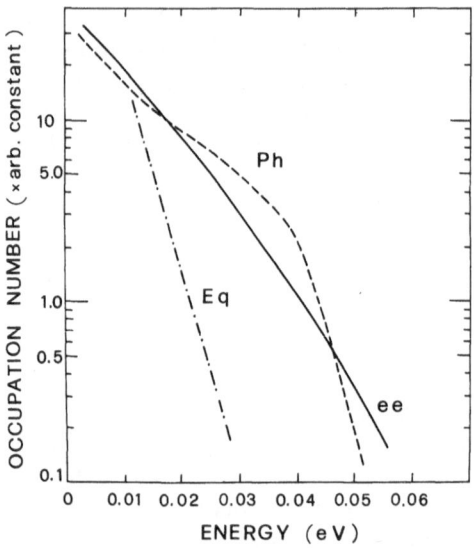

Fig. 2.16. Distribution function of electrons in a simple semiconductor modeled on Si obtained with Monte Carlo calculations at $T_0 = 45\,\mathrm{K}$ and $E = 300\,\mathrm{V/cm}$. (---) (Ph) has been obtained by considering only phonon scattering: (——) (e–e) includes electron-electron interaction with a concentration of $10^{17}\,\mathrm{cm^{-3}}$ electrons; (–·–) (Eq) indicates the equilibrium distribution [2.156]

2.13 Results for Drift, Diffusion, and Noise

We shall devote the first part of this section to presenting an analysis of the three main transport quantities, i.e., drift velocity, longitudinal diffusion coefficient, and equivalent noise temperature, as a function of field strength for a simple model semiconductor. Calculations have been performed with a Monte Carlo simulation which, by providing an exact solution of the Boltzmann equation, enables us to investigate the microscopic model in full detail with regard to the effect of scattering mechanisms on transport properties. The aim of such a model is to provide general background knowledge from which the correlation between the simple model and macroscopic parameters can be clearly understood. In this way, we shall also see how the failure of a simple theory to interpret experiments can be overcome by appropriate modifications of the model.

2.13.1 The Simple Model

The simple covalent semiconductor model consists of a single spherical and parabolic band with an effective mass $m = 0.35\,m_0$, and accounts for acoustic and nonpolar optical scattering as described in the deformation potential approach by single scattering coupling constant $E_1^0 = 4.6\,\mathrm{eV}$ and $d_0 = 40.3\,\mathrm{eV}$, respectively. Values characteristic of heavy holes in Ge are assumed (Table 2.6). This choice is motivated by the fact that transport properties of this material have been investigated widely from an experimental point of view and, in the first approach, the simple model used here is well suited to a realistic microscopic interpretation.

For the quantities of interest, the drift velocity is obtained from the estimator:

$$v_d = \frac{1}{eEt} \sum (\mathcal{E}_f - \mathcal{E}_i),$$
(2.216)

where the sum is evaluated over all carrier free flights.

The longitudinal diffusion coefficient is calculated from the second central moment as reported in Sect. 2.2.1. The equivalent noise temperature is calculated from the generalized Einstein relation (2.121). For the sake of completeness, the carrier mean energy measured in equivalent electron temperature through $T_e = 2 \langle \mathcal{E} \rangle / (3 K_B)$ has been reported together with T_n.

The results are reported as a function of field strength for the two temperatures of 8 and 300 K in Figs. 2.17, 18. As usual, for discussion purposes the electric field range can be divided into three regions, namely, the linear-response (i.e., $E \rightarrow 0$), an intermediate region where heating effects of carriers become evident, and the highest-field region (i.e., $E \rightarrow \infty$. The comparison between carrier mean energy and the value of the energy gap of the material sets a physically plausible upper limit on E).

To better analyze the behavior of different quantities in terms of single microscopic processes, we shall first discuss the low-temperature (8 K) case. In the linear-response region ($E < 0.2$ V/cm in Fig. 2.17), the following relations hold (Sect. 2.1):

$$v_d = \mu E,$$
(2.217)

$$D = \mu K_B T_0 / e$$
(2.218)

$$T_n = T_e = T_0.$$
(2.219)

The above equations, which are usually referred to as Ohm's law, Einstein's relation, and the equipartition of energy, respectively, are well reproduced by Monte Carlo results at the lowest field strengths (Fig. 2.17a).

The intermediate region starts at about 1 V/cm, and up to about 10 V/cm the acoustic scattering mechanism is the only active one (Fig. 2.17b). Owing to the quasi-elastic nature of this mechanism and since the scattering efficiency of acoustic modes increases with electron energy, the drift velocity is found to behave in a sublinear way. The diffusion coefficient is found to increase slightly above its thermal equilibrium value, and the noise temperature and the carrier temperature are found to increase both in a rather similar way.

In the region of field strengths between 10 and 10^3 V/cm, the optical scattering mechanism builds up until it predominates over acoustic scattering (Fig. 2.17b). Owing to strong optical scattering, the carrier returns to the bottom of the band after each emission process. Under this condition, the drift velocity tends to saturate and the diffusion coefficient exhibits a decreasing behavior due

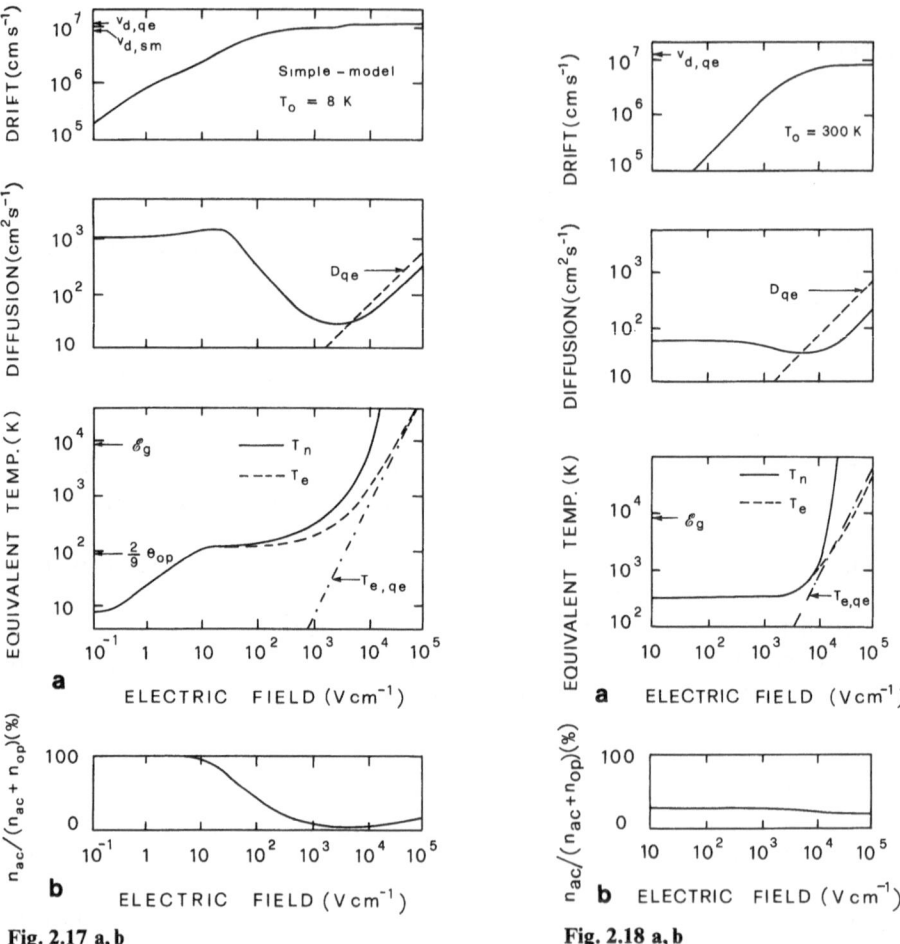

Fig. 2.17 a, b

Fig. 2.18 a, b

Fig. 2.17. (a) Drift velocity, longitudinal diffusion coefficient, and longitudinal noise and mean energy measured in equivalent temperature, as a function of the electric field strength for the simple model semiconductor at 8 K. The values of different quantities under the streaming motion (sm) and the quasi-elastic approximation (qe) are indicated for the sake of comparison. Here $\mathscr{E}_g$ is the value of the energy gap measured in equivalent temperature and θ_{op} is the optical-phonon equivalent temperature. (b) Ratio between acoustic scattering events n_{ac}, and total scattering events ($n_{ac}+n_{op}$), n_{op} being the number of optical scattering events, as a function of electric field strength for the simple model semiconductor at 8 K

Fig. 2.18 a, b. The same as in Fig. 2.17 at 300 K

to the rather deterministic nature of this motion which cannot cause much variance in the length traversed during a fixed time interval. Accordingly, both the carrier mean energy and the noise temperature attain an approximately constant value. These features follow what is predicted by the streaming motion approximation (Chap. 5) which, at this temperature, applies well to the field range and for which simple theoretical considerations lead to the relations

$$v_{\mathrm{d}} = \left(\frac{K_{\mathrm{B}}\theta_{\mathrm{op}}}{2m}\right)^{1/2},$$ (2.220)

$$T_{\mathrm{e}} = \frac{2}{9}\,\theta_{\mathrm{op}}.$$ (2.221)

For the sake of comparison, the limiting values given by (2.220, 221) are reported with Monte Carlo results in Fig. 2.17a.

In the region of field strengths above 10^3 V/cm, the optical scattering mechanism keeps its predominant role, but is no longer able to dissipate fully the energy gained by the carriers from the field. As a consequence, the drift velocity achieves a new saturation level higher than the previous one; the diffusion coefficient keeps increasing towards a limiting linear dependence with an increasing field; the carrier mean energy is found to increase more and more steeply with the field till it exhibits a limiting quadratic dependence while the noise temperature achieves asymptotically high values for fields above 20 kV/cm, where μ' is zero within the statistical uncertainty.

These features are predicted by an analytical solution of the Boltzmann equation obtained under the small anisotropy and quasi-elastic approximations [2.60,157]. This solution, which well applies to the highest-field region ($E \geq 10^4$ V/cm in the figure), considers only optical-phonon emission processes and leads to the following relations:

$$v_{\mathrm{d}} = \left(\frac{2^3\,K_{\mathrm{B}}\theta_{\mathrm{op}}}{3\pi m}\right)^{1/2},$$ (2.222)

$$D_{\mathrm{lo}} = v_{\mathrm{d}}\,l_{\mathrm{op}}\,\frac{E}{E_{\mathrm{c}}},$$ (2.223)

$$T_{\mathrm{e}} = \theta_{\mathrm{op}}\left(\frac{E}{E_{\mathrm{c}}}\right)^2.$$ (2.224)

Here E_{c} is a critical field given by

$$E_{\mathrm{c}} = \frac{3^{3/2}\,m^2 d_0^2}{2^2\,\pi e\hbar^2 a_0^2 \varrho_0},$$ (2.225)

which under $E/E_{\mathrm{c}} > 1$ it well justifies the above asymptotic solution, l_{op} is the mean free path due to optical-phonon scattering given by

$$l_{\mathrm{op}} = \frac{3^{1/2}\,K_{\mathrm{B}}\theta_{\mathrm{op}}}{eE_{\mathrm{c}}}.$$ (2.226)

The asymptotic prediction of (2.222–224), which within statistical uncertainty are found to agree with the Monte Carlo calculations, are reported in Fig. 2.17a.

Table 2.6. Parameters for hole-phonon scattering

	C [a]	Si [a]	Ge [a]	GaAs [b]
E_1^0 [eV]	5.5	5.0	4.6	3.5
d_0 [eV]	61.2	26.6	40.3	29.9

[a] [2.88]; [b] [2.86]

Table 2.7. Parameters for electron-phonon scattering[a]

	C	Si	Ge	GaAs
Intravalley				
$E_{1\Gamma}$ [eV]	—	—	5	7
E_{1L} [eV]	—	—	11	9.2
$E_{1\Delta}$ [eV]	8.7	9.0	9	9.3
$(D_t K)_L$ [10^8 eV/cm]	—	—	5.5	3
Intervalley $\Delta\Delta$				
θ_{g1} [K]	—	140 (TA)	—	—
D_{g1} [10^8 eV/cm]	—	0.5	—	—
θ_{g2} [K]	—	215 (LA)	100 (LA)	—
D_{g2} [10^8 eV/cm]	—	0.8	0.79	—
θ_{g3} [K]	1900 (LO)	720 (LO)	430 (LO)	345 (LO)
D_{g3} [10^8 eV/cm]	8.0	11.0	9.5	7
θ_{f1} [K]	—	220 (TA)	—	—
D_{f1} [10^8 eV/cm]	—	0.3	—	—
θ_{f2} [K]	1560 (LA)	550 (LA)	—	—
D_{f2} [10^8 eV/cm]	8.0	2.0	—	—
θ_{f3} [K]	1720 (TO)	685 (TO)	—	—
D_{f3} [10^8 eV/cm]	8.0	2.0	—	—
Intervalley LL				
θ_{LL1} [K]	—	—	320 (LA, LO)	340 (LA, LO)
D_{LL1} [10^8 eV/cm]	—	—	3.0	10
θ_{LL2} [K]	—	—	120 (TA)	—
D_{LL2} [10^8 eV/cm]	—	—	0.2	—
Intervalley $L\Delta$				
$\theta_{L\Delta}$ [K]	—	—	320 (LA)	340 (LA)
$D_{L\Delta}$ [10^8 eV/cm]	—	—	4.1	5
Intervalley $L\Gamma$				
$\theta_{L\Gamma}$ [K]	—	—	320 (LA)	325 (LA)
D_L [10^8 eV/cm]	—	—	2.0	10
Intervalley $\Delta\Gamma$				
$\theta_{\Delta\Gamma}$ [K]	—	—	320 (LA)	350 (LA)
$D_{\Delta\Gamma}$ [10^8 eV/cm]	—	—	10	10

[a] For references, see [2.43,86] (LA: Longitudinal acoustic; LO: Longitudinal optical; TA: Transverse acoustic; TO: Transverse optical)

In the high-temperature case $(T_0 = 300 \text{ K})$ (Fig. 2.18), owing to the increased efficiency of the scattering mechanisms in dissipating to the lattice the energy gained by the field, the region of field strengths for which the linear response region holds is extended ($E < 1 \text{ kV/cm}$ in the figure). Owing to the increased importance of acoustic-phonon scattering and optical absorption processes, the streaming motion region disappears and at the highest-field region ($E > 10 \text{ kV/cm}$), the results are no longer in close agreement with those of (2.222–224).

2.13.2 Real Cases

The simple model represents too crude a simplification to offer any deep microscopic physical insight or to permit a quantitative comparison between theory and experiments. Consequently, in the following we present a survey of the most significant results which, in our opinion, have been obtained for the case

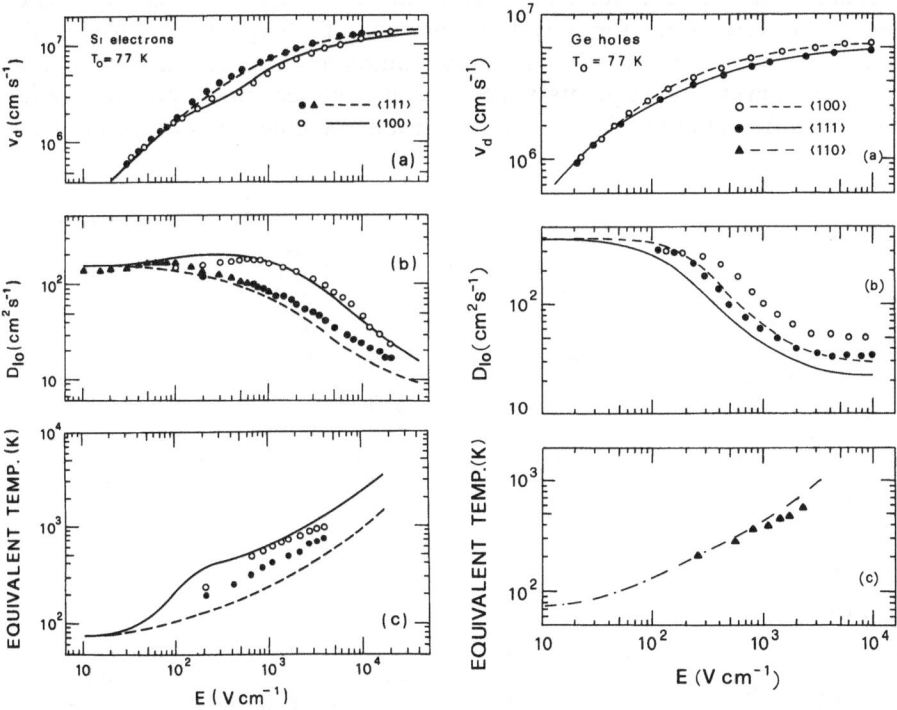

Fig. 2.19. (a) Drift velocity, (b) longitudinal diffusion coefficient, (c) longitudinal noise temperature as a function of electric field strength for the case of electrons in Si at 77 K. (O, ●) and (▲) refer to experiments, and (——) to Monte Carlo calculations [2.88]

Fig. 2.20. (a) Drift velocity, (b) longitudinal diffusion coefficient, (c) longitudinal noise temperature as a function of electric field strength for the case of holes in Ge at 77 K. (O, ●) and (▲) refer to experiments, and (——) to Monte Carlo calculations [2.75]

of electrons in Si, holes in Ge, and electrons in GaAs. In fact, these cases are well suited to illustrate the present knowledge of hot-electron phenomena when full details of the band structure and scattering mechanisms are taken into account; Chap. 3 will report a more thorough investigation. For completeness, in Tables 2.6 and 2.7 we summarize the values for the hole- and electron-phonon interaction in diamond, Si, Ge, and GaAs.

Figure 2.19 reports the results for the case of electrons in Si. The general behavior of drift, diffusion, and noise temperature resembles that of the simple model. Here the anisotropy exhibited by different quantities with respect to the orientation of the field is related to the peculiar many-valley structure of the conduction band.

Figure 2.20 reports the results for the case of holes in Ge. Even in this case, the results generally resemble those of the simple model. Here the anisotropy exhibited by different quantities is related to the warped shape of the heavy-hole equienergetic surfaces.

Figure 2.21 reports the results for the case of electrons in GaAs. For fields below threshold for Gunn effect ($\simeq 3.5\,\text{kV/cm}$) the polar scattering, with its peculiar streaming character and decreasing efficiency at increasing carrier energy, is responsible for a sudden increase of all quantities. Above the threshold, intervalley scattering from the minimum of the conduction band to upper minima at L and X points in the Brillouin zone becomes responsible for the negative differential mobility as well as for a marked decrease of the diffusion

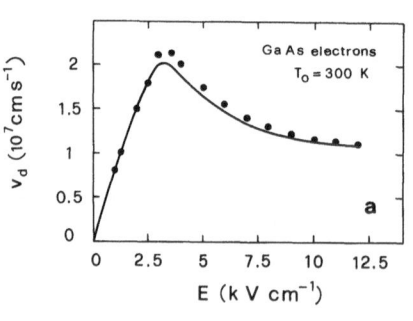

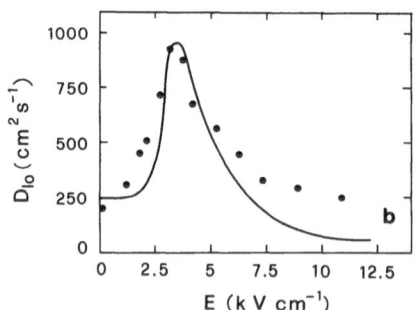

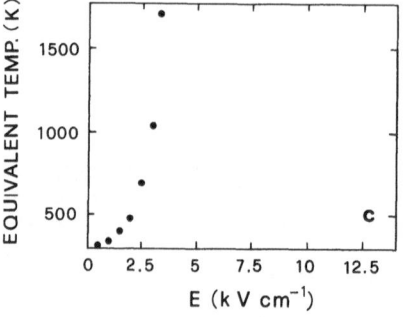

Fig. 2.21. (a) Drift velocity, (b) longitudinal diffusion coefficient, (c) longitudinal noise temperature as a function of electric field strength for the case of electrons in GaAs at 300 K. (●) refer to experiments [2.159] and (——) to Monte Carlo calculations [2.160]

coefficient. Furthermore, owing to a $\mu' < 0$, the system becomes electrically unstable with respect to fluctuations in charge density so that for field strengths approximately above threshold a noise temperature cannot be defined.

References

2.1   H.Fröhlich: Proc. Roy. Soc. **A188**, 521 and 532 (1947)
2.2   H.B.Callen, T.A.Welton: Phys. Rev. **83**, 34 (1951)
2.3   R.Kubo: Response, Relaxation and Fluctuations, in Lecture Notes Phys., Vol. 31, ed. by J.Ehlers, K.Hepp, H.A.Weidenmuller, (Springer, Berlin, Heidelberg, 1974) p. 74
2.4   K.M. Van Vliet, A. Van der Ziel: Solid State Electron. **20**, 931 (1977)
2.5   W.Shockley: Bell Syst. Techn. J. **30**, 990 (1951)
2.6   K.Okamoto, J.Nishizawa, K.Takahashi: J. Appl. Phys. **36**, 3716 (1965)
2.7   G.Ruch, G.S.Kino: Phys. Rev. **174**, 921 (1968)
2.8   E.Erlbach, J.B.Gunn: Phys. Rev. Lett. **8**, 280 (1962)
2.9   D.Gasquet, J.P.Nougier: Appl. Phys. Lett. **33**, 89 (1978)
2.10 G.Persky, D.J.Bartelink: Phys. Rev. **1B**, 1614 (1970)
2.11 R.Brunetti, C.Jacoboni, F.Nava, L.Reggiani, G.Bosman, R.J.J.Zijlstra: J. Appl. Phys. **52**, 6713 (1981)
2.12 V.Bareikis, K.Pozhela, I.B.Matulenene: Proc. 9th Intern. Conf. Phys. Semicon., ed. by S.M.Ryvkin (Nauka, Leningrad 1968) p. 760
2.13 W.Sasaki, M.Shibuya: J. Phys. Soc. Jpn. **11**, 1202 (1956)
2.14 C.Canali, F.Nava, G.Ottaviani, L.Reggiani: Proc. 13th Intern. Conf. Phys. Semicon., ed. by F.G.Fumi (Marves, Rome 1976) p. 1231
2.15 M.Rolland, J.P.Nougier: Proc. 13th Intern. Conf. Phys. Semicon., ed. by F.G.Fumi (Marves, Rome 1976) p. 1227
2.16 J.B.Gunn: Solid State Commun. **1** 88 (1963)
2.17 E.J.Ryder, I.M.Moss, D.A.Kleinman: Phys. Rev. **95**, 1342 (1954)
2.18 E.A.Davies: J. Phys. Chem. Sol. **25**, 201 (1964)
2.19 R.J.J.Zijlstra: *Noise in Physical Systems*, ed. by D.Wolf, Springer Ser. Electrophys., Vol. 2 (Springer, Berlin, Heidelberg, 1978)
2.20 G.Bauer: *Springer Tracts Mod. Phys.* 74, ed. by G.Hoehler (Springer, Berlin, Heidelberg 1974)
2.21 G.Bertolini, A.Coche (eds.): *Semiconductor Detectors* (North Holland, Amsterdam 1968) R.J.Keyes (ed.): *Optical and Infrared Detectors*, 2nd ed., Topics Appl. Phys., Vol. 19 (Springer, Berlin, Heidelberg, 1980)
2.22 D.K.Ferry, J.R.Barker, C.Jacoboni (eds.): *Physics of Nonlinear Transport in Semiconductors* (Plenum, New York 1980)
2.23 A.R.Hutson, A.Jayaraman, A.G.Chynoweth, A.S.Coriell, W. L.Feldman: Phys. Rev. Lett. **14**, 639 (1965)
2.24 S.Ashmontas, A.Olekas: Sov. Phys.–Semicond. **14**, 1301 (1980)
2.25 S.V.Gantsevich, V.L.Gurevich, R.Katilius: Rivista Nuovo Cimento **2**, 1 (1979)
2.26 J.J.Duderstadt, W.R.Martin: *Transport Theory* (Wiley, New York 1979)
2.27 P.Kocevar: In *Physics of Nonlinear Transport in Semiconductors* ed. by D.K.Ferry, J.R.Barker, C.Jacoboni (Plenum, New York 1980) p. 401
2.28 B.Vinter: Some Numerical Solutions of the Problem of High Field Transport in Semiconductors; Dissertation, The Technical University of Denmark, Lyngby (1973)
2.29 K.K.Thornber: IEEE Trans. EDL-3, 69 (1982)
2.30 A.Van der Ziel: *Solid State Physical Electronics* (Prentice Hall, Englewood Cliffs, NJ 1968)
2.31 H.Van Beijern: Rev. Mod. Phys. **54**, 195 (1982)
2.32 P.J.Price: *Semiconductors and Semimetals* 14, 249 (Academic, New York 1979)
2.33 A.Matulenis, Y.Pozhela, E.Starikov: Sov. Phys.-Semicond. **16**, 388 (1982)

2.34   A.Matulionis, Y.Pozhela, E.Starikov: Phys. Stat. Sol. **68**(a), K149 (1981)
2.35   L. Reggiani, R.Brunetti, C.Jacoboni: Proc. 3rd Intern. Conf. Hot Carriers in Semiconductors, J. Physique **42**, C7–111 (1981)
2.36   B.Boittiaux, E.Constant, L.Reggiani, R.Brunetti, C.Jacoboni: Appl. Phys. Lett. **40**, 407 (1982)
2.37   R.Zwanzig: Phys. Rev. **133**, A50 (1964)
2.38   R.G.Chambers: Proc. Phys. Soc. London **65A**, 458 (1952)
2.39   C.Jacoboni, L.Reggiani, R.Brunetti: Proc. 3rd Intern. Conf. Hot Carriers in Semiconductors, J. Physique **42**, C7–123 (1981)
        Similar results for the case of compound semiconductors were reported by M.Deb Roy, B.R.Nag: Appl. Phys. **A26**, 131 (1981); ibidem **A28**, 195 (1982); ibidem **A30**, 189 (1983)
2.40   C.Jacoboni: Phys. Stat. Sol. **65**(b), 61 (1974)
2.41   P.J.Price: *Fluctuation Phenomena in Solids*, ed. by R.E.Burgess (Academic, New York 1965) Chap. 8
2.42   W.A.Schlup: Phys. Kondens. Materie **13**, 307 (1971)
2.43   C.Jacoboni, L.Reggiani: Rev. Mod. Phys. **55**, 645 (1983).
        For a general treatment of the Monte Carlo method see: K.Binder (ed.): *Monte Carlo Methods*, Topics Current Phys., Vol. 7 (Springer, Berlin, Heidelberg 1979); Reuven Y.Rubinstein: *Simulation and the Monte Carlo Method* (Wiley, New York 1981); K.Binder: *Applications of the Monte Carlo Method*, Topics Current Phys., Vol. 36 (Springer, Berlin, Heidelberg 1984)
2.44   H.D.Rees: Phys. Lett. **26A**, 416 (1968)
2.45   H.D.Rees: J. Phys. Chem. Solids **30**, 643 (1969)
2.46   W.Fawcett, D.A.Boardman, S.Swain: J. Phys. Chem. Solids **31**, 1963 (1970)
2.47   P.J.Price: *Proc. 9th Intern. Conf. Phys. Semicon.*, ed. by S.M.Ryvkin (Nauka, Leningrad 1968) p. 753
2.48   J.H.Hammersley, D.C.Handscomb: In *Monte Carlo Methods*, ed. by M.S.Bartlett (Methuen, London 1964)
2.49   P.A.Lebwohl, P.J.Price: Solid State Commun. **9**, 1221 (1971a)
2.50   P.A.Lebwohl, P.J.Price: Appl. Phys. Lett. **19**, 530 (1971b)
2.51   K.Seeger: *Semiconductor Physics*, 2nd ed., Springer Ser. Solid-State Sci., Vol. 40 (Springer, Berlin, Heidelberg 1982)
2.52   G.Baccarani, C.Jacoboni, A.M.Mazzone: Solid State Electron **20**, 5 (1977)
2.53   J.Zimmermann, E.Constant: Solid State Electron. **23**, 915 (1980)
2.54   P.A.Lebwohl: J. Appl. Phys. **44**, 1744 (1973)
2.55   J.Zimmermann, Y.Leroy, E.Constant: J. Appl. Phys. **49**, 3378 (1978)
2.56   A.Reklaitis: Phys. Lett. **7**, 367 (1982)
2.57   M.Abramowitz, I.A.Stegun: *Handbook of Mathematical Function* (Dover, New York 1965)
2.58   W.Fawcett: In *Electrons in Crystalline Solids*, ed. by A.Salam (IAEA, Vienna 1973) p. 531
2.59   H.Budd: J. Phys. Soc. Jpn., Suppl. **21**, 420 (1966); and Phys. Rev. **158**, 798 (1967)
2.60   P.J.Price: IBM, J. Res. Dev. **14**, 12 (1970)
2.61   M.O.Vassel: J. Math. Phys. **11**, 408 (1970)
2.62   H.D.Rees: J. Phys. **C6**, 262 (1973)
2.63   D.A.McQuarrie: *Statistical Mechanics* (Harper and Row, New York 1976)
2.64   W.Shockley, J.A.Copeland, R.P.James: *Quantum Theory of Atoms, Molecules and Solid State* (Academic, New York 1966)
2.65   J.P.Nougier: Phys. Status Solidi **55B**, K43 (1973)
2.66   J.P.Nougier: Appl. Phys. Lett. **32**, 671 (1978)
2.67   J.P.Nougier: *Signaux et Bruit, Methodes de Calcul du Bruit de Fond* (Cours de D.E.A., University of Montpellier 1975)
2.68   K.M.Van Vliet, A.Friedman, R.J.J.Zijlstra, A.Gisolf, A.Van der Ziel: J. Appl. Phys. **46**, 1804, 1814 (1975)
2.69   J.P.Nougier, J.C.Vaissiere, D.Gasquet, A.Motadid: J. Appl. Phys. **52**, 5683 (1981)
2.70   S.Ramo: Proc. IRE **27**, 584 (1939)
2.71   W.Shockley: J. Appl. Phys. **9**, 635 (1938)

2.72   J.P.Nougier, M.Rolland: Phys. Rev. **8B**, 5728 (1973)
2.73   V.Bareikis, V.Viktoravichyus, A.Galdikas, R.Milyushite: Sov. Phys.–Solid State **20**, 85 (1978)
2.74   A.Gisolf, R.J.J.Zijlstra: J. Appl. Phys. **47**, 2727 (1976)
2.75   J.Zimmermann, S.Bonfis, Y.Leroy, E.Constant: Appl. Phys. Lett. **30**, 245 (1977)
2.76   G.Bosman, R.J.J.Zijlstra: Phys. Lett. **71A**, 464 (1979)
2.77   G.Bosman, R.J.J.Zijlstra, F.Nava: Solid State Electron. **24**, 5 (1981)
2.78   M.Bixon, R.Zwanzig: Phys. Rev. **187**, 267 (1969)
2.79   E.C.G.Stueckelberg: Helv. Phys. Acta **25**, 577 (1952)
2.80   S.V.Gantsevich, V.L.Gurevich, R.Katilius: Sov. Phys. – JEPT **30**, 276 (1970)
2.81   S.Chapman, T.G.Cowling: *The Mathematical Theory of Non-Uniform Gases* (Cambridge University Press 1970)
2.82   A.Ya.Shulman: Sov. Phys. – Solid State **12**, 922 (1970)
2.83   S.V.Gantsevich, V.L.Gurevich, R.Katilius: Phys. Kond. Matt. **18**, 106 (1974)
2.84   V.V.Paranjape: J. Phys. Chem. Solids **38**, 375 (1977)
2.85   G.L.Bir, G.E.Pikus: In *Symmetry and Strain-Induced Effects in Semiconductors*, ed. by D.Louvisch (Wiley, Jerusalem 1974)
2.86   C.Jacoboni, L.Reggiani: Advan. Phys. **28**, 493 (1979)
2.87   W.Van Haering, H.G.Junginger: Solid State Commun. **7**, 1135 (1969)
2.88   L.Reggiani: J. Phys. Soc. Jpn. **49**, 317 (1980)
2.89   G.Gagliani, L.Reggiani: unpublished results
2.90   J.Wiley: Semiconductors and Semimetals, Vol. 10, ed. by R.K.Willardson and A.C.Beer (Academic, New York 1974) p. 91
2.91   M.L.Cohen, T.K.Bergstresser: Phys. Rev. **141**, 789 (1966)
2.92   J.D.Wiley: Phys. Rev. **4B**, 2485 (1971)
2.93   D.Kranzer: Phys. Stat. Sol. **26**(a), 11 (1974)
2.94   G.Ottaviani, L.Reggiani, C.Canali, F.Nava, A.Alberigi-Quaranta: Phys. Rev. **12B**, 3318 (1975)
2.95   G.Dresselhaus, A.F.Kip, C.Kittel: Phys. Rev. **98**, 368 (1955)
2.96   E.M.Conwell, M.O.Vassel: Phys. Rev. **166**, 797 (1968)
2.97   E.G.S.Paige: In *The Electrical Conductivity of Germanium*, ed. by A.F.Gibson and R.E.Burgess (Heywood, London 1964)
2.98   E.O.Kane: J. Phys. Chem. Sol. **1**, 82 (1956)
2.99   G.Gagliani, L.Reggiani: Nuovo Cimento **30B**, 207 (1975)
2.100   R.G.Humphreys: J. Phys. **C14**, 2935 (1981)
2.101   C.Herring, E.Vogt: Phys. Rev. **101**, 944 (1956)
2.102   J.R.Barker, D.K.Ferry: Phys. Rev. Lett. **42**, 1779 (1979)
2.103   L.Reggiani, S.Bosi, C.Canali, F.Nava, S.F.Kozlov: Phys. Rev. **23B**, 3050 (1981a)
2.104   P.Agrain, M.Balkanski: *Selected Constants Relative to Semiconductors* (Pergamon Press, Oxford 1961)
2.105   S.S.Mitra, N.E.Massa: In *Handbook on Semiconductors*, Vol. 1, ed. by T.S.Moss (North-Holland, Amsterdam 1982) p. 81
2.106   D.L.Rode: Phys. Stat. Sol. **53**(b), 245 (1972)
2.107   B.K.Ridley: *Quantum Processes in Semiconductors* (Clarendon Press, Oxford 1982)
2.108   D.L.Rode: Phys. Rev.**B2**, 4036 (1970)
2.109   H. Ehrenreich: J. Phys. Chem. Sol. **9**, 129 (1959)
2.110   E.M.Conwell: *High Field Transport in Semiconductors*, Solid State Phys., Suppl., **9**, 105 (Academic, New York 1967)
2.111   D.L.Dexter: Phys. Rev. **85**, 936 (1952)
2.112   D.L.Dexter, F.Seitz: Phys. Rev. **86**, 964 (1952)
2.113   F.Duster, R.Labusch: Phys. Stat. Sol. **60**(b), 161 (1973)
2.114   H.Brooks, C.Herring: Phys. Rev. **83**, 879 (1951)
2.115   C.Erginsoy: Phys. Rev. **79**, 1013 (1950)
2.116   P.J.Price, R.L.Hartman: J. Phys. Chem. Sol. **25**, 219 (1964)
2.117   M.Asche, O.G.Sarbei: Phys. Stat. Sol. **33**(b), 9 (1969)

2.118 J.W.Harrison, J.H.Hauser: Phys. Rev. **B13**, 5347 (1976)
2.119 R.C.Curby, D.K.Ferry: Phys. Stat. Sol. **15**(a), 319 (1973)
2.120 J.M.Ziman: *Electrons and Phonons* (Oxford U. Press, London 1960)
2.121 N.Takenaka, M.Inoue, Y.Inuishi: J. Phys. Soc. Jpn. **47**, 861 (1979)
2.122 L.Reggiani, C.Calandra: Phys. Lett. **43A**, 339 (1973)
2.123 M.Costato, L.Reggiani: Phys. Stat. Sol. **58**(b), 47, 471 (1973)
2.124 C.Canali, C.Jacoboni, F.Nava, G.Ottaviani, A.Alberigi-Quaranta: Phys. Rev. **B12**, 2265 (1975)
2.125 L.Gherhardi, A.Pellacani, C.Jacoboni: Lett. Nuovo Cimento **14**, 225 (1975)
2.126 N.O.Gram, M.H.Joergensen: Phys. Rev. **8**, 3902 (1973)
2.127 G.L.Bir, G.E.Pikus: Sov. Phys. – Solid State **2**, 2039 (1961)
2.128 P.Lawaetz: Some Transport Properties of Holes in Germanium, Dissertation, The Technical University of Denmark, Lyngby (1967)
2.129 S.Bosi, C.Jacoboni, L.Reggiani: J. Phys. **C12**, 1525 (1979)
2.130 K.B.Tolpygo: Sov. Phys. – Solid State **4**, 1297 (1963)
2.131 P.Lawaetz: Phys. Rev. **183**, 730 (1969)
2.132 R.S.Crandal: Phys. Rev.**B1**, 730 (1970)
2.133 D.Chattopadhyay: J. Appl. Phys. **45**, 4931 (1974)
2.134 M.A.Renucci, J.B.Renucci, R.Zegber, M.Cardona: Phys. Rev. **10B**, 4309 (1974)
2.135 J.B.Renucci, R.N.Tyte, M.Cardona: Phys. Rev. **11B**, 3885 (1975)
2.136 J.M.Calleja, J.Kuhl, M.Cardona: Phys. Rev. **17B**, 876 (1978)
2.137 W.Poetz, P. Vogl: Phys. Rev. **24**, 2025 (1981)
2.138 D.K.Ferry: Phys. Rev. **14B**, 1605 (1976)
2.139 I.B.Levinson: Sov. Phys. – Solid State **7**, 1098 (1965)
2.140 W.A.Harrison: Phys. Rev. **104**, 1281 (1956)
2.141 H.W.Streitwolf: Phys. Stat. Sol. **37**, K47 (1970)
2.142 M.Lax, J.L.Birman: Phys. Stat. Sol. **49**(b), K153 (1972)
2.143 M.Asche, O.G.Sarbei: Phys. Stat. Sol. **103**(b), 11 (1981)
2.144 L.Evans, R.A.Hoult, R.A.Stradling, R.J.Tidey, J.C.Portal, S.Askenazy: J. Phys. **C8**, 1034 (1975)
2.145 W.T.Read, Jr.: Phyl. Magaz. **45**, 775 (1954)
2.146 E.M.Conwell, V.F.Weisskopf: Phys. Rev. **77**, 388 (1950)
2.147 D.Chattopadhyay, H.J.Queisser: Rev. Mod. Phys. **53**, 745 (1981)
2.148 P.Norton, H.Levinstein: Phys. Rev. **6B**, 470 (1972)
2.149 P.Norton, T.Braggins, H.Levinstein: Phys. Rev. **8B**, 5632 (1973)
2.150 T.C.McGill, R.Baron: Phys. Rev. **11**, 5208 (1975)
2.151 R.Baron, M.H.Young: In *Proc. 13th Intern. Conf. Phys. Semicon.*, ed. by F.G.Fumi (Marves, Rome 1976) p. 1158
2.152 N.Sclar: Phys. Rev. **104**, 1559 (1956)
2.153 G.Weinreich, T.M.Sanders Jr., H.G.White: Phys. Rev. **114**, 33 (1959)
2.154 H.M.A.El-Ghanem, B.K.Ridley: J. Phys. **C13**, 2041 (1980)
2.155 E.M.Conwell, M.O.Vassel: Phys. Rev. **166**, 797 (1968)
2.156 C.Jacoboni: In *Proc. 13th Intern. Conf. Phys. Semicon.*, ed. by F.G.Fumi (Marves, Roma 1976) p. 1195
2.157 I.B.Levinson: Sov. Phys. – Solid State **6**, 1665 (1965)
2.158 L.Reggiani, R.Brunetti, C.Jacoboni: Intern. Symp. on Noise in Physical Systems, ed. by P.H.E.Meijer, R.D.Mountain, R.J.Soulen (NBS Special Publications 614, Washington, DC 1981d) p. 414
2.159 V.Bareikis, A.Galdikas, R.Millusyte, V.Victorevicius: Abstracts 5th Intern. Conf. Noise in Physical Systems, Bad Nauheim (1978) p. 212 (unpublished)
2.160 J.Pozhela, A.Reklaitis: Solid State Commun. **27**, 1073 (1978)

3. Drift Velocity and Diffusion Coefficients from Time-of-Flight Measurements

Claudio Canali, Filippo Nava, and Lino Reggiani

With 29 Figures

The time-of-flight (ToF) technique has emerged over the last few years as the most reliable method of measuring drift velocity and longitudinal diffusion coefficient of hot carriers in semiconductors. In this respect it is worth noting that the results obtained with this technique have been supported by an accurate theoretical analysis so that, besides improving our knowledge on the transport coefficients of different materials, they can offer a good standard to validate alternative methods.

The aim of this chapter is to present the principles of this technique and then to report a comprehensive survey of the results which have appeared in the literature. Section 3.1 gives a brief historical outline on the development of this technique from the pioneering experiments of Haynes and Shockley to present. Sections 3.2, 3.3 describe the principles of the traditional time-of-flight technique and of the more recent microwave-time-of-flight technique. Section 3.4 reports the measurements of the drift velocity obtained for electrons and holes in different materials. Section 3.5 illustrates the method for measuring the longitudinal diffusion coefficient and reports the data obtained for electrons and holes in different materials.

3.1 Historical Survey

The basic ideas of the ToF technique as applied to the study of transport phenomena in semiconductors were first presented by *Haynes* and *Shockley* [3.1] in their pioneering experiments. The most important measurable quantity is the value of the transit time T_R, that is, the time taken by the charge carriers to travel across a given region of the sample under the influence of a known electric field. Thus, the ToF technique is usually synonymous with transit-time measurements [3.2].

Since the Haynes and Shockley experiment, in which minority carriers were injected through point contacts into a filamentary sample, other groups adopted this technique and several improvements were introduced. *Spear* [3.3, 4] produced electron-hole pairs by short pulses of low energy (about 10 keV) electrons in high-resistivity materials, thus enabling the analysis of drift velocity for carriers of each type to be carried out in the same sample.

By adopting the Spear's method, *Ruch* and *Kino* [3.5], and *Sigmon* and *Gibbons* [3.6] improved the technique in order to obtain the measurement of the longitudinal diffusion coefficient. *Neukermans* and *Kino* [3.7] succeeded in coupling the ToF technique with an external uniaxial pressure. The Modena group [3.8–16] completed a systematic analysis of drift and diffusion primarily in the elemental semiconductors of group four. *Evans* and *Robson* [3.17], by modulating the ionizing source at a microwave frequency (microwave-time-of-flight technique), obtained drift velocity measurements on very thin samples (2 to 10 μm), and for very high electric fields near the breakdown region.

3.2 Time-of-Flight Technique

3.2.1 Principles of the Method

The schematic principle of ToF is illustrated in Fig. 3.1. An ionizing radiation of sufficient energy to generate electron-hole pairs, and with an intensity low enough to avoid space-charge effects, hits the negative contact of a two-terminal semiconductor device, of length W, for a short time (Fig. 3.1a). Electron-hole pairs are generated near the bombarded contact and the presence of a uniform electric field (Fig. 3.1b), which is due to the voltage V_a applied to the device, enables carriers of one type (electrons in the case of Fig. 3.1) to travel across the sample, while the others are swept towards the opposite contact. Carriers traveling across the whole W region will induce a current transient signal (Fig. 3.1c) or a voltage transient signal (Fig. 3.1d) in the external circuit connected to the sample terminals. The duration of the signal T_R is exactly the same as the time carriers take to cross the W region.

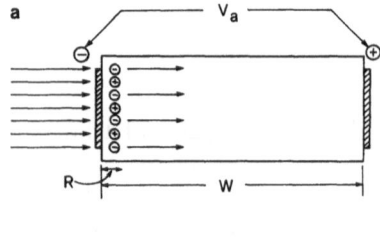
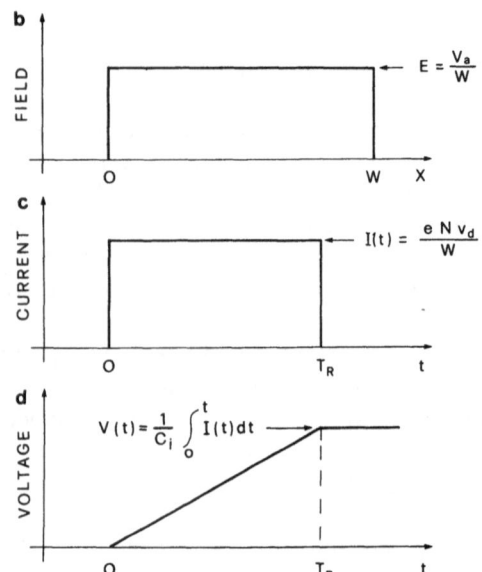

Fig. 3.1a–d. Schematic of the time-of-flight technique: (**a**) an ionizing radiation creating charge pairs in a narrow region R of the sample close to one contact; (**b**) electric field distribution inside the sample; (**c**) current and (**d**) voltage signals induced by the carriers drifting across the sample

Providing that the ionizing radiation generates charge pairs in a time τ_g which is much shorter than the transit time T_R, and in a range R which is much smaller than the sample length W, the current or voltage transient signal induced by the carriers' motion can be simply calculated under the following conditions: $\tau_L \gg T_R$ and $\tau_\varepsilon \gg T_R$, τ_L being the lifetime of the mobile carriers and $\tau_\varepsilon = \varrho \varepsilon \varepsilon_0$ the dielectric relaxation time of the material (ϱ being the resistivity, ε the free-space permittivity, and ε_0 the relative static dielectric constant of the material).

The condition $\tau_L \gg T_R$ ensures that drifting carriers do not recombine during their flight, while the condition $\tau_\varepsilon \gg T_R$ ensures that the local disturbance of the electrical neutrality due to the carrier layer drifting across the sample is not neutralized before T_R has elapsed.

According to the Ramo-Shockley theorem [3.18, 19] a single carrier moving a distance Δx between two parallel contacts, separated by a distance W, in a direction parallel to the electric field will induce, at the electrodes, a charge ΔQ given by

$$\Delta Q = e \frac{\Delta x}{W} \tag{3.1}$$

(e being the electron charge) regardless of the dependence of the electric field E on x.

Therefore, for noninteracting electrons, the amplitude of the current pulse $I(t)$ induced by carriers drifting across the sample is obtained by making the time derivative of (3.1) and by summing over the number N of drifting carriers:

$$I(t) = \begin{cases} \dfrac{Nev_d}{W} = \dfrac{Ne}{T_R} & 0 \le t \le T_R \\[2mm] 0 & t > T_R, \end{cases} \tag{3.2}$$

where v_d is the drift velocity. The duration of the current transient signal is exactly equal to the time required for carriers to drift across the whole W region, and the drift velocity v_d is simply given by $v_d = W/T_R$.

It is also worth noting that (3.1) proves that virtually all the charge induced at the sample contacts, and consequently the current signal, is only due to the motion of the carriers which cross the whole sample width (electrons in the case shown in Fig. 3.1). In fact, the other type of carriers (holes in the case shown in Fig. 3.1) only travel a negligible fraction of the sample length before being collected and give a negligible contribution to the induced charge and to the transient signal. It is also evident that, in principle, this technique allows the observation of the charge pulses induced by the drifting of each type of carrier separately in the same sample. This is easily obtained either by reversing the polarity of the applied voltage V_a or by bombarding the other contact.

The current waveforms observed in experiments can be different from the ideal ones predicted by (3.2), and provided that the above discussed conditions are fulfilled, the following events can be identified:

1) a nonuniform electric field is present in the sample, particularly when the examined devices are Schottky or p–n junctions;
2) space-charge effects are present when the density of the carriers generated by the ionizing radiation is so high that the electric field inside the sample is strongly disturbed;
3) trapping and detrapping effects are present;
4) thermal diffusion phenomena are present.

3.2.2 Ionizing Sources

Three main different ionizing sources have been used in ToF experiments: heavily charged particles of low energy [3.20], short light flashes [3.21], and electron bursts [3.8]. Their main characteristics and advantages are briefly recalled in what follows.

Low-energy charged particles, such as 5 MeV α particles emitted by a natural Am^{241} α source, find their main advantage in experimental simplicity. However, each α particle generates a small number of electron-hole pairs in a very small volume, a few μm^3, with a consequent high local density. Owing to the small number of total pairs generated, an amplifier is required to observe the current or voltage transient signal while, owing to the high density of pairs generated, the space-charge effect cannot be neglected.

Short-duration light pulses emitted by light-emitting diodes and/or by lasers [3.22] are used again primarily for their experimental simplicity. However, the wavelength of the source must be matched with the energy gap of the material under study and difficulties may arise particularly with materials with a high-energy gap. Furthermore, the contact of the bombarded sample must be transparent to light.

A pulsed electron gun which generates electron bursts is the most used, advantageous, and versatile ionizing radiation source. In fact, 5 to 30 keV electrons have enough energy to pass through the sample contact and penetrate a few microns in the material.

With either of the last two sources, the bombarded area is large, about 10 mm^2, so that the total number of generated carriers can be high enough to monitor the induced signal directly without an amplifier, while the density can be kept low enough to neglect space-charge effects. Furthermore, with these sources the arrival time of the ionizing radiation can be controlled, and pulsed voltages can be applied to the sample to avoid breakdown and Joule heating.

3.2.3 Materials and Samples

In order to obtain an experimental determination of T_R, the material studied must be characterized by values of the lifetime τ and mean free-drift time before trapping, τ^+, of the drifting charge carriers longer than T_R. Furthermore, the dielectric relaxation time should always be longer than T_R. This last condition is clearly satisfied with semi-insulating materials ($\varrho > 10^8\ \Omega$cm), while for the case of materials with medium-low resistivities, like Si and Ge, it implies the use of surface-barrier diodes or $p^+ - v - n^+$ and $n^+ - \pi - p^+$ junctions where p^+ and n^+ are heavily doped thin layers of semiconductor material transparent to bombarding radiation. In fact, in this case the space-charge region, which arises by reversely biasing the junction, is practically depleted of free carriers so that it presents an ideally infinite value of ϱ and thus of τ_ε, and constitutes the drifting region W.

The preparation of samples to be used in ToF experiments requires an appropriate choice of contact characteristics and sample dimensions. In order to minimize both the induced noise and the noise-to-signal ratio, and to avoid breakdown and Joule's heating of the sample, a small or negligible *dc* current should flow across the device when the bias is applied. This goal can be simply achieved by using ohmic contacts in the case of semi-insulating materials, whereas for materials like Si and Ge, in which the resistivity is not high enough to assure a negligible low dc current, a reverse-bias surface-barrier diode or a p–n junction must be used.

The contacts used most frequently are obtained by evaporating thin metal films which give ohmic or rectifying contacts, in particular their barrier height at the metal-semiconductor interface should be high enough to prevent carrier injection and/or give satisfactory low values of the reverse current. Typically, gold or aluminium is used. Moreover, films of these metals a few hundreds Å thick are transparent enough to ionizing radiation.

Sample dimensions should be selected according to the following criteria. Large capacitance (> 1 to $2\ \text{pF}$) of the devices due to an area which is too large and/or a thickness which is too small should be avoided; the requirement $W \gg R$ must be satisfied. Even considering the availability of a fast sampling oscilloscope, transit times measurable with sufficient accuracy ($T_R \gg$ a few hundred ps) should be observed. Either transit times longer than τ^+, or excessive sample thicknesses causing the amplitude of the current signal (which is proportional to $1/W$) to become too small, should be avoided. Typical dimensions of samples used in ToF experiments are thickness 0.2 to 1 mm, and contact area $\leq 10\ \text{mm}^2$.

3.2.4 Electric Field Distribution

When reverse-bias surface-barrier diodes or p–n junction structures are used, the field distribution in the space-charge region, where carriers drift, is not uniform and its space dependence must be considered in evaluating the drift velocity $v_d(E)$ curves. For both structures, surfaces, surface-barrier or $p^+ - v - n^+$ (and $n^+ - \pi - p^+$) diodes, the field distribution can be calculated as for a one-dimensional abrupt junction where the space-charge region extends only in the lightly doped semiconductor material.

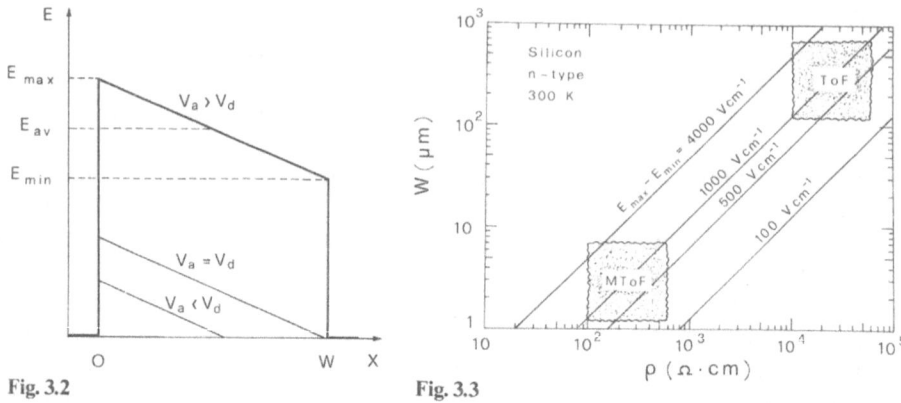

Fig. 3.2   Fig. 3.3

Fig. 3.2. Electric field distribution in the space-charge region of a reverse-bias abrupt p-n junction or of a surface-barrier diode. For $V_a < V_d$ the field strength decreases down to zero linearly with the distance. For $V_a > V_d$ the field distribution is a superposition of a constant field, $E = E_{min}$, with the linearly decreasing one

Fig. 3.3. Length versus resistivity of the active region of the sample. The lines show the maximum deviation of the electric field inside a reverse-bias abrupt p-n junction or inside a surface-barrier diode. Data refer to n-Si at room temperature. Dotted areas represent typical thickness and resistivity ranges of the devices used with ToF and MToF techniques, respectively

Accordingly, from Poisson's equation the field distribution depends on the net density, $N_I = |N_d - N_a|$, of ionized centers (N_d, N_a being the donors' and acceptors' concentrations). The width of the space-charge region increases proportionally to the root square of the reverse applied voltage up to a depletion of the whole sample thickness W; this occurs for $V_a = V_d = eW^2 N_I/(2\varepsilon\varepsilon_0)$ (Fig. 3.2). For $V_a > V_d$ the maximum deviation of the electric field inside the sample from the average value, $E_{av} = V_a/W$, is found to be

$$\frac{E_{max} - E_{min}}{E_{av}} = \frac{2v_d}{V_a} = \frac{eW^2 N_I}{\varepsilon\varepsilon_0 V_a}. \tag{3.3}$$

Thus the value of $2V_d/V_a$ increases with increasing W and N_I, while it decreases with a higher applied voltage V_a, as shown in Fig. 3.2.

Figure 3.3 illustrates, for n-type Si, how the maximum deviation of the electric field inside a device is correlated to the resistivity and thickness of the sample. It can be seen that, in the case of a high-resistivity material (10 to 60 kΩcm), the value of $(E_{max} - E_{min})$ can be kept lower than 1 kV/cm even in samples with thicknesses ranging from 100 to 600 μm, as normally occurs in ToF. On the other hand, in the case of a material of medium-low resistivity (0.1 to 0.6 kΩcm), in order to keep the same value of $(E_{max} - E_{min})$ the thickness of the sample used must be reduced down to 1 to 7 μm, as is normally done with the microwave-time-of-flight technique (Sect. 3.3).

3.2.5 Current and Voltage Transients

From the electrical point of view, the sample excited by the ionizing radiation can be considered the equivalent of the circuit shown in Fig. 3.4. It consists of a current generator $I(t)$ with the device capacitance C_d and a stray (or parasitic) capacitance C_s in parallel. The series resistance R_s accounts for the resistances of the back contact of the sample or of the interconnections. When device fabrication and holder are optimized, both C_s and R_s can be neglected.

The transient pulse induced by the drifting of charge carriers can be observed as a current waveform $I(t)$, when a low-resistance impedance system (current amplifier) is used, or as a voltage transient $V(t)$, when an amplifier with a capacitive input impedance is employed (Fig. 3.4). To measure the current signal, a fast and thus widely used system is obtained by taking the signal with a $50\,\Omega$ coaxial cable directly connected to a low input impedance, $R_0 = 50\,\Omega$, of a sampling oscilloscope. The rise time of this system is given by $R_0 C'$, where C' is the parallel of the device capacitance C_d and the parasitic capacitance C_s.

The voltage signal is just the integral of the current pulse:

$$V(t) = \frac{1}{C_i} \int_0^t I(t)\,dt, \tag{3.4}$$

which, for the case of a uniform electric field, leads to

$$V(t) = \begin{cases} \dfrac{eN}{C_i} \dfrac{t}{T_R} & 0 \le t \le T_R \\[2ex] \dfrac{eN}{C_i} & t > T_R, \end{cases} \tag{3.5}$$

where C_i, the input capacitance of the amplifier, is much greater than C_d and C_s. The voltage signal is reported and compared to the current signal in Figs. 3.1, 4.

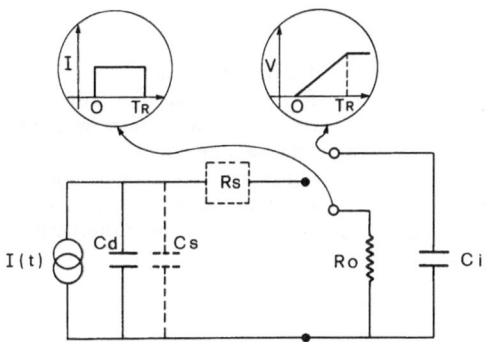

Fig. 3.4. Equivalent circuit of excited sample in ToF experiments. Inserts show current and voltage signals observed when a resistive (R_0) or a capacitive (C_i) impedance is connected to the device terminals

3.2.6 Trapping and Detrapping

When a uniform distribution of traps is present inside the material, trapping and detrapping (capture and release) processes can be properly described by the mean free-drift time τ^+ (i.e., the average time between two subsequent trapping events), and the detrapping time τ_D (i.e., the time spent in the trap by a captured carrier before its release), which are respectively defined as

$$\tau^+ = \frac{1}{N_T \sigma_c v_{th}}, \tag{3.6}$$

$$\tau_D = \frac{1}{N_T \sigma_e v_{th}} \exp\left(\mathscr{E}_T / K_B T_0\right). \tag{3.7}$$

Here N_T is the density of the trap centers; $\mathscr{E}_T$ their energy in the band gap; σ_c and σ_e the cross sections for capture and emission of a charge carrier; v_{th} the thermal velocity of free carriers; and T_0 and K_B the absolute temperature and the Boltzmann constant, respectively.

When trapping and detrapping processes occur during the carrier flight, the current or voltage waveforms observed experimentally may differ from those predicted by (3.2) or (3.5). Without giving a detailed and exhaustive discussion, we recall here only the three main effects produced by trapping and detrapping on current waveforms (Fig. 3.5).

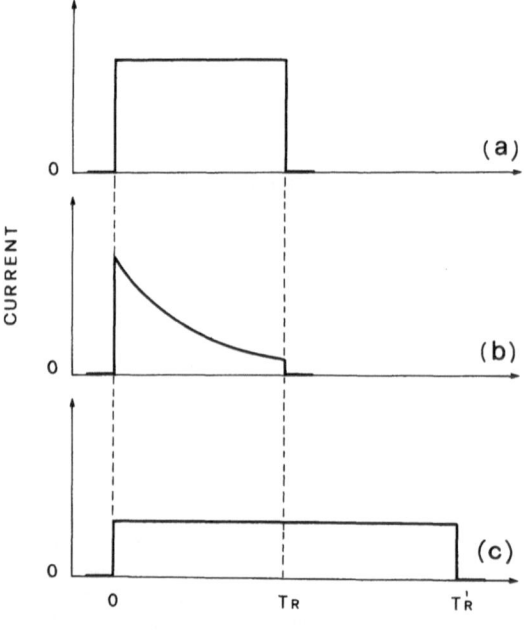

Fig 3.5a–c. Possible current waveforms obtained with ToF measurements: (a) no trapping is present; (b) trapping but no detrapping occurs; (c) trapping and detrapping are present with characteristic times much shorter than the transit time

When $\tau^+ \gg T_R$, trapping phenomena do not affect the carrier-drift velocity and consequently the current signal is equal to that reported in (3.2) (Fig. 3.5a).

When τ^+ is comparable with T_R, but the conditions $\tau_D \gg \tau^+$ and $\tau_D \gg T_R$ are fulfilled, trapping occurs but no detrapping takes place during the transit time, and the current signal is given by

$$I(t) = \begin{cases} \dfrac{eN}{T_R} \exp\left(-t/\tau^+\right) & 0 \le t \le T_R \\ \\ 0 & t > T_R, \end{cases} \tag{3.8}$$

as shown in Fig. 3.5b. From this waveform it is possible to determine both T_R and τ^+. In fact, τ^+ can be obtained from the exponential decay of the current signal.

When trapped charges are released in a time comparable with the transit time $(\tau^+ \le T_R, \tau_D \le T_R)$, the detrapping process severely degrades the fall down to zero of the current signal at T_R, and thus the transit time is no longer easily determinable. A simple case which leads to a constant amplitude current signal is observed when multiple capture and release events occur during the flight of carriers, i.e., when τ^+ and τ_D are much shorter than T_R. In this case the charge cloud will cross the sample in an extended transit time (Fig. 3.5c) given by

$$T_R' = T_R + \frac{T_R}{\tau^+}\tau_D, \tag{3.9}$$

where the term $(T_R/\tau^+)\tau_D$ accounts for the total time spent by carriers in the traps. In other words, carriers drift with a reduced velocity W/T_R', or, as is usually said, with a reduced mobility.

A detailed analysis of the current waveform for different values of the ratios τ^+/T_R and τ_D/τ^+ can be found in [3.23–25].

3.2.7 Space-Charge Effects

In the previous sections an ideal model which neglects the electrostatic forces among drifting carriers has been considered. This model is physically plausible if the created charge density is small enough so that the associated self-field can be considered as negligible with respect to the applied field. When this assumption is no longer fulfilled, space-charge effects must be accounted for and the current signals observed in experiments differ from those given in (3.2). The two following cases are briefly considered.

Broadening of the Carrier Sheet. With an increasing density of the created carriers we first observe a broadening effect of the carrier sheet during its flight. This occurs when the density of the carriers remains low enough so that after being generated by the ionizing radiation they immediately drift under the influence of the electric field, but, during the flight, the repulsion due to space charge spreads the drifting cloud. Thus arriving at the opposite electrode as an enlarged sheet, the carriers are collected at different times and give rise to a long fall time in the current pulse. This case will be discussed in more detail in connection with the diffusion measurements (Sect. 3.5).

Space-Charge-Limited Current. For high density of electron-hole plasma, the field in the sample may be shielded inside the plasma itself. Therefore, electrons and holes cannot be immediately separated, but the plasma they form is slowly eroded by the electric field. If the number of carriers created near the bombarded contact is so high as to be comparable to an infinite reservoir, space-charge-limited-current conditions are satisfied.

In this case the current pulse exhibits a characteristic cusp with a time position approximately equal to $0.8\,T_R$, irrespective of the density, as illustrated in Fig. 3.6. As can be seen in this figure, for a low density of carriers the current signal shows a rectangular shape with a width at half maximum corresponding to the transient time through the W region. When the density of carriers is increased, the current signal exhibits a shape similar to the one predicted by the theory of space-charge-limited current, and a cusp appears at about $0.8\,T_R$. Although the position of the cusp is found to be independent of charge density, the shape and amplitude of the signal depend on the density.

The physical reason for this pulse shape is the following. Electrons advance into the sample, and the entire space charge between the cathode and their leading front remains filled with electron space charge because of the electron reservoir at the cathode. As a consequence, the electric field inside the sample is strongly perturbed and, in particular, the electrons near the leading front

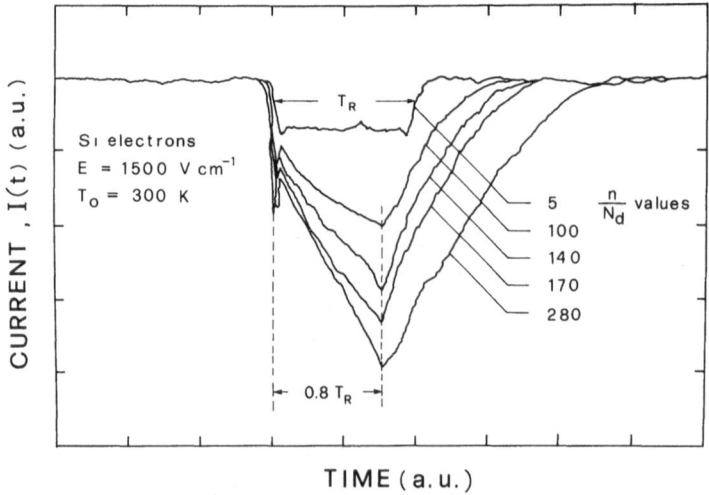

Fig. 3.6. Current signals due to the motion of electrons for five initial densities of created charge carriers. Signals were obtained from a Si surface-barrier diode, $W = 390\,\mu\text{m}$, $\varrho = 185\,\text{k}\Omega\text{cm}$

undergo a field greater than the average value V_a/W, so they arrive at the anode in a time $(0.8\,T_R)$, which is shorter than that necessary with a constant electric field, $E = V_a/W$.

3.3 Microwave-Time-of-Flight Technique

A major limitation of ToF lies in the practical impossibility to measure, even with a sampling oscilloscope, transit times shorter than a few hundred ps. In the length scale, this means that the drift region cannot become smaller than about one hundred μm. This limitation is undesired for two reasons, as can be seen in Fig. 3.3. In fact, a reduction of the sample thickness can help to minimize the electric field variation inside the drift region as well as to apply the ToF to materials of medium-low resistivity ($\leq 10^2\,\Omega$cm). Furthermore, epitaxial layers (either of Si or of III–V compound semiconductors) with dimensions in the order of 1 to 10 μm are widely used in devices, and consequently a large interest exists for characterizing their transport properties.

This limitation can be overcome by employing the microwave-time-of-flight (MToF) technique which, in principle, is based on the same assumptions of ToF; however, it introduces a sinusoidal modulation of the ionizing radiation, so that, in place of measuring a transit time, an amplitude and a phase measurement of the modulated induced current is performed.

When, during the carrier flight, thermal diffusion, trapping-detrapping, and avalanche multiplication effects are negligible, the induced microwave current $\tilde{I}$, with the device terminals short-circuited, can be written as [3.17]

$$\tilde{I} = \tilde{I}_0 \, \frac{\sin\,[\omega W/(2\,v_d)]}{\omega W/(2\,v_d)} \, \exp\left[\,i\left(\omega t - \frac{W}{2\,v_d}\right)\right]. \tag{3.10}$$

In (3.10) all ac quantities are assumed to vary as $\exp(i\omega t)$, ω being the angular frequency, while $\tilde{I}_0$ is the microwave current induced immediately under the bombarded contact. It can be seen from (3.10) that when the drift velocity and, in turn, the applied voltage change, the amplitude of the microwave current signal varies as $\sin\,[\omega W/(2\,v_d)]/\omega W(2\,v_d)$ while the phase varies as $\omega W/(2\,v_d)$. As a consequence, from the knowledge of ω and W, the drift velocity can be obtained by measuring the amplitude and/or the phase of the induced current signal. Unfortunately, in the experiments only relative changes in phase or in amplitude can be conveniently measured as a function of the applied voltage. Thus when the velocity-field curve is to be determined from the change in amplitude (or phase) of the output current with the bias voltage, it will be necessary:

1) to have "a priori" knowledge of at least one point in the curve (for example, from the low-field mobility value), or
2) to choose the sample thickness so that a null or maximum in the microwave current output can be observed for given values of sample bias.

In the latter case, since the current amplitude takes the functional form $\sin x/x$, its extremes are given by the condition

$$\omega W/(2 v_d) = m\pi. \qquad (3.11)$$

The amplitude is thus zero for $m = 1, 2, 3, \ldots$, while realtive maxima occur at $m = 1.430, 2.459, 3.471, \ldots$. The experimental observation of an extreme leads directly, by guessing the value of m, to the carrier-drift velocity through (3.11).

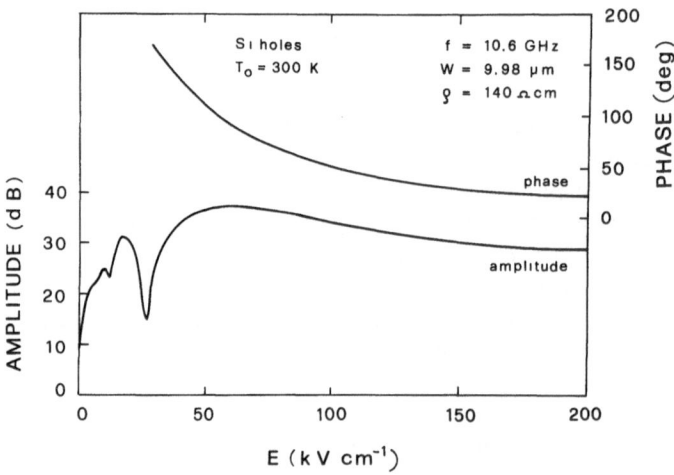

Fig. 3.7. Relative amplitude and phase of test signal measured with MToF technique as a function of the applied field [3.26]

Figure 3.7 shows the experimental measurements of the amplitude and phase as a function of the applied voltage when holes drift in a Si sample at room temperature. Two minima and three maxima are observed in the amplitude, from which a total of five points of velocity-field characteristics are obtained. Furthermore, the analysis of phase data provides a continuous $v_d(E)$ curve covering fields ranging from 30 to 200 kV/cm. Since only relative-phase variations can be measured, the $v_d(E)$ curve is determined using as reference the absolute points obtained from the amplitude extremes.

Recently the experimental apparatus has been improved [3.27] so that it can operate over a range of microwave frequencies (variable frequencies microwave time of flight). In this case, for the same sample a large number of minima and maxima in the current output can be obtained by changing ω; these extremes correspond to different absolute values in drift velocity according to (3.11), so that a complete $v_d(E)$ curve can be more easily built using directly measured values of v_d.

3.4 Drift Velocity Measurements

In the following, the experimental data of drift velocity for electrons and holes will be surveyed. The uncertainty in the measurements is estimated to be within 5% [3.9]. When available, the theoretical calculations which are based on the microscopic models developed in Chap. 2 have been included for the sake of completeness.

3.4.1 Electrons

Figures 3.8,9 report the results for the case of diamond and Si. The dependence on the field of drift velocity is very similar for these materials because of their similar band structure. In fact, anisotropy with respect to the direction of the

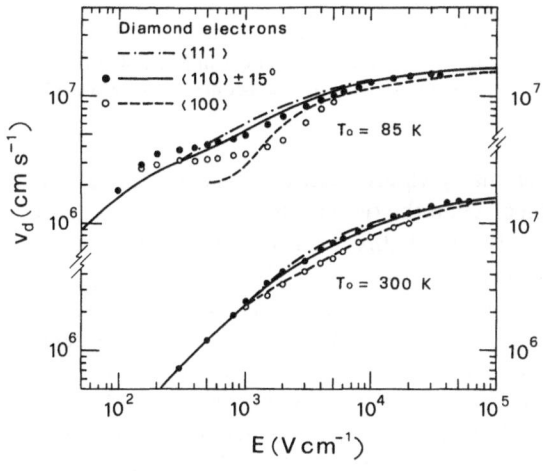

Fig. 3.8. Drift velocity of electrons as a function of electric field strength in diamond. ($\bullet$, O) and (——) refer to experimental and Monte Carlo theoretical results, respectively [3.28]

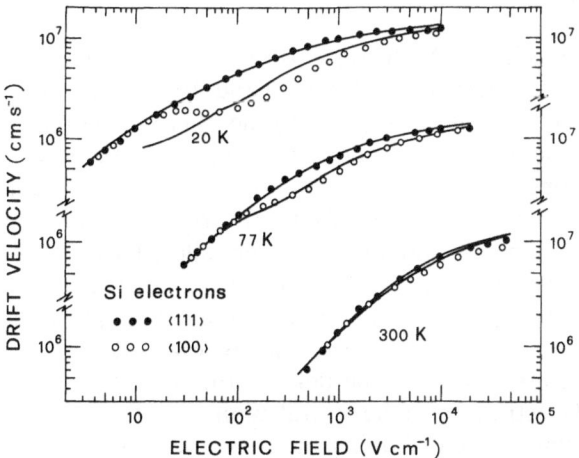

Fig. 3.9. Drift velocity of electrons as a function of electric field strength in Si. ($\bullet$, O) and (——) refer to experimental and Monte Carlo theoretical results, respectively [3.9]

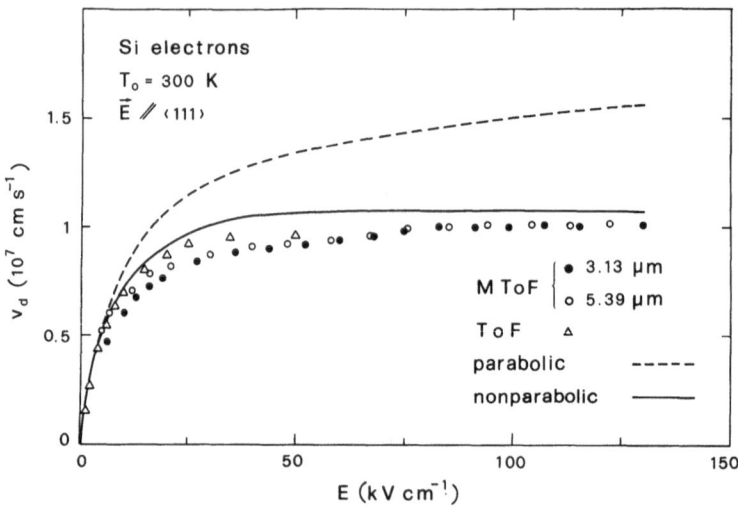

Fig. 3.10. Drift velocity of electrons as a function of electric field strength in Si as obtained with MToF ($\circ$, $\bullet$) and with ToF ($\triangle$) [3.29]; ($---$, ———) report theoretical Monte Carlo results [3.30]

applied field is correlated to the many-valley characteristic of the electron conduction band. As a consequence, when the field is along a $\langle 100 \rangle$ direction, a repopulation of the valleys along the field direction at the expense of those perpendicular to the field direction produces the net effect of a $v_{d111} \geq v_{d100}$. Saturation effects at the highest fields due to the emission of high-energy phonons and to the nonparabolicity of the conduction band are better evidenced by lowering the temperature, as predicted in Sect. 2.13. Figure 3.10 shows MToF measurements for electrons in Si. Data are found to be in reasonably good agreement with ToF results in the common range of electric fields and extend ToF results up to about 130 kV/cm. At the highest-field strengths the nonparabolicity of the band is shown to be crucial for interpreting the experiments.

Figure 3.11 refers to electrons in Ge. Anisotropy effects are again correlated to the many-valley structure of the electron conduction band. It is worth noting that since the minima of the conduction band lie along the $\langle 111 \rangle$ direction, as opposed to the case of diamond and Si, one finds $v_{d100} \geq v_{d111}$. At the lowest temperature and for fields along the $\langle 100 \rangle$ direction above about 1 kV/cm, a region of negative differential mobility has been found. It originates from the scattering between absolute minima at L points and upper valleys located at Γ and near X points in the Brillouin zone. This effect is better evidenced by lowering the energy separation between L and X minima through the application of an external uniaxial pressure, as shown by the results of Fig. 3.12.

Figure 3.13 reports the results of GaAs. The polar nature of the material coupled with the scattering of electrons form the central minimum at Γ to higher minima located at L and X points in the Brillouin zone are at the basis of a very

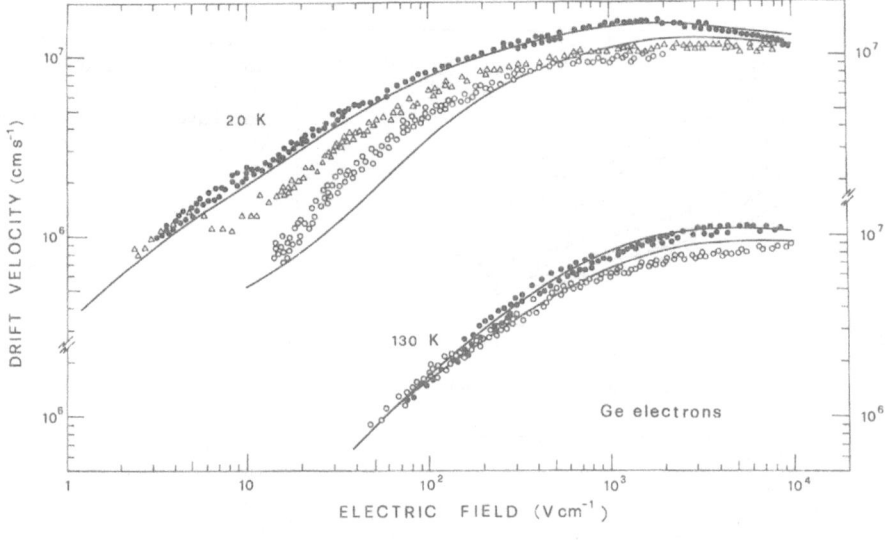

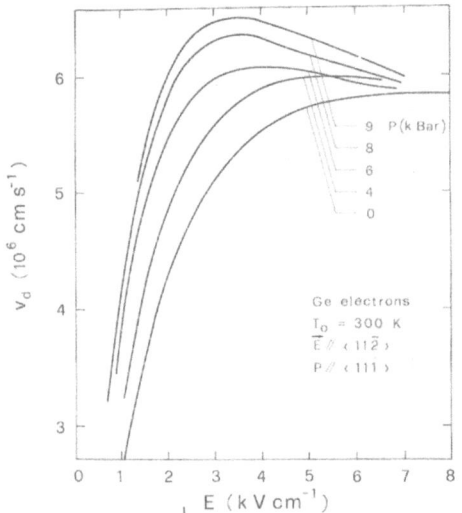

Fig. 3.11. Drift velocity of electrons as a function of electric field strength in Ge. (——) refers to Monte Carlo theoretical results, symbols to experiments. (O) refers to $E \parallel \langle 111 \rangle$, (●) to $E \parallel \langle 100 \rangle$ both with $n = 10^{12}$ cm^{-3}; (△) refers to $E \parallel \langle 111 \rangle$ with $n = 3\ 10^{13}$ cm^{-3} [3.15]

Fig. 3.12. Drift velocity of electrons in Ge as a function of electric field strength for different values of pressure as indicated in figure. The current flows in the $\langle 11\bar{2} \rangle$ direction and the uniaxial pressure is applied along the $\langle 111 \rangle$ direction [3.31]

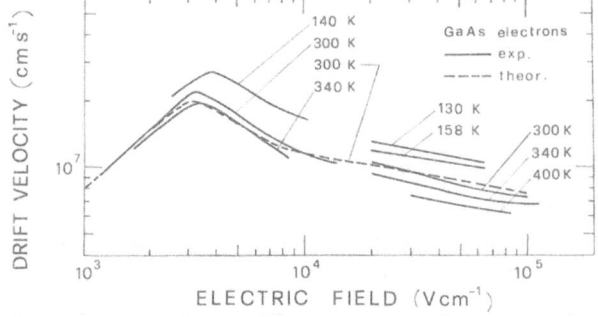

Fig. 3.13. Drift velocity of electrons as a function of electric field strength in GaAs. (——) represent experimental data [3.5,32]; (– – –) indicates the Monte Carlo theoretical results [3.33]

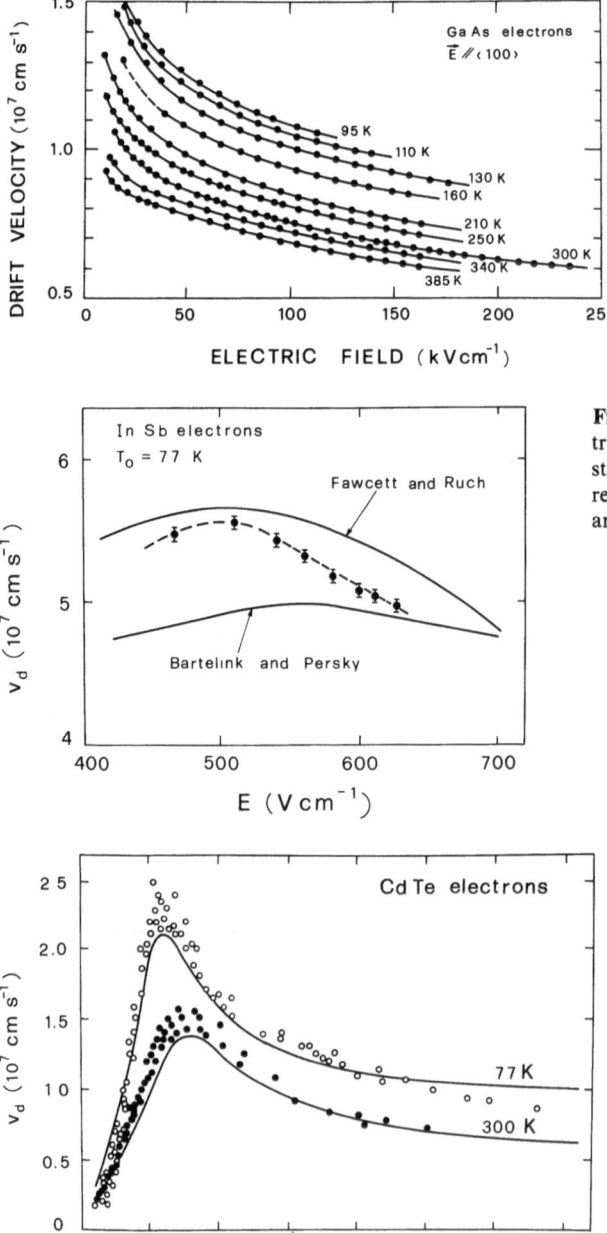

Fig. 3.14. Drift velocity of electrons as a function of electric field strength in GaAs. Data have been obtained with MToF [3.34]

Fig. 3.15. Drift velocity of electrons as a function of electric field strength in InSb [3.35]. The theoretical calculations from [3.36,37] are also indicated

Fig. 3.16. Drift velocity of electrons as a function of electric field strength in CdTe. (●, ○) and (——) refer to experimental and Monte Carlo theoretical results [3.38]

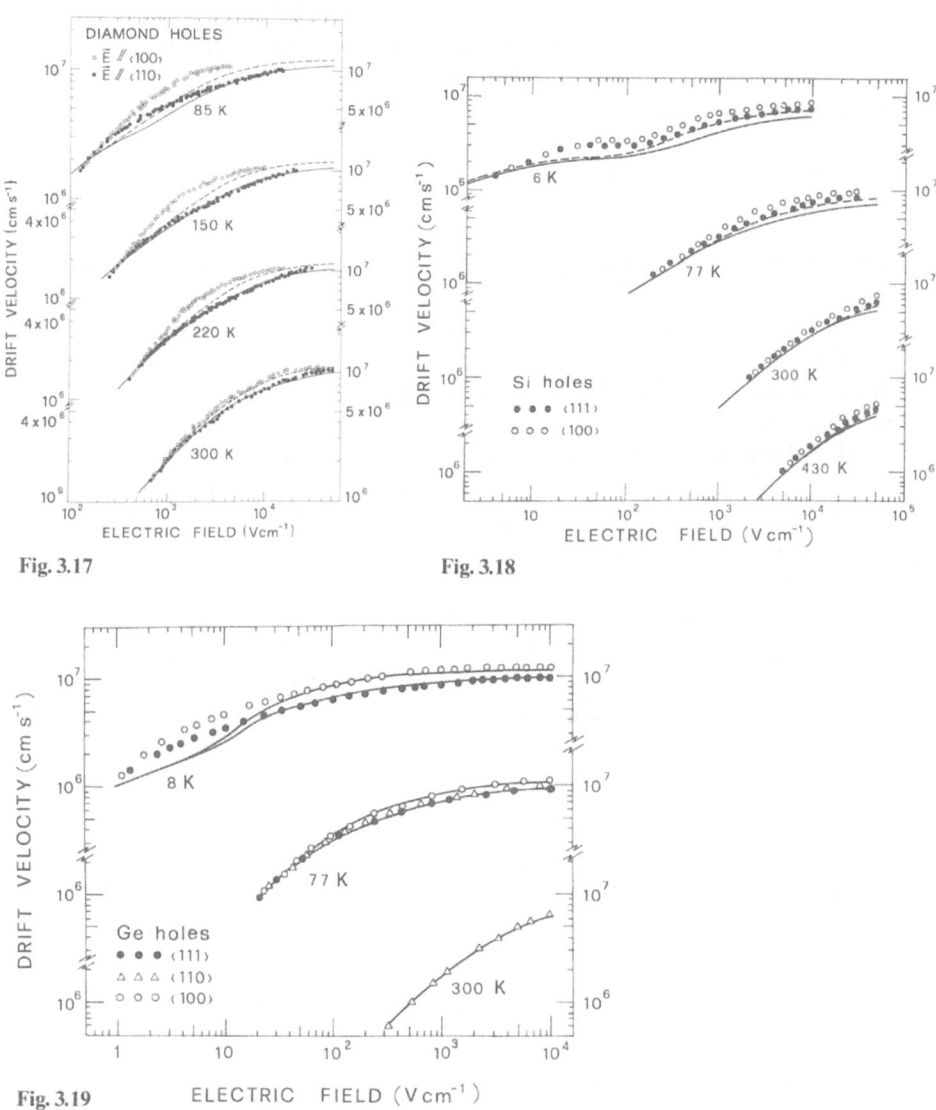

Fig. 3.17

Fig. 3.18

Fig. 3.19   ELECTRIC   FIELD $(V\,cm^{-1})$

Fig. 3.17. Drift velocity of holes as a function of electric field strength in diamond. (●, ○) and (——, ---) refer to experimental and Monte Carlo theoretical results, respectively [3.39]

Fig. 3.18. Drift velocity of holes as a function of electric field strength in Si. (●, ○) and (——, ---) refer to experimental and Monte Carlo theoretical results, respectively [3.39]

Fig. 3.19. Drift velocity of holes as a function of electric field strength in Ge. (●, ○, △) refer to experimental data; (——) indicate the Monte Carlo theoretical results [3.40]

pronounced effect of a negative differential mobility region for fields above a threshold value of about 3.5 kV/cm. The highest-field region has been investigated mainly with the MToF technique and results are summarized in Fig. 3.14. Analogous results for the case of InSb and CdTe are reported in Figs. 3.15 and 16, respectively.

3.4.2 Holes

Figures 3.17, 18, and 19 report the results for the cases of diamond, Si, and Ge, respectively. Anisotropy with respect to field direction is correlated to the warped shape of the heavy-hole band (Sect. 2.7), which is found to give the main contribution for the determination of the total drift velocity. Accordingly, the higher values of the drift velocity when the field is applied along a ⟨100⟩ direction with respect to a ⟨111⟩ direction are ascribed to an effective mass exhibited by heavy holes along a ⟨100⟩ direction smaller than that exhibited along a ⟨111⟩ direction. It is worth noting that the effect of nonparabolicity due to spin-orbit interaction (Sect. 2.7.3) is maximum in diamond, for which, in the whole region of fields and temperatures, the asymptotic value of the heavy-hole effective mass $m_h/m_0 = 1.08$ (m_0 being the free-electron mass) is needed to interpret the experiments. The same effect is intermediate in Si, for which it should be responsible for an approximately saturated region of v_d at 6 K and for electric fields ranging from 20 to 150 V/cm, and minimum in Ge, for which a 10% decrease of drift velocity values at the highest fields is theoretically predicted.

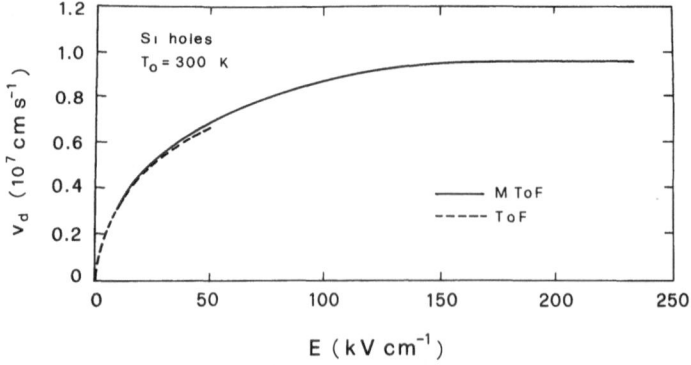

Fig. 3.20. Drift velocity of holes as a function of electric field strength in Si obtained with MToF and ToF techniques [3.26]

Figure 3.20 shows, for the case of Si, MToF results which are found to be in reasonable agreement with ToF results in the common range of fields and which extend previous ToF results up to about 250 kV/cm.

3.5 Longitudinal Diffusion Coefficient Measurements

The application of ToF technique to determine the longitudinal diffusion coefficient D_{lo} requires several assumptions and a high degree of accuracy; therefore the principles of the method have been briefly surveyed in the following. It is worth mentioning that, through an ingeneous modification of ToF technique, *Bartelink* and *Persky* [3.41] succeeded in measuring the transverse diffusion coefficient D_{tr} of electrons in Si. However, the difficulty in the preparation of a suitable sample has forbidden their "beam spreading technique" to be systematically applied to other materials.

3.5.1 Principles of the Method

During its motion between the contacts of the sample, the space distribution of carriers evolves in space and time because of thermal diffusion and Coulomb repulsion (space-charge) phenomena. Normally these phenomena lead to a broadening of the initial distribution which, in turn, produces primarily a value of the fall time τ_f of the induced current pulse longer than the rise time τ_r. Figure 3.21 illustrates the major modifications in the current waveform as expected from diffusion phenomena.

Apart from the contribution of the electronic rise time, which is common to both τ_f and τ_r (for normalization purposes here measured between 0.05 and 0.95 of the pulse height), τ_f can be analyzed in terms of the times τ_{diff} and τ_{fsc}, which indicate the contribution to the current signal rise time given by diffusion and space-charge phenomena, respectively. To do so we assume that the initial thickness x_0 of the electron layer is given by $x_0 = \tau_g v_d + R$, where τ_g is the excitation time (i.e., the time the ionizing source is active). If three conditions are satisfied – (i) $x_0 \ll x(T_R)$, where $x(T_R)$ is the final thickness of the electron layer;

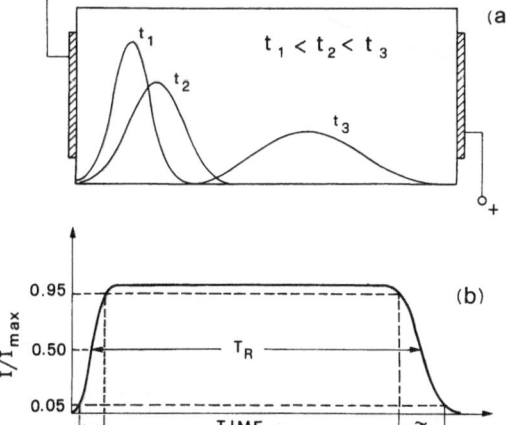

Fig. 3.21a, b. Schematic of the ToF technique used for the determination of the longitudinal diffusion coefficient. **(a)** Space distribution of electrons at three successive times as they travel across the sample towards the positive contact. **(b)** Current pulse induced at the sample contacts by the drifting carriers

(ii) $\tau_r < \tau_f$; and (iii) the electric field is uniform through the sample – τ_{diff} can be calculated in terms of macroscopic quantities as [3.5]

$$\tau_{diff} = (2.16 \, D_{lo} W / v_d^3)^{1/2}. \tag{3.12}$$

Furthermore, τ_{fsc}, which refers to the spreading of the drifting charge layer due to electrostatic forces, can be evaluated in the dielectric relaxation approach [3.42] as

$$\tau_{fsc} = (x_0 / v_d) \exp (T_R / \tau_\varepsilon') = (x_0 / v_d) \exp (I / I_c), \tag{3.13}$$

$$I_c = (\varepsilon \varepsilon_0 \mathcal{E}_e v_d^2 S) / (\mu' W \mathcal{E}_p). \tag{3.14}$$

Where the ratio between T_R and the differential dielectric relaxation time τ_ε' of the electron layer has been expressed in terms of the current of the impinging electrons originated by the ionizing source I, through a normalized current I_c, defined by the (3.14); thus the condition $I \ll I_c$ corresponds to a negligible space-charge effect. In (3.14) $\mathcal{E}_e$ is the energy of the impinging electrons, $\mathcal{E}_p$ is the energy necessary to create an electron-hole pair in the material, S the useful area of the sample, and μ' the differential mobility at the applied electric field.

When both diffusion and space-charge effects are present, τ_f can be expressed as

$$\tau_f = [(x_0 / v_d)^2 \exp (2 \, I / I_c) + \tau_{diff}^2 + \tau_r^2]^{1/2}, \tag{3.15}$$

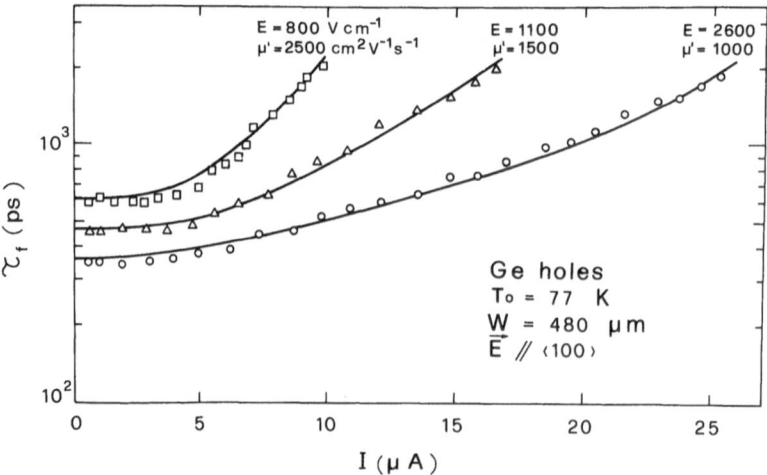

Fig. 3.22. Fall time τ_f of the current pulse as a function of the impinging electron current I for three different applied fields. (O, △, □) refer to experiments and (——) to theoretical calculations making use of (3.15) [3.12]

where it has been implicitly assumed that the current rate of each contribution is Gaussian (the reliability of this assumption has been analyzed [3.43–45]). Eq. (3.15) has been supported by experimental findings, and Fig. 3.22 reports typical results for the case of holes in Ge.

In view of the foregoing, this procedure can be adopted for the systematic collection of data on D_{lo}. To avoid space-charge effects, experimental measurements can be performed at a low current of impinging electrons (e.g., $I \simeq 5\,\mu A$). Then, providing the experimental conditions satisfy the requirements for the validity of (3.12), reasonably accurate data on D_{lo} can be obtained by substituting in (3.12) $\tau_{diff} = (\tau_f^2 - \tau_r^2)^{1/2}$, where τ_f and τ_r are deduced from the experimental current waveform (Fig. 3.21).

The total experimental error on D_{lo} due to the experimental uncertainty regarding v_d, τ_f, τ_r, and W can be estimated to be about 25% [3.12].

3.5.2 Results

In the following the experimental data on the longitudinal diffusion coefficient for electrons and holes will be surveyed. The theoretical calculations which are based on the microscopic models developed in Chap. 2 have been included for the sake of completeness.

3.5.3 Electrons

Figure 3.23 reports the results for the case of Si. Here data obtained with the noise-conductivity technique are compared with those obtained with ToF technique to show the reliability and complementarity of these two techniques. Anisotropy of the longitudinal diffusion coefficient with $D_{lo\,100} \geq D_{lo\,111}$ is related to an additional intervalley diffusion which arises owing to the location of the minima in the conduction band along the $\langle 100 \rangle$ directions. In fact, when the field is oriented along a $\langle 100 \rangle$ direction, the electrons, by exhibiting different mean velocities in different valleys, give rise to a further spreading in space which is responsible for the intervalley diffusion. In Fig. 3.24 the transverse diffusion coefficient measurements of *Persky* and *Bartelink* [3.46] are reported together with the longitudinal diffusion coefficient for E along the $\langle 111 \rangle$ direction at 300 K. The results are found to satisfy the condition $D_{tr} \geq D_{lo}$ in the whole range of field strengths.

Figure 3.25 reports the results for the case of Ge. In analogy with the case of Si, anisotropic effects are evident but, in this case, $D_{lo\,111} \geq D_{lo\,100}$ owing to minima in the conduction band located at L points along the $\langle 111 \rangle$ directions. It is worth noting that in Si and Ge the anisotropy effects on the longitudinal diffusion coefficient are reversed with respect to those on the drift velocity.

Figure 3.26 reports the results for the case of GaAs. The diffusion coefficient has an initial sharp increase; then it exhibits a maximum value at the threshold

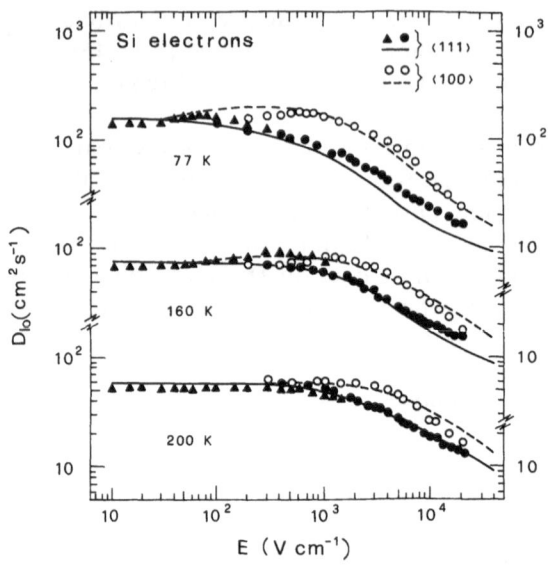

Fig. 3.23. Longitudinal diffusion coefficient of electrons in Si as a function of electric field strength. (●, ○) indicate ToF data, (▲) noise-conductivity data, (---, ——) the results of the Monte Carlo theoretical calculations [3.16]

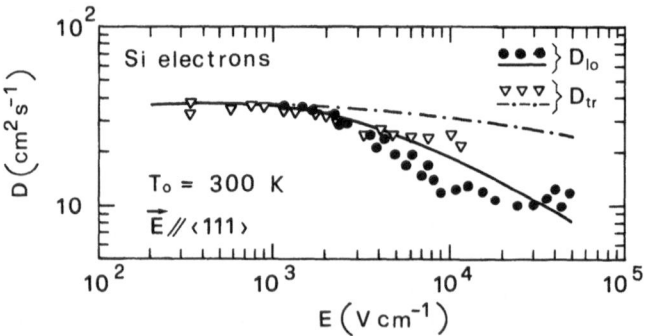

Fig. 3.24. Longitudinal and transverse diffusion coefficient of electrons in Si as a function of electric field strength along the ⟨111⟩ direction at 300 K. (●) and (▽) refer to experiments and (-·-, ——) to Monte Carlo theoretical calculations [3.16]

for negative differential mobility, finally decreasing systematically. The dominant role of polar optical interaction, the scattering efficiency of which decreases with increasing carrier energy, thereby leading to a net increase in the spatial spreading, is responsible for the initial increase of D_{lo}. Then the setup of intervalley interaction between the minimum of the conduction band and upper valleys, located at L and X points in the Brillouin zone, leads to a strong increase of the scattering efficency and is responsible for the maximum and the final decrease in diffusivity. To interpret the experiments, several theoretical models have been proposed. Three examples are shown in Fig. 3.26 by different lines.

Figure 3.27 reports the results for electrons in CdTe, which are quite similar to those of GaAs.

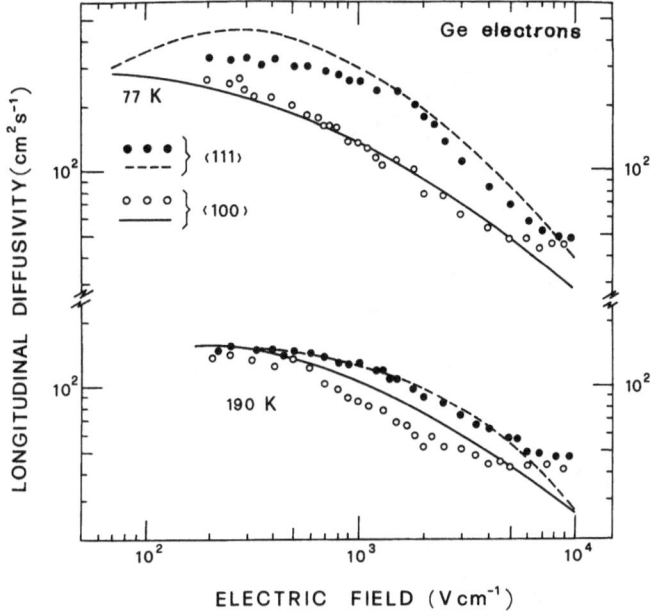

Fig. 3.25. Longitudinal diffusion coefficient of electrons in Ge as a function of electric field strength. (●,○) and (---, ——) refer to experiments and Monte Carlo theoretical calculations, respectively [3.15]

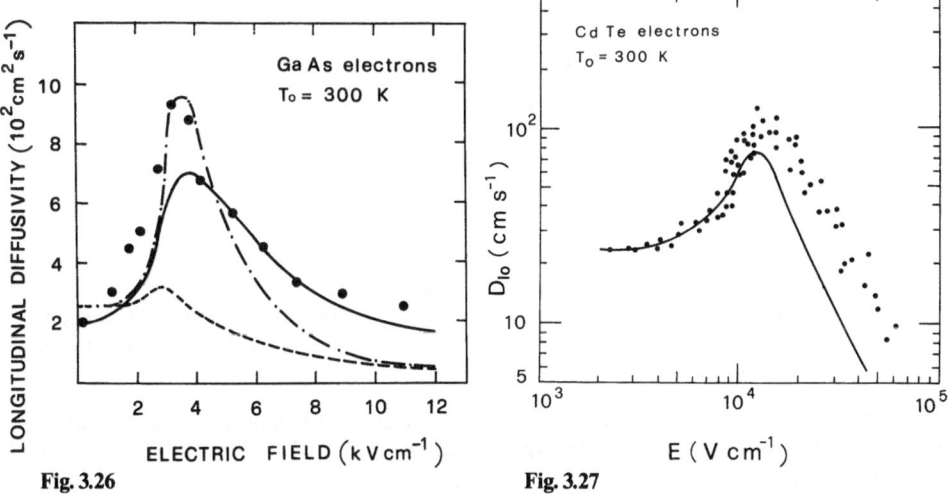

Fig. 3.26   Fig. 3.27

Fig. 3.26. Longitudinal diffusion coefficient of electrons in GaAs as a function of electric field strength. (●) indicate experimental results of *Ruch* and *Kino* [3.5]; the curves show the different theoretical results of *Fawcett* and *Rees* [3.47] (---), *Ohmi* and *Hasuo* [3.48] (——), and *Pozhela* and *Reklaitis* [3.33] (-·-)

Fig. 3.27. Longitudinal diffusion coefficient of electrons in CdTe as a function of electric field strength. (●) and (——) refer to experimental and Monte Carlo theoretical results, respectively [3.49]

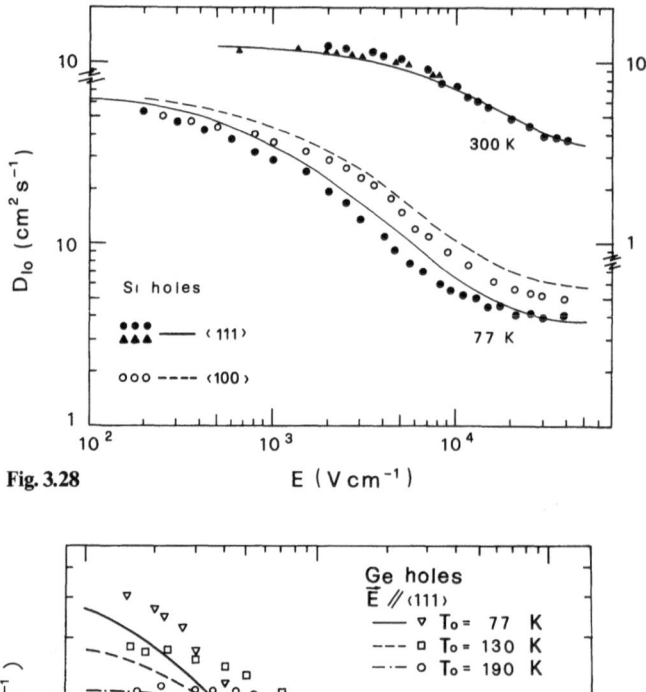

Fig. 3.28

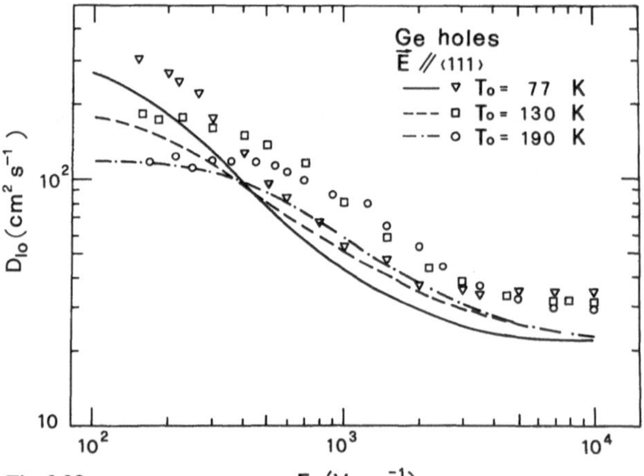

E (V cm⁻¹)

Fig. 3.29

E (V cm⁻¹)

Fig. 3.28. Longitudinal diffusion coefficient of holes in Si as a function of electric field strength. (●, ○) present experimental data obtained with ToF, (▲) with noise-conductivity technique; (---, ——) report Monte Carlo theoretical results [3.39]

Fig. 3.29. Longitudinal diffusion coefficient of holes in Ge as a function of electric field strength. (∇, □, ○) and (——, ---, —·—) refer to experimental and Monte Carlo theoretical results, respectively [3.12]

3.5.4 Holes

Figure 3.28 reports the results for the case of Si. At 300 K, data obtained with the noise-conductivity technique are compared with those obtained with ToF to confirm the reliability and complementarity of these techniques. Anisotropy of the longitudinal diffusion coefficient with $D_{lo\,100} \geq D_{lo\,111}$ is related to the warped shape of the heavy-hole band. In analogy with the anisotropy of the drift velocity, the smaller value of the effective mass when the electric field is oriented along a $\langle 100 \rangle$ direction, compared to a $\langle 111 \rangle$ direction, is at the origin of this effect.

Figure 3.29 reports the results for the case of Ge. It is worth noting that anisotropic effects analogous to the case of Si have been evidenced even for this material (Sect. 2.13.2).

3.6 Conclusions

This chapter has reviewed the working principles and the main results obtained with the time-of-flight techniques for hot carriers in covalent and compound semiconductors. The ToF technique has emerged as the most direct and reliable experimental technique for investigating both drift velocity and longitudinal diffusion coefficients. Even if limited to drift velocity, the MToF technique offers the possibility to perform measurements at very high electric fields (up to the limit of electrical breakdown) in materials of medium-low resistivity and very small thicknesses (down to the micrometer region), which are of growing interest in both basic and applied research.

Both drift velocity and diffusivity have been systematically analyzed in covalent semiconductors, while data on compound semiconductors are still somewhat scarce, apart from the case of electrons in GaAs.

In the near future some novel results for high-field drift velocity are to be expected by using the MToF technique, in particular, in the areas of ballistic transport and compound semiconductor heterolayers.

References

3.1 J.R.Haynes, W.Shockley: Phys. Rev. **81**, 835 (1951)
3.2 For a general review of the possibilities of the ToF technique as applied to semiconductors, see L.Reggiani: In *Physics of Nonlinear Transport in Semiconductors*, ed. by D.K.Ferry, J.R.Barker, C. Jacoboni (Plenum, New York 1980) p. 243
3.3 W.E.Spear: Proc. Phys. Soc. **70**, 669 (1957)
3.4 W.E.Spear: J. Non-Crystall. Sol. **1**, 197 (1969)
3.5 J.G.Ruch, G.S.Kino: J. Appl. Phys. **174**, 921 (1968)
3.6 T.W.Sigmon, J.F.Gibbons: Appl. Phys. Lett. **10**, 320 (1969)
3.7 A.Neukermans, G.S.Kino: Solid State Commun. **8**, 987 (1970)
3.8 A.Alberigi-Quaranta, C.Canali, G.Ottaviani: Rev. Scient. Instrum. **41**, 1205 (1970)
3.9 C.Canali, C.Jacoboni, G.Ottaviani, A.Alberigi-Quaranta: Phys. Rev. **12B**, 2265 (1975)
3.10 G.Ottaviani, L.Reggiani, C.Canali, F.Nava, A.Alberigi-Quaranta: Phys. Rev. **12B**, 3318 (1975)
3.11 L.Reggiani, C.Canali, F.Nava, G.Ottaviani: Phys. Rev. **16B**, 2781 (1977)
3.12 L.Reggiani, C.Canali, F.Nava, A.Alberigi-Quaranta: J. Appl. Phys. **49**, 4446 (1978)
3.13 F.Nava, C.Canali, L.Reggiani, D.Gasquet, J.C.Vaissiere, J.P.Nougier: J. Appl. Phys. **50**, 922 (1979)

3.14 L.Reggiani, S.Bosi, C.Canali, F.Nava, S.F.Kozlov: Phys. Rev. **23B**, 3050 (1981)

3.15 C.Jacoboni, F.Nava, C.Canali, G.Ottaviani: Phys. Rev. **24B**, 1014 (1981)

3.16 R.Brunetti, C.Jacoboni, F.Nava, L.Reggiani, G.Bosman, R.J.J.Zijlstra: J. Appl. Phys. **52**, 6713 (1981)

3.17 A.G.R.Evans, P.N.Robson: Solid State Electron. **17**, 805 (1974)

3.18 S.Ramo: proc. IRE **27**, 584 (1939)

3.19 W.Shockley: J. Appl. Phys. **9**, 635 (1938)

3.20 A.Alberigi-Quaranta, F.Cipolla, M.Martini: Phys. Lett. **17**, 102 (1965)

3.21 R.G.Kepler: Phys. Rev. **119**, 1226 (1960)

3.22 M.H.Evanno, J.L.Vaterkowski: Electron. Lett. **18**, 417 (1982)

3.23 J.W.Mayer: In *Semiconductor Detector*, ed. by G.Bertolini, A.Coche (North-Holland, Amsterdam 1968) Chap. 5

3.24 M.Martini, J.W.Mayer, K.R.Zanio: In *Applied Solid State Science*, Vol. 3 ed. by R.Wolf (Academic, New York 1972)

3.25 G.Ottaviani, C.Canali, C.Jacoboni, A.Alberigi-Quaranta, K.R.Zanio: J. Appl. Phys. **44**, 360 (1973)

3.26 P.M.Smith, J.Frey, P.Chatterjee: Appl. Phys. Lett. **39**, 333 (1981)

3.27 G.Hill, P.N.Robson: Solid State Electron. **25**, 589 (1982)

3.28 F.Nava, C.Canali, C.Jacoboni, L.Reggiani: Solid State Commun. **33**, 475 (1980)

3.29 P.M.Smith, M.Inoue, J.Frey: Appl. Phys. Lett. **37**, 797 (1980)

3.30 R.Brunetti, C.Jacoboni, L.Reggiani: unpublished results

3.31 A.Neukermans, G.S.Kino: Phys. Rev. **7B**, 2693 (1973)

3.32 P.A.Houston, A.G.R. Evans: Solid State Electron. **20**, 197 (1977)

3.33 J.Pozhela, A.Reklaitis: Solid State Commun. **27**, 1073 (1978)

3.34 T.H.Windhorn, T.J.Roth, L.M.Zinkiewicz, O.L.Gaddy, G.E.Stillman: Appl. Phys. Lett. **40**, 513 (1982)

3.35 A.Neukermans, G.S.Kino: Phys. Rev. **7B**, 2703 (1973)

3.36 G.Persky, D.Bartelink: IBM J. Res. Develop. **13**, 607 (1969)

3.37 W.Fawcett, J.Ruch: Appl. Phys. Lett. **15**, 368 (1969)

3.38 V.Borsari, C.Jacoboni: Phys. Status Solidi **54(B)**, 649 (1972)

3.39 L.Reggiani: J. Phys. Soc. Jpn. **49**, Suppl. A, 317 (1980)

3.40 C.Jacoboni, L.Reggiani: Adv. Phys. **28**, 493 (1979)

3.41 D.J.Bartelink, G.Persky: Appl. Phys. Lett. **16**, 191 (1970)

3.42 J.E.Carroll: *Hot Electron Microwave Generators* (Arnold, London 1970)

3.43 T.H.Glisson, R.A.Sadler, J.R.Hauser, M.A.Littlejohn: Solid State Electron. **23**, 627 (1980)

3.44 G.Duggan, F.Berz: J. Appl. Phys. **53**, 470 (1982)

3.45 D.F.Nelson: Phys. Rev. **25B**, 5267 (1982)

3.46 G.Persky, D.J.Bartelink: J. Appl. Phys. **42**, 4414 (1971)

3.47 W.Fawcett, H.D.Rees: Phys. Lett. **A29**, 578 (1969)

3.48 T.Ohmi, S.Hasuo: Proc. 10th Intern. Conf. on the Physics of Semiconductors (Cambridge USAEC, 1970) p. 60

3.49 C.Canali, F.Catellani, C.Jacoboni, R.Minder, G.Ottaviani, A.Alberigi-Quaranta: Solid State Commun. **17**, 1443 (1975)

4. Transport Parameters from Microwave Conductivity and Noise Measurements

Yuras K. Pozhela

With 26 Figures

The present chapter will analyze hot-carrier transport and fluctuation phenomena at microwave frequencies. The hot-carrier investigations by use of microwaves have been under way for more than 25 years. At the beginning, the microwave carrier heating was proposed to overcome the difficulties associated with the non-ohmic behavior of contacts at high fields. This enabled an average conductivity to be measured as a function of the microwave field amplitude from which the dc conductivity at field E, $\sigma(E)$, could be obtained by some numerical procedure. Later on, the microwave technique was successfully applied to materials exhibiting negative differential conductivity [i.e., $\sigma_d(E) < 0$], since it became possible to ensure conditions under which the space-charge redistribution in the semiconductor volume did not take place. Furthermore, as an alternative to the time-of-flight technique, the microwave technique was found to be especially suited for measuring the drift velocity versus field characteristics, $v_d(E)$, in high conductivity samples, or even under carrier avalanche multiplication conditions.

The frequency range covered by microwaves is confined from 10^9 to 10^{12} Hz, and the technique appears especially valuable for investigating various dynamical aspects of hot-electron behavior in elemental semiconductors, such as Si and Ge, and compound semiconductors, such as GaAs. In these materials the frequencies associated with the energy relaxation time and the momentum relaxation time lie just in this range of values. It should be pointed out that, in the last two decades, the progress in the microwave solid-state electronics has been largely due to the successful results achieved in the hot-electron studies. The knowledge of the material parameters at microwave frequencies is also of basic importance in optimizing the performance of the devices.

On the experimental side, the microwave technique, in which small dc voltages and currents can be measured in the presence of large microwave fields, has led to the discovery of a number of new effects, e.g., various electromotive forces induced by nonuniform microwave carrier heating.

On the theoretical side, the problem of nonequilibrium microwave current fluctuations has emerged as a fundamental one in hot-electron physics since the electron temperature and the relations between the diffusion coefficient, mobility, and the current-fluctuation spectral density, so useful in thermodynamic equilibrium, are no longer valid. It should be pointed out that $1/f$ and generation-recombination noises are practically absent at microwave frequencies. This allows one to separate experimentally the noise due to hot-electron velocity fluctuations from other sources of noise, such as $1/f$, generation-recombination, etc.

During the last few years, the theoretical and experimental studies of the microwave current fluctuations have brought about some clarity in the problem. In particular, under hot-electron conditions exact relationships between the diffusion coefficient and the current-fluctuation spectral density at microwave frequencies have been established.

The hot-carrier microwave investigations can be classified into the following groups: (i) determination of the microwave conductivity and related relaxation effects; (ii) analysis of the effects associated with nonhomogeneous carrier heating; (iii) analysis of the nonequilibrium hot-electron noise and diffusion. Accordingly, it is the aim of this chapter to report on these three subjects.

4.1 Microwave Conductivity

The microwave technique allows us to measure the dependence of the conductivity on the electric field; it also gives the conductivity relaxation time in the range 10^{-9} to 10^{-12} s. In this section we describe the application of the microwave technique to obtain hot-electron conditions and then to measure the drift velocity as a function of the electric field (Sect. 4.1.1), as well as the hot-electron energy and momentum relaxation times (Sect. 4.1.2).

4.1.1 Electron Heating by Microwave Fields

A typical experimental setup for measuring the conductivity at high microwave fields is shown in Fig. 4.1. The sample is placed inside a waveguide in the region where the electric field is maximum. With the aid of ohmic contacts, the sample is connected in series with a constant voltage source (a pulse generator or a battery) and a dc current meter. The microwave currents leaving the sample are shunted off by the capacitance formed between the upper and the lower guide walls and the sample. Thus, the dc and microwave currents are decoupled outside the

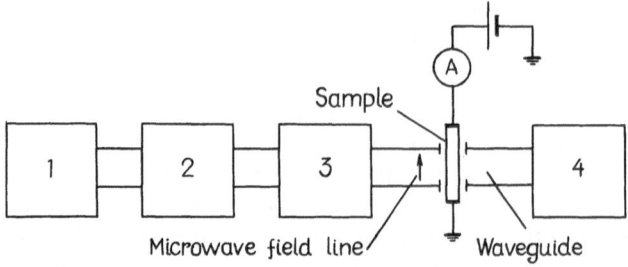

Fig. 4.1. Schematic of the setup for the measurement of the microwave conductivity: (*1*) microwave generator, (*2*) tuning elements, (*3*) incident and reflective power meters, (*4*) transmitted power meter

waveguide. The microwave circuit consists of a microwave generator, usually of a magnetron type (Block 1 in the figure); tuning elements (Block 2); and meters for measuring the incident and reflected power (Block 3), and transmitted power (Block 4).

Two kinds of fields exist in the semiconductor sample, the dc field E_{dc}, and the ac field $E_{ac} = E_1 \sin \omega t$. Depending on which of the above fields heats the electrons, two alternative techniques are possible. In the former, the *differential microwave technique*, the condition $E_{dc} \gg E_1$ is required; while in the latter, the *integral microwave technique*, the opposite condition $E_1 \gg E_{dc}$ should be satisfied.

In its essence, the differential microwave technique does not differ from the well-known microwave methods for measuring the complex dielectric constant. It has been used to investigate the small signal conductivity and the dielectric constant anisotropy as a function of the dc electric field strength and orientation [4.1, 2]. The experimental results, obtained by this differential technique, have been partly described in [4.3]; thus only the new results, concerning the energy relaxation time in semiconductors, will be discussed in Sect. 4.1.2.

The integral microwave technique has been introduced in [4.4–7] and a detailed description of the technique can be found in [4.3–9]. A number of substantial improvements and interesting modifications added during the last decade will be analyzed below.

The dc current density which flows through the sample under the simultaneous action of the homogeneous dc and microwave fields is

$$j_{dc} = \frac{1}{2\pi} \int_0^{2\pi} j(E_{dc} + E_1 \sin \omega t) d(\omega t), \tag{4.1}$$

with $E_{dc} = V_{dc}/L$, and where V_{dc} is the external dc voltage over the sample of length L. In order to find the current density versus field characteristics $j(E)$ from the experimentally measured values of $j_{dc}(E_{dc}, E_1)$, the integral equation (4.1) must be solved with respect to $j(E)$. Assuming that j is an odd function of E and that the current follows the microwave field instantaneously, under the condition $E_{dc} \ll E_1$, (4.1) reduces to

$$j_{dc} \approx \frac{2 E_{dc}}{\pi} \int_0^{\pi/2} \sigma' d(\omega t), \quad \text{where} \tag{4.2}$$

$$\sigma' = \frac{dj(E)}{dE}\bigg|_{E = E_1 \sin \omega t}$$

is the differential conductivity at field $E_1 \sin \omega t$ [4.3]. Equation (4.2) can be inverted with the aid of the Schlömilch's integral theorem [4.10]:

$$j(E_1) = \sigma_0 \int_0^{\pi/2} E_1 \sin \omega t \frac{j_{dc}(E)}{j_0}\bigg|_{E = E_1 \sin \omega t} d(\omega t). \tag{4.3}$$

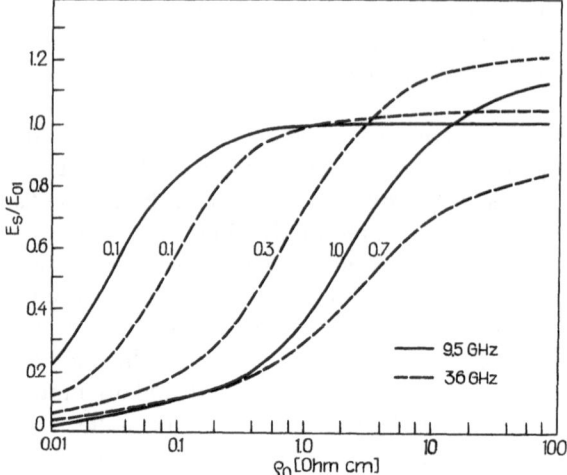

Fig. 4.2. Electric field E_s on the surface of a cylindrically shaped InSb sample, placed in the center of the X-band guide (——) and K-band guide (– – –), as a function of the resistivity. The values of the field are normalized to the field in the empty guide E_{01} and the numbers on the curves refer to the sample diameters in millimeters [4.14]

After a change of variables, one obtains

$$\frac{\sigma'(E_1)}{\sigma_0} = \int_0^{E_1} \frac{j_{dc}(E)}{j_0} \frac{E}{E_1} \frac{1}{\sqrt{1-(E/E_1)^2}} \frac{dE}{E_1}. \tag{4.4}$$

Here, $j_0 = \sigma_0 E_{dc}$ and $\sigma_0 = \lim_{E \to 0} \sigma'(E)$.

Equation (4.4) shows that to obtain the values of the differential conductivity at field E_1, one has to integrate the experimentally measured function j_{dc}/j_0 with the kernel $x/(1-x^2)^{1/2}$.

The main problems in applying the microwave integral technique are associated with the two following points: (1) the difficulty in determining the amplitude E_1 of the microwave field in the sample, (2) the presence of spurious effects due to higher harmonics, which are radiated either by the microwave source, or by the sample itself and reflected back to the sample from the discontinuities in the guide.

The determination of the value of E_1 in the semiconductor sample is rather involved because the sample itself may bring about severe distortions in the waveguide field pattern. The problem of the field distribution in a rod-shaped semiconductor sample was solved by many workers [4.9, 11–14], in most cases with the aid of computers. To simplify the problem, the experimenters often use the condition that the sample impedance Z_s is much greater than the guide impedance Z_g, so that the field in the sample E_1 coincides practically with the field E_{01} in the empty guide at the place where the sample is located. This condition can be satisfied if the sample diameter is small enough when compared to the microwave wavelength and the skin depth of the semiconductors. Figure 4.2 shows the field on the surface of a rod-shaped InSb sample, normalized to the field in the empty guide, as a function of the resistivity for various sample diameters [4.14]. It can be seen that in a high-resistivity material the field can be higher than in the empty guide while in a low-resistivity material it can be lower. Within an uncertainty of 5%, one can assume that, for $\varrho_0 > 5\,\Omega$cm, the field in the sample coincides with that in the empty guide if the sample diameter is smaller than 0.4 mm for the X band and 0.1 mm for the K band.

When the values of the impedances Z_s and Z_g are of the same order, the field in the sample is usually deduced from the measurement of the power absorbed by the sample, as was suggested by *Seeger* [4.4]. This is a rather common procedure when the width of the sample is smaller than that of the guide as, for example, happens while performing measurements on epitaxial layers deposited on conducting substrates. In this case, one measures the dependence of j_0/j_{dc} on the average microwave power absorbed by the sample, P_s, the field E_1 in the sample is thus obtained by integrating the experimental curve as

$$E_1^2 = \frac{2}{V_s \sigma_0} \int_0^{P_S(E_1)} \frac{j_0}{j_{dc}(P_s)} \, dP_s, \tag{4.5}$$

where V_s is the active volume of the sample [4.14].

To find a conductivity-field characteristic, some authors [4.15–17] have used the dependence of the reflection or transmission coefficient on the power incident onto the sample. However, because of its derivation from a linear theoretical approach, this method is in general unsuitable to be applied to hot-electron conditions where strong nonlinear current-field characteristics usually occur.

In the following the effect of a nonsinusoidal behavior of the microwave field on the current-field characteristic will be considered. Since in a strong microwave field a semiconductor sample behaves as a nonlinear element, the presence of the higher frequency harmonics in addition to the fundamental harmonic, can give rise to a mixing dc voltage in the semiconductor. The magnitude and the sign of this voltage depend on the relative phase of the harmonics. An example is the case of the simplest nonlinearity (warm-electron) $j = \sigma_0 E(1 + \beta E^2)$. In presence of the fundamental $E_1 \sin(\omega t + \varphi)$ and second harmonic $E_2 \sin 2\omega t$, there will appear the dc current density

$$j_h = \frac{3}{4} \sigma_0 \beta E_1^2 E_2 \cos 2\varphi. \tag{4.6}$$

This current will give rise to the harmonic mixing field $E_h = 0.75\, \beta E_1^2 E_2 \cos 2\varphi$.

Experimentally, the mixing voltage $V_h = E_h \cdot L$ was observed for the first time independently by *Carlin* and *Pozhela* [4.18] and *Schneider* and *Seeger* [4.19]. Its value can be as large as a few volts and, therefore, the mixing voltage introduces substantial errors if special caution is not taken to eliminate them [4.20]. To this end, for example, one can insert harmonic filters before and after the sample, or subtract j_h from j_{dc} at small values of j_h.

It should be noted that during the last decade the microwave techniques have been mainly used to investigate the drift velocity in the region of negative differential conductivity. In fact, there is no alternative technique to obtain the drift velocity in this region for materials with an electron concentration higher than 10^{14} cm^{-3}. Furthermore, while measuring the drift velocity, one must also be sure that the high-field-domain formation is suppressed since this may lead to a nonhomogeneous field distribution in the sample (in the case of GaAs, this occurs provided the condition $\varrho_0 \cdot f \geq 35\,\Omega$cm GHz is satisfied [4.21]). If this is not the case, the resistance of the sample may become absolutely negative. In other words, a positive external dc voltage will drive a negative current through the sample and through the external circuit, as shown in Fig. 4.3 [4.22]. If the dc circuit containing the absolute negative resistor is broken up, dc voltage values as high as hundreds of volts will appear between its opposite free terminals. The absolute negative resistance effect is explained by the appearance of the microwave rectification, after a dc bias has been applied to the sample [4.22].

In the methods discussed above, it has been tacitly assumed that the carrier concentration is kept constant during the experiment so that the drift velocity is proportional to the current flowing through the sample. However, especially in narrow-gap semiconductors, at high electric field carrier multiplication due to band-band impact ionization may set in. This certainly leads to a change in the carrier concentration during the experiment. In these cases, the microwave techniques described above cannot be applied directly for measuring the drift velocity $v_d(E)$ or the mobility $\mu(E)$.

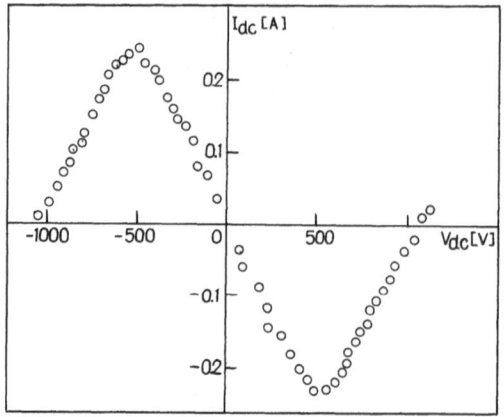

Fig. 4.3. Direct current versus voltage showing the absolute negative resistance for a n-GaAs sample subject to a microwave field $E_1 = 1.4$ kV/cm of 10 GHz, $n_0 = 2.56 \times 10^{15}$ cm^{-3}, $T_0 = 300$ K

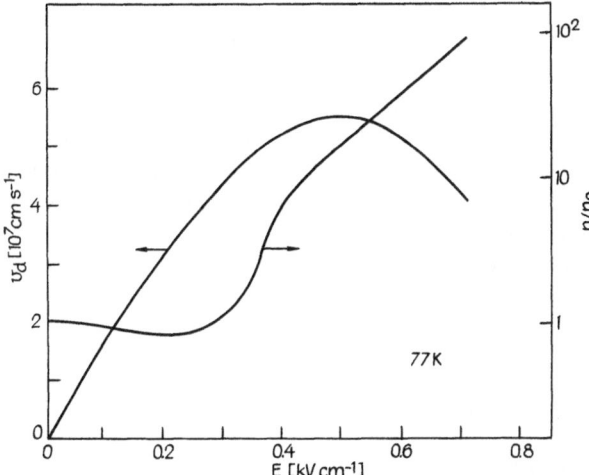

Fig. 4.4. Drift velocity (*left scale*) and free-carrier concentration normalized to the equilibrium value n_0 (*right scale*) for a n-InSb:Cr sample at $T_0 = 77$ K as a function of the electric field. The results are obtained by the microwave double-pulse technique [4.23]

To overcome these difficulties, a modified microwave technique has been proposed [4.23] which allows one to obtain simultaneously the dependence on the electric field of both the mobility and the carrier concentration. This modified technique makes use of doubled microwave pulses with very steep fronts. The time interval between the pulses is much shorter than the excess carrier lifetime. The dc current is measured both during the high microwave pulses (when the excess carriers are generated and their mobility is changed due to heating effect) and in between the pulses (when their mobility is ohmic because of the absence of the heating effect). From these data, the drift velocity and the carrier concentration can be deduced as functions of the electric field [4.23]. Figure 4.4 shows the drift velocity and the electron concentration as obtained with the modified microwave technique in a sample of InSb, doped with chromium.

4.1.2 Microwave Conductivity and Relaxation Times

In the previous section the drift velocity was assumed to be an instantaneous function of the electric field. As mentioned above, in this limit the microwave techniques allow one to obtain the steady-state current-field characteristics of the semiconductor sample. These investigations were carried out mainly on materials exhibiting negative differential conductivity at high electric field, e.g., GaAs [4.21, 24–26], InP [4.27], InSb [4.23], $Al_xGa_{1-x}As$ [4.28], $Ga_xIn_{1-x}Sb$ [4.29], $GaSb_xAs_{1-x}$, $In_xGa_{1-x}As$ [4.30], p-Te [4.31]. However, since the time constants characterizing the momentum, the energy, and the intervalley relaxation processes are usually in the range between 10^{-9} and 10^{-13} s, the frequency of a microwave electromagnetic field may become comparable with the inverse of these time constants. As a consequence, strong relaxation effects arise and a direct observation of the inertness of the physical processes involved becomes possible. In this context, during the last decade a large number of experiments and theoretical analyses have been performed on Ge, Si and III–V compounds. These have provided rapid progress in the understanding of the semiconductor behavior in a strong electric field of high frequency. In the following, we shall describe briefly some of the experimental and theoretical results but, before doing this, we shall give a simple idea concerning how the relaxation processes can affect the microwave transport.

In principle, the time dependence of a given parameter y can be estimated assuming the following time behavior for y:

$$\frac{dy}{dt} = \frac{y_0 - y}{\tau_y}. \tag{4.7}$$

Here y_0 is the steady-state value of y and τ_y is the relaxation time, characterizing the inertness of the physical process under consideration. In the small-signal approximation one can assume $y \sim \exp(i\omega t)$, so that the solution of (4.7) is

$$y = y_0 (1 + i\omega\tau_y)^{-1}. \tag{4.8}$$

From (4.8) it can be seen that both the real and the imaginary parts of y are frequency dependent.

In this approximation the real part of the small signal conductivity is given by

$$\sigma_d = \sigma_0 (1 + \omega^2 \tau_\sigma^2)^{-1}, \tag{4.9}$$

where τ_σ is the conductivity relaxation time.

It is well known that, under low-field conditions $(E \to 0)$, τ_σ coincides with the momentum relaxation time τ_m. When the electric field is sufficiently high so that non-ohmic conditions are established, the conductivity relaxation is determined, at least, by two physical processes: the momentum and the energy dissipation. In this case, an equation like (4.7) has to be written for each process and, in order to obtain the frequency dependence of σ_d, the resulting system of equations is to be solved. This problem was considered in the literature rather extensively as soon as it was realized that the microwave experiments could give information about the energy relaxation times [4.3, 8, 32].

To perform the measurement of the energy relaxation time many microwave techniques were proposed. Accordingly, harmonic mixing has been used [4.33] and the harmonic generation method has been used [4.34]. Both of them make use of the fact that the inertness of the carrier heating causes a delay in time between the electric field and the ac component of conductivity.

In the harmonic mixing experiment, this results in a phase shift $\Delta\varphi$ in the relation of the mixing signal on the relative phase shift of the harmonics, i.e., instead of (4.6) one has $E_h \sim \cos 2(\varphi + \Delta\varphi)$. Once the value of $\Delta\varphi$ is measured, the energy relaxation time $\tau_{\mathscr{E}}$ is obtained from the relation

$$\tan 2\Delta\varphi = \frac{2\omega^3\tau_{\mathscr{E}}^3}{1+3\omega^2\tau_{\mathscr{E}}^2}.\tag{4.10}$$

In the harmonic generation method proposed in [4.34], the power P_k and the phase φ_k of the kth harmonic generated by the sample are measured. Both P_k and φ_k are functions of $\tau_{\mathscr{E}}$ and therefore this type of experiment can be used also for investigations of the energy dissipation in semiconductors.

Recently similar investigations have been carried out in a wide range of frequencies [4.35, 36]. This has enabled one to examine a frequency dependence of the effective warm-electron coefficient, α^*. The definition of this coefficient can be obtained from (3.1) by inserting the relation $j(E) = \sigma_0(1+\beta E^2)E$, which describes the dependence on a field of the current density in the warm-electron region. After some calculations one obtains

$$j_{\mathrm{dc}} = j_0(1+\alpha^*P_{\mathrm{m}}/\sigma_0),\tag{4.11}$$

where $P_{\mathrm{m}} = \sigma_0 E_1^2(1+\omega^2\tau_{\mathrm{m}}^2)^{-1}$ is the microwave power, absorbed in a unit volume of the sample.

The dependence of α^* on the frequency for the case of n-Si is shown in Fig. 4.5. The experimental results agree well with the relation

$$\alpha^*(\omega) = \alpha^*(0)\,\frac{1}{3}\left(1+\frac{2+\omega^2\tau_{\mathrm{m}}\tau_{\mathscr{E}}}{1+\omega^2\tau_{\mathscr{E}}^2}\right),\tag{4.12}$$

obtained by solving the system of balance equations for the average momentum and energy in analogy with (4.7).

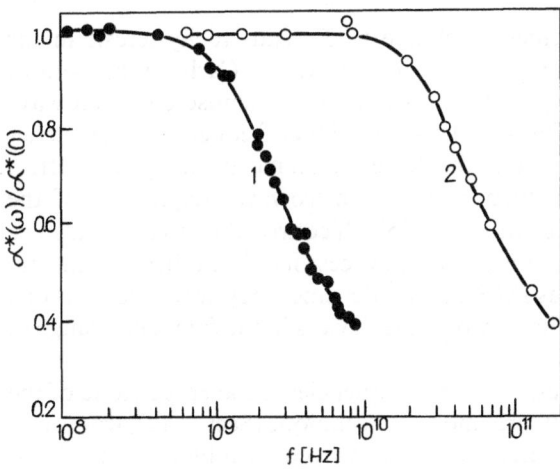

Fig. 4.5. Frequency dependence of the coefficient α^* normalized to its zero frequency value for n-Si with $n = 2\times10^{13}\,\mathrm{cm}^{-3}$, and $E\|\langle111\rangle$. Solid curves represent the best fit of the experimental data as calculated from (4.12): $1 - T_0 = 80\,\mathrm{K}$, $\tau_{\mathrm{m}} = 30\,\mathrm{ps}$, $\tau_{\mathscr{E}} = 580\,\mathrm{ps}$; $2 - T_0 = 290\,\mathrm{K}$, $\tau_{\mathrm{m}} = 0.2\,\mathrm{ps}$, $\tau_{\mathscr{E}} = 2\,\mathrm{ps}$

In the range of intermediate electric field strengths the decrease of $\tau_{\mathscr{E}}$ at increasing electric field and lattice temperature was observed, for example, in n- and p-Ge and Si [4.33, 34, 38], n-GaAs [4.37, 39–43], n-GaSb [4.45], n-InAs [4.7b, 44, 48], p-Te [4.46], n-InP [4.40, 43], n-InSb [4.47, 48], etc. This type of dependence is due to the fact that optical and intervalley scatterings, which are responsible for the energy dissipation in these materials, increase both with the lattice and the electron temperature. Furthermore, the stronger the coupling between the free carriers and the optical vibration modes, the shorter the mean energy and conductivity relaxation times are.

The dc heating field cannot be used for the investigations of the hot-electron dynamics in the negative differential conductivity region because of the domain formation. To suppress the domain formation, the bias high electric field has been created by applying a 10 GHz microwave field [4.37]. Furthermore, to obtain the small signal conductivity as a function of the bias field, a weak probe field of 50 GHz has been applied simultaneously.

The simple relaxation time approximation remains a qualitative and rough approach, especially at high electric fields. A more rigorous theory which analyzes the high field microwave conductivity should be based on the solution of the time-dependent Boltzmann equation. During the last decade several numerical methods have been proposed for the solution of this problem. They are based either on some modification of Monte Carlo method [4.50–54] or make use of an iterative procedure [4.39]. A time response of the high field electron distribution function is usually calculated, and these results are later used for evaluating the microwave conductivity. Calculations were carried out for GaAs, InP, InSb, and other semiconductors. As a matter of fact, such calculations give an explanation to all qualitative features of the corresponding microwave experiments.

At large microwave fields and for frequencies of the order of τ^{-1}, where τ is some characteristic relaxation time (energy, intervalley, etc.), the change of drift velocity and mean energy of hot electrons is delayed in time with respect to the change in the microwave field.

The phase shift of drift velocity, relative to E, leads to hysteresis in the dependence of $v_d(E)$. As an example, the results of Monte Carlo simulations of the dependence of drift velocity upon the strength of a sinusoidal microwave field in n-InSb are shown in Fig. 4.6 [4.54]. At higher frequencies, the carrier inertness leads also to the decrease of the hot-electron mean energy. This effect reveals itself through the dependence on the microwave frequencies of the threshold field for avalanche breakdown [4.55]. Of course, this effect can also be due to the inertness of the impact ionization process itself. All these results are very important in devices such as IMPATT diodes since they are at the basis of a considerable decrease in the efficiency of these devices in the frequency range of about 100 GHz.

In spite of its strong hysteresis, at high frequencies the average value of the drift velocity, obtained by the integral mobility technique (Sect. 4.11), was found to depend only slightly on frequency. This was proved both by theoretical and

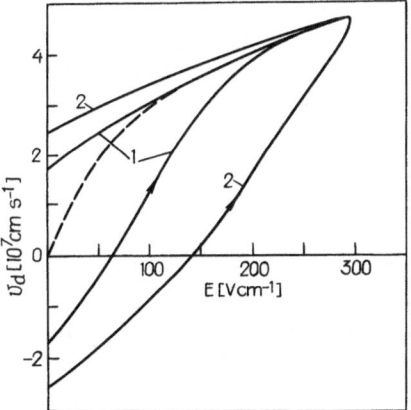

Fig. 4.6. Electron drift velocity dependence on momentum microwave field $E(t) = E_1 \sin \omega t$ during the positive half-period at the frequencies of 10 GHz (*1*) and 35 GHz (*2*) in InSb at $T_0 = 77$ K. (–––) represents the steady-state dependence

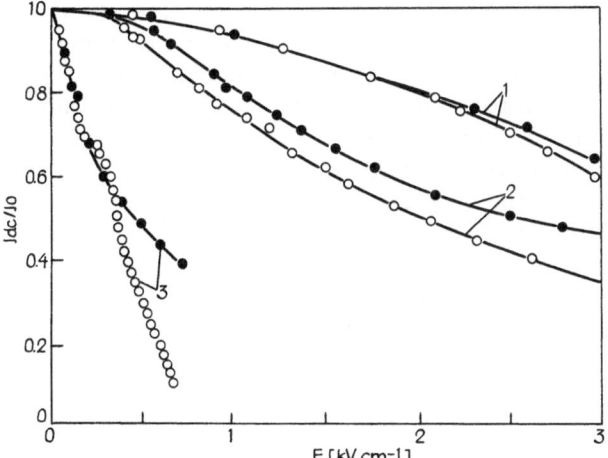

Fig. 4.7. Normalized average dc current density as a function of the microwave field amplitude at 9–10 GHz (●) and 35–37 GHz (○) for (*1*) GaAs [4.42], (*2*) n-Ge [4.56], (*3*) InSb [4.54]; (*1,2*) $T_0 = 300$ K, (*3*) $T_0 = 77$ K. (——) are a guide to the eye

experimental investigations, and some experimental data for n-Ge [4.56], n-InSb [4.54], and n-GaAs [4.42] are shown in Fig. 4.7.

Let us note that for the case of n-GaAs, calculations of σ_d predict that at frequencies as high as 70 GHz the negative differential conductivity should disappear [4.49, 50]. Moreover, the inertness of the intervalley repopulation effect should be observed already at 30 GHz. Nevertheless, the intervalley-transfer diodes made of n-GaAs are known to be functioning at 100 GHz [4.57]. This disagreement between the theoretical predictions and the experiment is caused by the fact that the dimensions of the active regions of the microwave devices operating at the highest frequency are so small that the electron transit time becomes comparable with the energy and even the momentum relaxation time. In this situation the microwave conductivity becomes essentially dependent on the dimensions of the device and its doping profile (Chap. 8).

4.2 Nonuniform Microwave Heating

Nonuniform heating of the electron gas can be easily achieved with microwaves by placing the whole semiconductor sample inside the guide where the microwave field is nonuniform or by placing only a part of the semiconductor sample through the wall into the guide, as shown in Fig. 4.8.

The nonuniform electron heating due to nonuniform fields causes an electron flux from the hot side to the cold one. Under the open circuit conditions, this flux yields a compensating potential difference V_{dc} at the terminals of the semiconductor sample. V_{dc} is caused by the same physical processes as in the usual thermoelectric Seebeck effect, the main difference being the fact that now the crystal lattice remains at uniform temperature while the gradient of the electron temperature is the only driving force responsible for the effect. The electron temperature may be much higher than that of the lattice, and, as a rule, the electron gas is hotter at lower lattice temperatures.

It can be easily seen that all the ordinary thermoelectric effects, e.g., the Nernst effect, the Dember effect, et al., can also be observed under nonuniform electron heating. In the ohmic region the thermoelectric effects can be used to obtain information on various scattering processes, find the diffusion coefficient, and measure the temperatures of the sample. The hot-carrier thermoelectric effects can be useful tools as well to obtain information on the temperature and diffusion coefficient of hot electrons but, in this case, it should be noted that the above effects are caused by highly nonequilibrium electron gas.

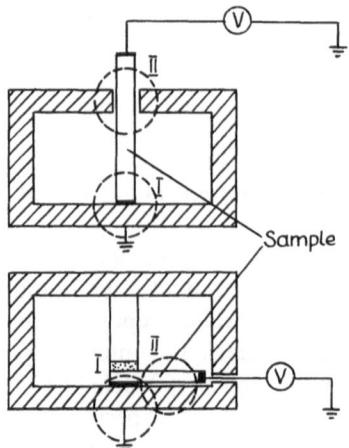

Fig. 4.8. Sample mountage in the waveguide suitable for the measurement of the hot electron thermoelectric voltage. (○) show the regions where the microwave electron heating is nonhomogeneous: (*I*) a nonhomogeneously doped region, (*II*) a nonhomogeneous microwave field region

The hot-carrier thermoelectric effects were experimentally observed by many workers. However, under the nonuniform electron heating, new specific effects, such as negative differential conductivity and rectification in a homogeneously doped semiconductor, were also observed. Therefore the present section will illustrate these new effects and their application to the measurements of the energy relaxation time and the diffusion coefficient as a function of the electric field strength. Accordingly, Sect. 4.2.1 will discuss the effects due to nonuniform heating, and then Sects. 4.2.2, 3 will present the real-space-transfer effects.

4.2.1 Hot-Carrier Thermoelectric Effects

First, the Seebeck effect due to hot electrons will be considered. This effect was experimentally observed at high microwave fields in Ge, Si, InSb, and analyzed theoretically by many workers [4.3, 8, 58–63]. Various analytical models were proposed to explain it, and a simple phenomenological model incorporating all salient features of the effect will be presented below.

If a circuit is made of two differently doped semiconductors and one of its contacts is heated to temperature T_1 while the other contact is left at T_0, a thermoelectric current (or a thermoelectric voltage if the circuit is open) appears. Under the open-circuit conditions, the thermoelectric current is compensated by the drift one:

$$eD \frac{\partial n}{\partial x} = en \, \mu E_x, \tag{4.13}$$

where D is the diffusion coefficient. If the temperature gradient is assumed to exist only in the bulk of the semiconductor where the charge carrier concentration is constant, (4.13) can be easily integrated over the entire circuit. The integration yields the following expression for thermoelectric voltage:

$$U_T = \frac{K_B T_0}{e} \ln \left(\frac{n_1}{n_2} \right) \left(\frac{D(T_1)}{D(T_0)} \frac{\mu(T_0)}{\mu(T_1)} - 1 \right). \tag{4.14}$$

After application of the Einstein's relation, from (4.14) we obtain

$$U_T = V_{k0} \left(\frac{T_1}{T_0} - 1 \right), \tag{4.15}$$

where

$$V_{k0} = \frac{K_B T_0}{e} \ln \left(\frac{n_1}{n_2} \right). \tag{4.16}$$

This is the usual expression for the ordinary Seebeck thermoelectric voltage.

Now, let us consider the sample with nonhomogeneity in the form of a gradual n^+–n junction placed in a strong microwave field. If the microwave frequency is so high that the displacement currents are much larger than the conduction currents, then the microwave field in the n^+–n junction may be assumed to be uniform. There is no need to resort to the concept of electron temperature, and one can use (4.14) in order to write down the carrier thermoelectric voltage in the form

$$U_T = V_{k0} \left(\frac{D(E)}{D(0)} \frac{\mu(0)}{\mu(E)} - 1 \right). \tag{4.17}$$

This formula contains only that part of the thermoelectric voltage which is caused by the heating of the nonhomogeneous region of a semiconductor sample (Region I in Fig. 4.8). The contribution to the thermoelectric voltage due to the energy gradient in the homogeneous region of the sample (Region II in Fig. 4.8) is $eV_{k0}/K_B T$ times smaller than U_T (4.17). Experimentally, the observed values of U_T range from tens to hundreds of millivolts at the microwave fields of 1 kV/cm.

By measuring U_T one can find $D(E)$ rather directly if $\mu(E)$ is known. The results obtained in this way are shown in Fig. 4.9. Since the measurements were performed by using microwaves, some averaged values $\bar{D}(E_1)$ were found. In order to deduce the instantaneous value of $D(E)$, an integral transformation similar to that used in the calculations of $\sigma_d(E)$ from the microwave measurements should be performed (Sect. 4.1.1). It should be noted that, qualitatively in n-Si, the magnitude and the character of $D(E)$, found from hot-electron thermoelectric voltage, coincides with that obtained by noise measurements and by the time-of-flight technique. For other materials there are some discrepancies.

These discrepancies are related to the phenomenological approximation used to obtain (4.17). It does not take into account the anisotropy and the valley

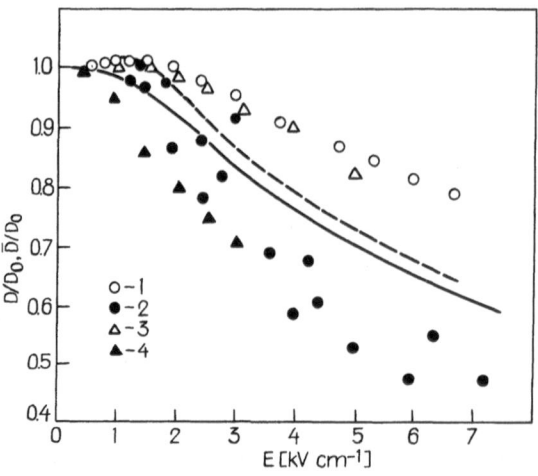

Fig. 4.9. Longitudinal diffusion coefficient of electrons in Si for the $\langle 111 \rangle$ crystallographic direction at $T_0 = 300$ K. Here $(1, 3)$ refer to experimental values of average diffusion coefficient, $\bar{D}(E_1)$, as determined from the thermoelectric voltage U_T [4.61]; (2) refers to $D(E)$ as measured with the time-of-flight technique [4.66]; (4) refers to $D(E)$ as deduced from noise temperature T_n [4.65]. (——) refers to the Monte Carlo calculations [4.66]. (---) refers to instantaneous $D(E)$ values, calculated from the experimental values of $\bar{D}(E_1)$ [4.64]

nonparabolicity as well as the tensor character of the diffusion coefficient and mobility for hot electrons.

In the approximation considered above, the ratio $D(E)/\mu(E)$ characterizes qualitatively the energy of the electron gas, and from the experimentally deduced values of this ratio, the phenomenological time of the energy relaxation could be determined [4.67]. The values of $\tau_{\mathscr{E}}$, obtained for silicon, were found in good agreement with those deduced from the harmonic mixing experiments, as described in Sect. 4.1.2.

The investigation of the thermoelectric voltage in n-InSb leads to the discovery of the "absolute" cooling effect of an electron gas in which the electron energy in the presence of an electric field becomes lower than the equilibrium energy corresponding to the lattice temperature. This effect, predicted by *Gribnikov* and *Kochelap* [4.68], was observed by *Ashmontas* et al. [4.63]. The cooling effect manifested itself as a change of the sign of the hot-electron thermoelectric voltage at some value of the microwave electric field, as shown in Fig. 4.10. It was observed at low temperatures in highly compensated samples where the electron-electron scattering is negligible and the prevailing momentum scattering mechanism comes from impurities.

The electron cooling effect is determined by nonelastic optical-phonon scattering. It occurs in the range of electric fields at which electrons can already reach the energy of the optical phonon, $\hbar\omega_0$, but remain below the energy $\hbar\omega_0 + K_B T_0$, where T_0 is the lattice temperature. Thus, after the nonelastic optical scattering, the electrons are found in the region of energies lower than the lattice energy, $\mathscr{E} < K_B T_0$.

The thermomagnetic hot-electron effect, analogous to the ordinary Nernst effect, was experimentally observed in n-InSb [4.69] and the theory of the effect was presented in [4.70]. In the experiments, the semiconductor plate, with a thickness comparable to or larger than the skin depth, filled the window of the guide. With the electron temperature gradient being formed in the direction of wave propagation, the thermoelectric current sets up in the same direction. In a

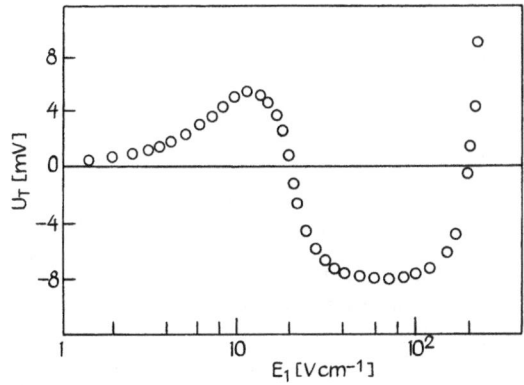

Fig. 4.10. Hot-electron thermoelectric voltage in a n-InSb:Cr sample as a function of the microwave field amplitude: $n = 9 \times 10^{10}\,\text{cm}^{-3}$, $n^+ = 1.5 \times 10^{12}\,\text{cm}^{-3}$, $T_0 = 78\,\text{K}$ [4.63]. At the fields $E = 20$ to $200\,\text{V/cm}$, the electron temperature is lower than the lattice temperature

magnetic field, applied perpendicularly to the direction of the wave propagation, the current is deflected and, as a result, the thermomagnetic voltage appears.

Instead of a $n - n^+$ junction, a nonuniform illumination could be used to obtain the gradient of concentration in the crystal. The hot-electron thermoelectric voltage, caused by the nonuniform illumination, was observed in [4.71] and named the photothermogradient voltage.

In the case of an intrinsic semiconductor, the flux of electrons and holes gives rise to a voltage in the direction of the temperature gradient which is due to different electron and hole mobilities and diffusion coefficients (Dember effect). The hot-electron Dember effect was observed in [4.72]. Thus, a number of new effects, which collectively may be named electrogradient effects in semiconductors [4.8], have been observed under nonuniform carrier heating by electric field.

4.2.2 Microwave Studies of Hot-Electron Real-Space Transfer Effects

Real-space transfer effects have been observed in uniformly doped semiconductors, graded-gap semiconductors, and layered heterostructures.

The carrier flux from a hot to a cold region due to nonuniform carrier heating leads, naturally, to a carrier transport in real space. If, at the same time, parameters like effective mass, momentum and energy relaxation times and lifetime are space dependent, the carrier transfer in real space will lead to a change of dc and microwave impedances of the sample. This effect is named hot-electron real-space transfer. It is especially important in thin semiconductor layers.

This effect was first observed by *Kalvenas* and *Pozhela* [4.73] in intrinsic germanium at room temperature under a high microwave field. As the consequence of an electron heating at the sample surface weaker than in its volume, a flux of hot electron-hole pairs occurred towards the surface. At the same time, the carrier recombination rate was considerably higher on the surface

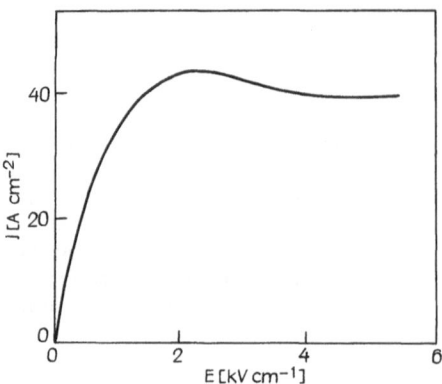

Fig. 4.11. Current density as a function of electric field as obtained from the microwave measurement of $\sigma'(E)$ in Ge ($\varrho = 20\,\Omega\,\text{cm}$, $T_0 = 300\,\text{K}$). The negative slope of the curve is attributed to the real-space-transfer effect (see text)

than in the volume of the semiconductor and, for an etched surface, it was even increasing with the field increase; thus the flux brought about a change of the mean concentration in the volume.

In the case of a sample with thickness comparable to or less than the diffusion length, the concentration change was so considerable that the current-field characteristic exhibited a negative slope (Fig. 4.11).

Another type of real-space transfer effect can be observed in graded-gap semiconductors when a high electric field is applied in the direction normal to the energy gap gradient.

One-valley graded-gap semiconductors can have a gradient of effective mass; thus moving electrons are shifted along the mass gradient [4.74]. Experiments in graded-gap $n\text{-}Cd_xHg_{1-x}Te$ show that an alternating current (normal to the gradient) produces an alternating voltage of doubled frequency in the direction of the gradient [4.75].

In many-valley semiconductors, e.g., $Al_xGa_{1-x}As$, hot-electron kinetic energy in the lower high mobility valley is limited because of the strong scattering at energies exceeding the intervalley energy gap. This latter is coordinate dependent in a graded-gap semiconductor; therefore, electron heating leads to a kinetic energy gradient. The electron motion due to this gradient produces an electromotive force and also changes the conductivity in the direction of the electric field used for heating [4.76]. Microwave investigations of graded-gap $Al_xGa_{1-x}As$ show that the real-space transfer enhances the Gunn effect and negative differential conductivity [4.28].

Recently hot-electron real-space transfer effects have been observed in layered heterostructures. They are considered to a wider extent by Hess and Iafrate (Chap. 7).

4.2.3 Bigradient Effect

A sample made from a uniformly doped semiconductor in a geometric shape reminding one of the conventional symbol of a diode (see insets in Fig. 4.12), effectively works as a diode at microwave frequencies. This effect, observed in [4.77] and shown in Fig. 4.12, is called the bigradient effect. Several phenomena contribute to this effect.

First, the electron kinetic energy gradients, due to the nonuniformity of the electric field, cause hot-electron electromotive forces. They appear to be gradient dependent and do not compensate each other if the gradients are not very small, i.e., when the neck dimensions are comparable to the electron energy relaxation length $\sim (D\tau_\mathscr{E})^{1/2}$.

Next, because of the occurrence of the hot-electron velocity saturation (in Ge, Si, etc.), the space charge is accumulated in the neck. Figure 4.13 reports the field distribution as obtained from the solution of the Poisson equation for the sample geometry illustrated under the assumption that the velocity saturation takes place at $E = 1.7\,\text{kV/cm}$ [4.22]. The field distribution changes when the

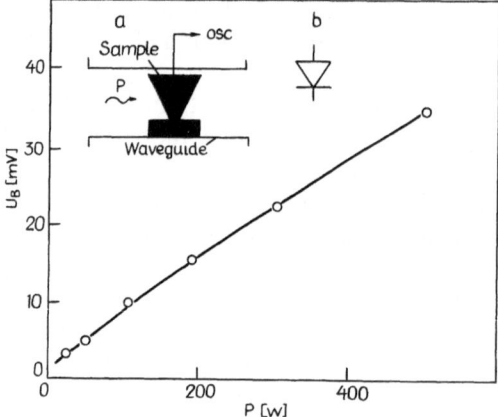

Fig. 4.12. Bigradient voltage in n-Ge as a function of the microwave power in the waveguide ($T_0 = 300$ K) [4.78]. (O) are experimental results and (——) is a guide to the eye. The insets show the shape of the sample in the waveguide (*a*) and a sign of a diode (*b*)

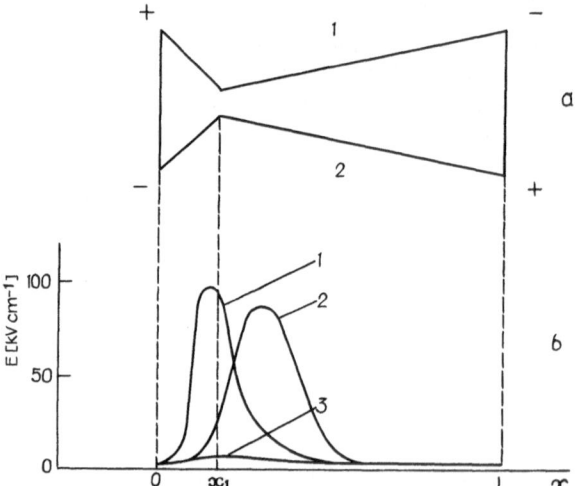

Fig. 4.13. Bigradient effect: (*a*) sample configuration with cross-sectional area: $S(0) = S(L) = 6.6 \times 10^{-2}$ cm², $S(x_1) = 8 \times 10^{-5}$ cm², $x_1 = 0.14$ cm, $L = 0.8$ cm. $n_0 = 2 \times 10^{14}$ cm⁻³; (*b*) electric field distribution at $V_{dc} = 500$ V; (*1*) current from the left to the right; (*2*) current from the right to the left; (*3*) electric field in the sample under ohmic conditions, i.e., $\mu = 3900$ cm² V⁻¹ s⁻¹

current is reversed; therefore different voltages are required to maintain the same current, and the bigradient diode exhibits a rectification function.

When the velocity saturation, assisted by the charge accumulation, takes place, the electric field in the neck is much higher than that expected from the sample configuration under Ohm's law (Fig. 4.13). Thus, at relatively low voltage applied to the sample, the field in the neck may be so high that impact ionization becomes possible. The generated drifting holes compensate the negative space charge, the field in the neck decreases, and the impact ionization stops. The process is repeated when the holes leave the neck area. It becomes periodic and is accompanied by current oscillations at the microwave frequency [4.78].

Thus the bigradient diode can perform two functions at microwave frequencies: current rectification and power generation.

4.3 Microwave Noise and Diffusion Coefficient

Current fluctuations in semiconductors result from the microscopic random processes; therefore, their measurement enables us to achieve information about these processes.

Investigations of the current-fluctuation spectrum and of its dependence on the strength of the electric field enable the energy and momentum relaxation times as well as the intervalley scattering rate to be determined. In this respect it is worth noting that the current-fluctuation spectral density, as will be shown below, is more sensitive to the details of scattering mechanisms and band structure of the material, as compared to the microwave conductivity.

When carrier-carrier interaction is negligible the current-fluctuation spectral density (in steady-state conditions) can be simply related to the diffusion coefficient of hot electrons (Chap. 2). Thus it provides the possibility to determine the diffusion coefficient by means of a hot-electron noise and small signal ac conductivity measurement in homogeneous materials without producing any concentration gradient. Furthermore, this noise-conductivity technique, as an alternative to the time-of-flight technique, enables the diffusion coefficient in low- as well as in high-resistivity materials to be measured.

In this section, after a brief description of the experimental technique for the microwave noise measurement, the dependence of the current-fluctuation spectral density upon frequency as well as upon the strength of electric field is considered. Finally, the dependence of the diffusion coefficient on time will be discussed.

4.3.1 Experimental Technique for the Microwave Noise Measurements

The spectral density of current fluctuations $S_{j\alpha}(\omega, E)$ along the α direction in the presence of steady electric field E is defined by the equation analogous to that of Nyquist:

$$S_{j\alpha}(\omega, E) = \frac{1}{V_0} \, 4 \, K_B T_{n\alpha}(\omega, E) \sigma'_\alpha(\omega, E), \tag{4.18}$$

where K_B is the Boltzmann constant; V_0 is the sample volume; and $\sigma'_\alpha(\omega, E)$ and $T_{n\alpha}(\omega, E)$ are the small signal ac conductivity and the noise temperature at angular frequency ω, respectively.

The noise temperature has no microscopic meaning but it is determined by (4.18) as a quantity which has the dimension of a temperature. However, $T_{n\alpha}(\omega, E)$ may be experimentally determined by measuring the maximum noise power dissipated by the noise source on the matched output impedance at frequency $\omega = 2\pi f$ per unit bandwidth Δf:

$$P_{n\alpha}(\omega, E) = K_B T_{n\alpha}(\omega, E) \Delta f. \tag{4.19}$$

Thus, for the determination of $S_{j\alpha}(\omega, E)$, it is necessary to measure the power of the noise source and the small signal ac conductivity $\sigma'_\alpha(\omega, E)$.

In order to measure $S_{j\alpha}(\omega, E)$, use is made of a sample mountage in the waveguide as for the measurement of the microwave conductivity, already described in Sect. 4.1.1. A strong electric field E, applied to the sample, heats the carriers, and a weak microwave field is used for the measurement of $\sigma'_\alpha(\omega, E)$. In absence of the weak field, the noise power, generated by the sample, is determined. Thus, the experimental apparatus for the measurement of $S_{j\alpha}(\omega, E)$ consists of two parts: the former for the determination of $\sigma'_\alpha(\omega, E)$, and the latter for the measurement of $T_{n\alpha}(\omega, E)$.

The technique for the $\sigma'_\alpha(\omega, E)$ measurement has been described in Sect. 4.1.1. The noise temperature is determined by comparing the noise power, radiated by the semiconductor sample, with that of a standard noise generator. To get the desired noise temperature $T_{n\alpha}(\omega, E)$ from the measured noise power, both the reflection and the losses are taken into account. Since the impedance of the sample varies according to the value of the applied electric field E, the matching of the sample impedance in the waveguide must be controlled for each value of E. The ordinary microwave techniques which are described in [4.79, 80] are used for this kind of measurement.

At high electric fields in order to avoid thermal heating of the sample, all the experiments are performed using voltage pulses as short as a few microseconds with low repetition rate. Due to these short pulses, a high-sensitivity gated radiometer is used.

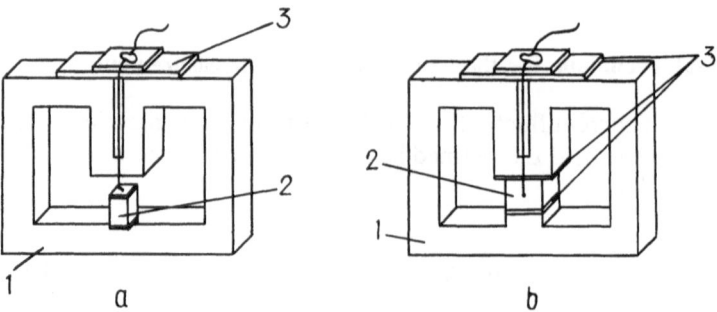

Fig. 4.14a, b. Schematic picture of the H-type waveguide cross-section with the samples mounted for the longitudinal ($E_0 \| E_n$) (a) and the transverse ($E_0 \perp E_n$) (b) noise temperature measurements: (1) waveguide section, (2) sample, (3) isolators, E_0 heating dc field, E_n microwave noise field excited in the waveguide

Information of relevant interest about the microscopic processes is obtained by measuring the microwave spectral density of the current fluctuations along $S_{j\parallel}$ and transverse $S_{j\perp}$ to the electric field direction. Accordingly, for these measurements either the position of the sample in the waveguide or the direction of the electric field in the sample is changed. A schematic diagram of an experimental setup for the longitudinal and the transverse noise measurements is shown in Fig. 4.14. The sample is mounted in an H-shaped waveguide where the microwave noise electric field vector, E_n, parallel to the narrow wall of the guide, is excited and detected. When the direction of the microwave noise electric field is colinear or perpendicular to that of the heating dc field E_0, the longitudinal $T_{n\parallel}$ and the transverse $T_{n\perp}$ noise temperatures are measured, respectively. The same holds for the longitudinal and the transverse small signal ac conductivity measurements.

The technique for the $S_{j\alpha}(\omega, E)$ measurement is rather complicated and, consequently, the experimental uncertainty may amount to 10–20%.

4.3.2 Dependence of the Microwave Current-Fluctuation Spectral Density on Frequency and Electric Field

In best known semiconductors, e. g., Ge, Si, A^3B^5, except for rare specific case, the noise temperature increases while $\sigma_{d\alpha}(\omega, E)$ decreases with an increasing electric field. Therefore, as follows from (4.18), the competition of these two tendencies determines the field dependence of $S_{j\alpha}(\omega, E)$. Consequently, $S_{j\alpha}(\omega, E)$ may either increase or decrease with the electric field.

For fixed E, the spectrum of $S_{j\alpha}(\omega, E)$ in the microwave region may be expressed schematically by the following sum

$$S_{j\alpha}(\omega, E) = \sum_i [S_{j\alpha}(0, E)]_i \frac{1}{1 + (\omega\tau_i)^2}, \qquad (4.20)$$

where τ_i can represent a momentum, or an energy or an intervalley relaxation time. Such a hypothetical spectrum of $S_{j\alpha}(\omega, E)$ is plotted in Fig. 4.15. There the generation-recombination as well as the $1/f$ noise are neglected.

Inspection of Fig. 4.15 shows that in the microwave region, when $\omega\tau_m \geq 1$, $S_{j\alpha}(\omega, E)$ sharply decreases with frequency. In the region where $\omega\tau_\mathscr{E} < 1$, $\tau_\mathscr{E}$ being the energy relaxation time, along with the thermal fluctuations of hot electrons there occurs an additional convective noise. This is caused by fluctuations of average values of drift velocity, as a result of random variations of the mean energy. This type of noise occurs only in the presence of a current flow across the semiconductor, and its contribution to the spectral density of the current fluctuations may be either positive or negative.

In fact, fluctuations of the mean energy are caused by a change in the power absorbed from the external supply in the presence of a thermal current fluctuation $\delta\mathscr{E} \sim (\delta j \cdot E)$; then the contribution to the total fluctuation depends

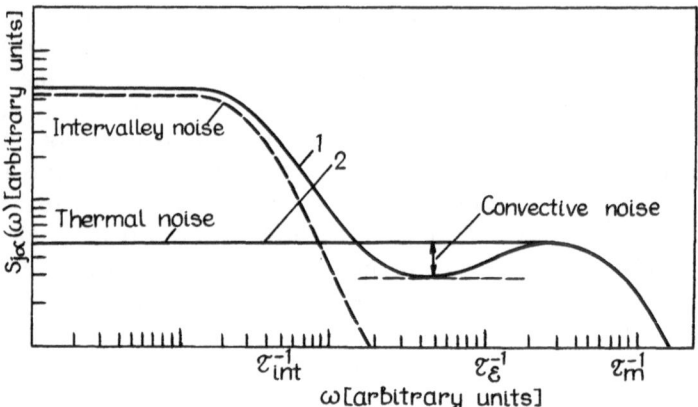

Fig. 4.15. Schematic dependence of the spectral density of the current fluctuations on frequency parallel (*1*) and transverse (*2*) to the electric field direction. Steps on the curves correspond to $\omega\tau_i \approx 1$ for different fluctuation sources

on the value of the differential mobility. In semiconductors with sublinear current-voltage characteristic, the current fluctuation is suppressed by the decrease of mobility. Thus, in such semiconductors the fluctuations of absorbed power reduce the total spectral density of current fluctuations, as shown in Fig. 4.15. Accordingly, semiconductors with sublinear current-voltage characteristic, the contribution of convective noise is positive. The occurrence of an intervalley noise, due to the presence of the intervalley scattering, increases $S_{j\alpha}(\omega, E)$ in the frequency region $\omega\tau_{\text{int}} \leq 1$.

Owing to the field dependence of the relaxation times and $\sigma'_{\parallel}(\omega, E)$, both the amplitude and position of the steps of Fig. 4.15 will depend upon the electric field strength. Consequently, the $S_{j\alpha}(\omega, E)$ spectrum will, in general, depend on the electric field strength.

In materials with a simple spherical and parabolic conduction-band minimum, the intervalley and the convective noises are present only in the longitudinal case. In the transverse case, since the noise arises from the thermal motion of carriers only, $S_{j\perp}(\omega)$ does not depend on frequency under $\omega\tau_{\text{m}} \ll 1$ condition, as it is shown in Fig. 4.15.

4.3.3 Microwave Noise in One-Valley Semiconductors

In the following, the experimental results and the Monte Carlo calculations of $S_{j\alpha}(\omega, E)$, when the intervalley scattering is neglected, are presented. According to the schematic representation of $S_{j\alpha}(\omega, E)$ in Fig. 4.15, in the case when $\omega\tau_{\mathscr{E}} < 1$, only one peculiarity should be noted: for semiconductors with sublinear drift velocity versus field characteristic, the spectral density of the transverse current fluctuations exceeds the longitudinal one [4.81].

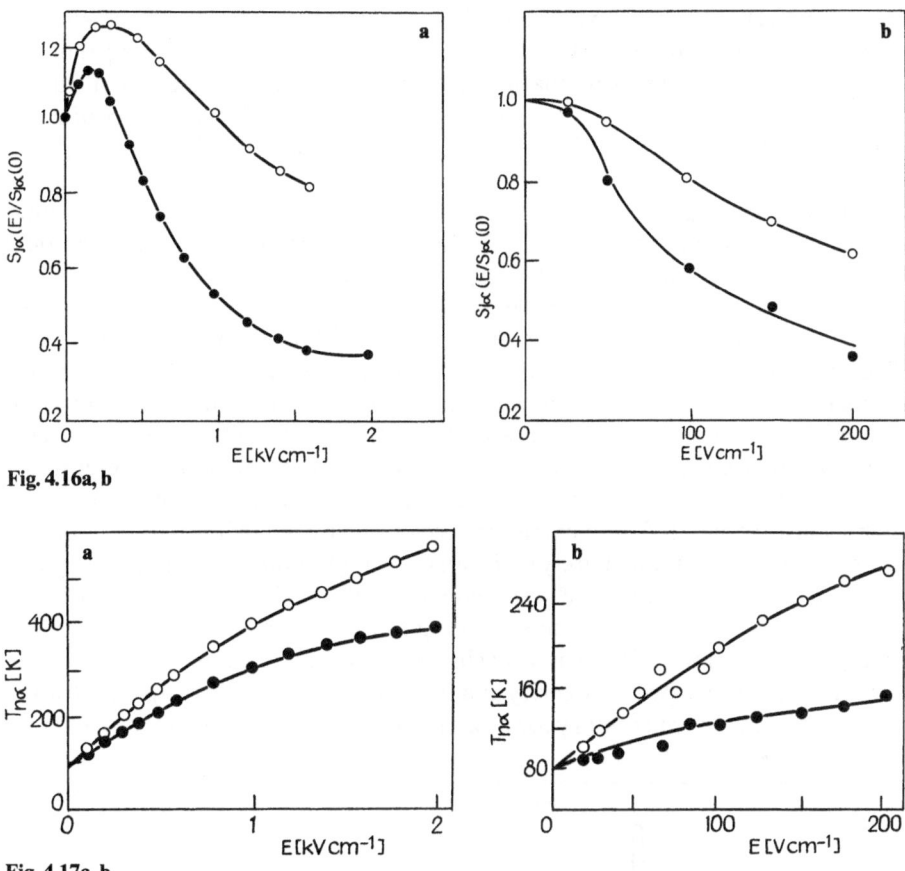

Fig. 4.16a, b

Fig. 4.17a, b

Fig. 4.16a, b. Experimental dependence of $S_{j\alpha}(\omega, E)/S_{j\alpha}(\omega, 0)$ versus electric field $E \| \langle 110 \rangle$ for p-Ge [4.84] (a), and n-InSb [4.85] (b); $T_0 = 78$ K, $f = 10$ GHz. ($\circ$, $\bullet$) refer to $S_{j\perp}$ and $S_{j\|}$, respectively. (———) are a guide to the eye

Fig. 4.17a, b. Experimental dependence of the noise temperature $T_{n\alpha}$ versus electric field $E \| \langle 110 \rangle$ for p-Ge [4.84] (a), and n-InSb [4.85] (b); $T_0 = 78$ K, $f = 10$ GHz. ($\bullet$, $\circ$) refer to $T_{n\perp}$ and $T_{n\|}$, respectively. (———) are a guide to the eye

The first microwave experiment on the hot-carrier current-fluctuation spectral density in Ge by *Bareikis* et al. [4.82] confirmed that indeed $S_{j\|}(\omega, E) < S_{j\perp}(\omega, E)$. Later on, the presence of a convective noise was confirmed experimentally not only for p-Ge [4.83], but also for p-Si, n-InSb, and other materials for which the drift velocity exhibits a sublinear dependence on the electric field [4.84,85]. The experimental results for the case of p-Ge and n-InSb are reported in Fig. 4.16.

The noise temperature as a function of the electric field, in accordance with the materials mentioned above, is reported in Fig. 4.17. The anisotropy with respect to the field density exhibited in Fig. 4.17 results from both the anisotropy of $\sigma'_\alpha(\omega, E)$ and $S_{j\alpha}(\omega, E)$. Therefore, the assumption made in [4.86] that the anisotropy of the current-fluctuation spectral density as well as of the diffusion coefficient is due only to the anisotropy of $\sigma'_\alpha(\omega, E)$ is not confirmed experimentally. Probably, this assumption may be justified only in some special cases [4.87].

In a series of experiments and of Monte Carlo calculations, it has been found that in the same material, within certain limits of frequency and electric field strength, the inequality $S_{j\parallel}(\omega, E) < S_{j\perp}(\omega, E)$ may change to the opposite one $S_{j\parallel}(\omega, E) > S_{j\perp}(\omega, E)$. According to the scheme shown in Fig. 4.15, this is not allowed for materials where the drift velocity is sublinear with the electric field. Theoretically, it has been shown that this inversion of anisotropy in the current-fluctuation spectral density is possible in materials with strong optical-phonon scattering due to a resonant increase of $S_{j\parallel}(\omega, E)$ [4.88, 89]. At low temperatures, when weak acoustical and strong optical scatterings are present, the momentum distribution function (as it is shown in Chap. 6 of this volume) is highly anisotropic. In fact, a carrier is accelerated by the field up to the optical-phonon energy and immediately emits an optical phonon. The carrier is thereby scattered to its ground state $\mathscr{E} = 0$, dissipating all its kinetic energy. Then it is accelerated again, and the same streaming motion is repeated. The associated repeated fluctuations of the carrier velocity result in a resonant increase of the current-fluctuation spectral density at a frequency equal to the reciprocal transit time, that is, the time necessary for a carrier to reach the optical-phonon energy.

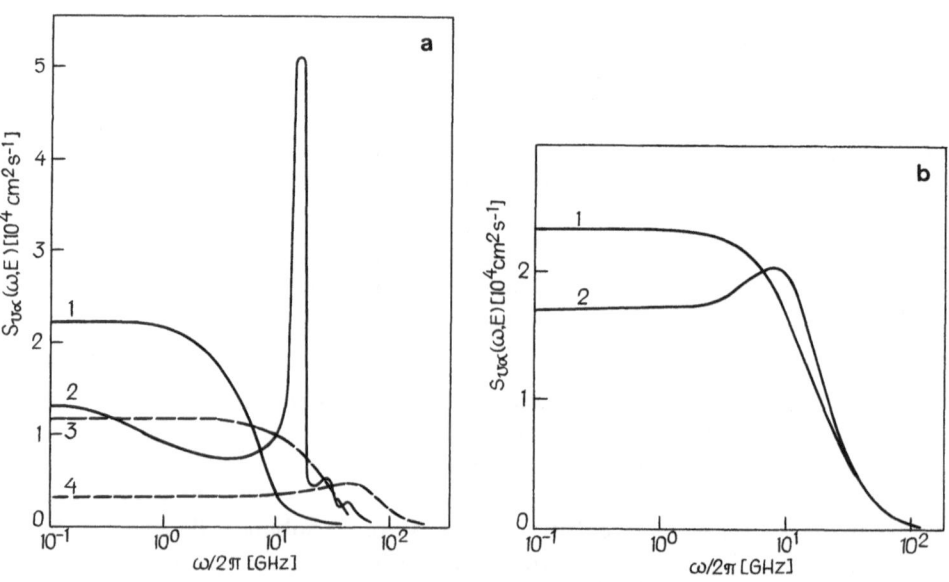

Fig. 4.18a, b. Calculated spectra of velocity-fluctuation spectral density, $S_{v\alpha}(\omega, E)$, for n-InSb, $T_0 = 10\,\mathrm{K}$: $(1, 3)\,S_{v\perp}(\omega, E)$ and $(2, 4)\,S_{v\parallel}(\omega, E)$. (——) refer to $E = 10\,\mathrm{V/cm}$, (———) to $E = 50\,\mathrm{V/cm}$: (a) $N_I = 0$, (b) $N_I = 10^{14}\,\mathrm{cm}^{-3}$

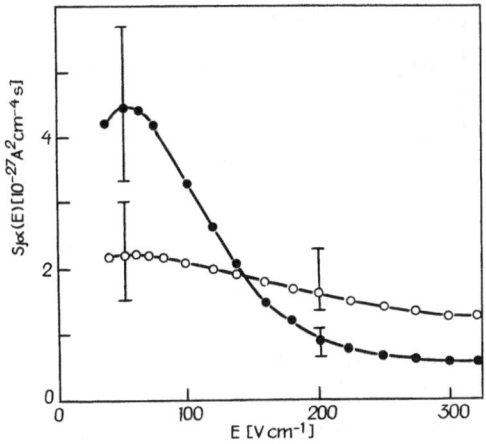

Fig. 4.19. Experimental dependence of the current-fluctuation spectral density $S_{j\alpha}(\omega, E)$ versus electric field $E \| \langle 100 \rangle$ for p-Ge; $T_0 = 10$ K, $f = 10$ GHz, $n = 8.5 \times 10^{13}$ cm^{-3}. (O, ●) refer to $S_{j\perp}$ and $S_{j\|}$, respectively. (——) are a guide to the eye. The error bars indicate the uncertainty of the measured points

The frequency dependence of the velocity-fluctuation spectral density, calculated by the Monte Carlo technique for the case of n-InSb, is reported in Fig. 4.18 [4.85]. Provided that the scattering by ionized impurities is negligible, $S_{v\|}(\omega, E)$ has its maximum at the transit frequencies (see Curve 2 in Fig. 4.18a. Note that for a frequency of 10 GHz, $S_{v\|} > S_{v\perp}$ at 10 V/cm, but $S_{v\|} < S_{v\perp}$ at 50 V/cm). This effect and the peak of $S_{v\|}(\omega)$ is strongly smoothed when scattering by ionized impurities is taken into account (see Curve 2 in Fig. 4.18b). Analogous spectra for p-Ge have been also obtained [4.90], and the experimental data of $S_{j\|}(\omega, E)$ vs E are shown in Fig. 4.19. It is seen that $S_{j\|}(\omega, E) > S_{j\perp}(\omega, E)$ for $E \leq 140$ V/cm, and $S_{j\|}(\omega, E) < S_{j\perp}(\omega, E)$ for $E > 140$ V/cm. The analogous effect was observed experimentally in n-InSb [4.91].

4.3.4 Microwave Noise in Many-Valley Semiconductors

The electron transfer from one valley to another usually leads to a change in the value of the mobility and, as a consequence, to a change in the value of the current. Due to the random character of these transitions, a current noise, often referred to as the intervalley noise, is produced [4.92]. The spectral density of these intervalley-current-density fluctuations $S_{j\,\text{int}}(\omega, E)$ can be written as

$$S_{j\,\text{int}}(\omega, E) = \frac{4 e^2}{V_0} \frac{n_1 n_2}{n} (\langle v \rangle_1 - \langle v \rangle_2)^2 \frac{\tau_{\text{int}}}{1 + (\omega \tau_{\text{int}})^2}, \tag{4.21}$$

where n_1, $\langle v \rangle_1$ and n_2, $\langle v \rangle_2$ are the average carrier concentrations and drift velocities in valleys of type 1 and 2, respectively.

When the average drift velocities $\langle v \rangle_1$ and $\langle v \rangle_2$ are assumed to be parallel to the electric field, the intervalley noise occurs only in the field direction. However, when the drift velocities $\langle v \rangle_1$ and $\langle v \rangle_2$ are not parallel to the electric field, as it

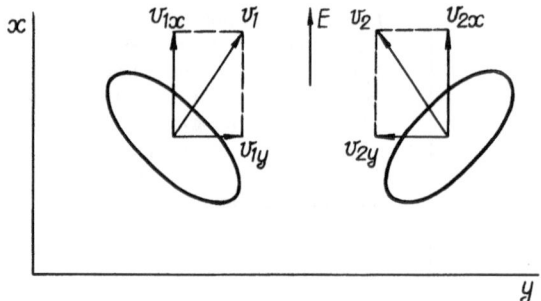

Fig. 4.20. Schematic representation of the drift velocity of electrons in a two-ellipsoidal-valley system

may occur in n-Ge and n-Si, the intervalley transitions may increase the current noise in the transverse directions as well [4.93]. In fact, an increase in the current-noise spectral density in the transverse direction may occur even if the transverse average drift velocities do not contribute to the longitudinal one. Let us take, as a simple example, two ellipsoidal valleys located symmetrically with respect to the electric field direction (Fig. 4.20). Due to the effective-mass anisotropy, the direction of the drift velocity of each valley does not coincide with the field direction, that is, $\langle v \rangle_{1x} = \langle v \rangle_{2x}$, but $\langle v \rangle_{1y} = -\langle v \rangle_{2y}$. The substitution of the corresponding components in (4.21) gives for $S_{j\,\mathrm{int}\,\|}(\omega, E) \sim (\langle v \rangle_{1x} - \langle v \rangle_{2x})^2 = 0$, and for $S_{j\,\mathrm{int}\perp} \sim \langle v^2 \rangle_{1y} \neq 0$.

The intervalley noise due to the transitions between the ellipsoidal equivalent valleys was observed experimentally in n-Ge [4.94, 95] and n-Si [4.95, 96].

According to (4.21), the intervalley noise reveals itself if $\omega \tau_{\mathrm{int}} < 1$. In n-Si, τ_{int} decreases with an increasing electric field, and for $E \simeq 1 \,\mathrm{kV/cm}$ at the frequency of $10\,\mathrm{GHz}$, $\omega \tau_{\mathrm{int}} \sim 1$ is expected [4.97]. This suggests that at the frequency of $10\,\mathrm{GHz}$, the intervalley noise in n-Si reveals itself at $E > 1 \,\mathrm{kV/cm}$. The data of the current-fluctuation spectral density in n-Si are reported in Fig. 4.21 [4.95]. The

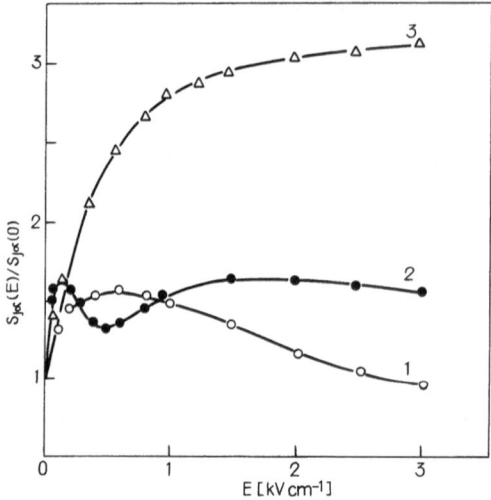

Fig. 4.21. Measured field dependence of $S_{j\alpha}(E)/S_{j\alpha}(0)$ for n-Si, $T_0 = 85\,\mathrm{K}$, and $f = 10\,\mathrm{GHz}$: $(1, 2)\, S_{j\|}$ along the $\langle 111 \rangle$ (O) and $\langle 100 \rangle$ (●) directions, respectively, $(3)\, S_{j\perp}$ along the $\langle 110 \rangle$ direction with $E \| \langle 001 \rangle$ (△). (———) are a guide to the eye

minimum, observed for the case of Curve 2 in Fig. 4.21, should be due to an electron concentration increase in cold valleys established, in this material, at $E \simeq 300\text{--}400$ V/cm [4.8]. At higher values of E, the intervalley noise sets up and thus the current-fluctuation spectral density for $E \| \langle 100 \rangle$ is found to exceed that for $E \| \langle 111 \rangle$. As it is seen from Fig. 4.21, the spectral density of the transverse current fluctuations is much higher than that of the longitudinal ones. This effect is associated with the electron transfer from hot into the cold valleys, where carrier mobility in the direction transverse to E is high. The transverse component of the current-fluctuation spectral density is further increased due to the absence of the convective noise, whose contribution to $S_{j\alpha}(\omega, E)$ is negative in this case.

The frequency dependence of the longitudinal velocity-fluctuation spectral density in n-Si at $E = 200$ V/cm, as calculated by the Monte Carlo technique and measured for the two directions of E with respect to the crystallographic directions ($E \| \langle 100 \rangle$ and $E \| \langle 111 \rangle$), is reported in Fig. 4.22 [4.98]. When the measurements are performed in the $\langle 100 \rangle$ direction, the intervalley noise is present. Therefore, at frequencies below about 4 GHz, the values of $S_{v\|}$ in the $\langle 100 \rangle$ direction exceed those in the $\langle 111 \rangle$ direction. In the frequency range above about 4 GHz, the intervalley noise tends to disappear due to the inertia of the intervalley transitions ($\omega \tau_{\text{int}} \gtrsim 1$, $\tau_{\text{int}} \approx 50$ ps), and an inversion of the anisotropy, $S_{v\langle 100 \rangle} < S_{v\langle 111 \rangle}$, is observed. The reason for this final anisotropy is in the different heating of electrons when the field is in the $\langle 100 \rangle$ and $\langle 111 \rangle$ directions. Along the $\langle 100 \rangle$ direction in the frequency range above 10 GHz, and along the $\langle 111 \rangle$ direction at all the frequencies used, only thermal and convective noises are present. The increase of $S_{v\|}(\omega, E)$ at frequencies $f > 10$ GHz is associated with the disappearance of the convective noise which occurs since $\omega \tau_{\mathscr{E}} \gtrsim 1$ (compare Fig. 4.15). At the highest frequencies, $S_{v\|}(\omega, E)$ decreases due to the condition $\omega \tau_{\text{m}} \gtrsim 1$.

The intervalley noise in semiconductors with nonequivalent valleys is due to the transitions from the lower valley with high mobility to the upper one

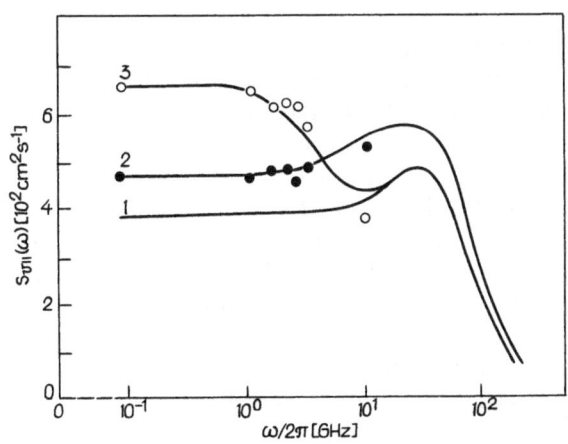

Fig. 4.22. Frequency dependence of the longitudinal velocity-fluctuation spectral density, $S_{v\|}(\omega)$, for n-Si [4.98]; $T_0 = 78$ K, $E = 200$ V/cm. Experimental $S_{j\|}(\omega)/(e^2 n)$: (○) for $E \| \langle 100 \rangle$, (●) for $E \| \langle 111 \rangle$. (——) refer to the Monte Carlo calculations: (1) for $E \| \langle 100 \rangle$, and intervalley transitions neglected; (2) for $E \| \langle 111 \rangle$; (3) for $E \| \langle 100 \rangle$ and intervalley transitions included. The scattering parameters used are the same as in [4.99]

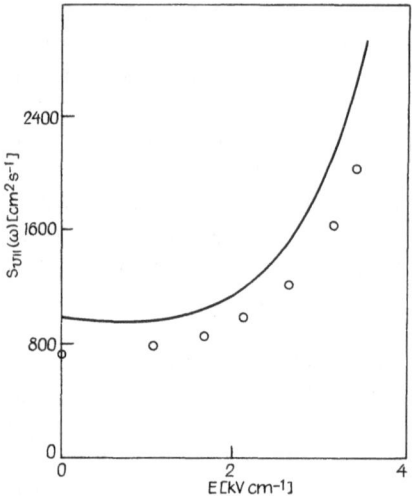

Fig. 4.23. Field dependence of the longitudinal velocity-fluctuation spectral density, $S_{v\parallel}(\omega, E)$, for n-GaAs; $T_0 = 300$ K, $f = 10$ GHz. (O) refer to experiments [4.65]; curve refers to calculations [4.100]

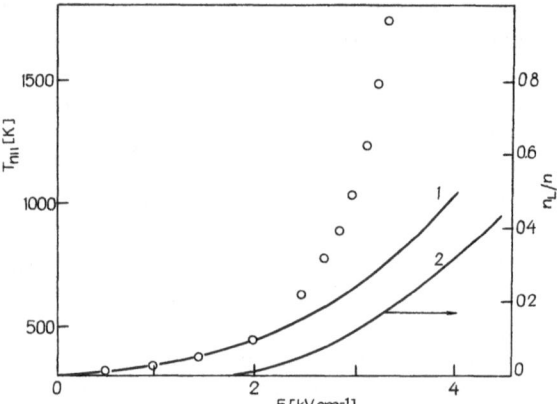

Fig. 4.24. Dependence with electric field of the experimental microwave (10 GHz) noise temperature (O) in n-GaAs at 300 K. The electron temperature in the Γ valley (*1*), and the relative electron concentration n_L/n in the L valley (*2*), as calculated in [4.100], are reported in the same figure

with low mobility. Figure 4.23 reports the experimental data on the field dependence of the spectral density of longitudinal velocity fluctuations $S_{v\parallel}(\omega, E) = 4K_B T_{n\parallel}\sigma'_{\parallel}(\omega, E)/(e^2 n)$ in GaAs at the frequency of 10 GHz. The appreciable increase in $S_{v\parallel}$, observed when the electric field exceeds 2 kV/cm, is explained by the occurrence of intervalley noise due to the electron transition from the Γ minimum to the upper valleys located at L points [4.100, 101]. The curves reported in Fig. 4.24 support the above interpretation by showing that in GaAs, the number of electrons in the L valley increases for $E > 2$ kV/cm; therefore, at these field values, the noise temperature is found to exceed the electron temperature in the Γ valley.

It should be noted that the field dependence of the noise temperature is a more sensitive indicator for electron transitions into the L valley as compared to the field dependence of conductivity.

4.3.5 Microwave Noise and Time-Dependent Diffusion Coefficient

Microwave noise experiments can be used to obtain in a rather direct way data on the diffusion coefficient according to the well-known relation (Chap. 2)

$$S_j(0) = \frac{4\,e^2 n D_0}{V_0}, \tag{4.22}$$

where $S_j(0)$ is the spectral density of current noise at "zero" frequency ("zero" should be intended as a frequency low enough when compared with the value of some inverse relaxation times, e.g., momentum, energy) and D_0 is the diffusion coefficient entering Fick's law. It is worth noting that the first experimental evidence of a field-dependent diffusion coefficient in semiconductors has been determined from noise measurements [4.82,102]. The data on the diffusion coefficient can then be used to describe the spreading of hot-electron packets at times that are long as compared with the above-mentioned relaxation times.

An analysis of fast devices often requires consideration of the packet spreading at very short times when Fick's law is not valid. Then the above-mentioned diffusion coefficient is not useful, but a time-dependent diffusion coefficient (Chap. 2) can be helpful to make an estimate of the effective dimensions of the packet. Below, we shall consider theoretically conditions under which the spectral characteristics at microwave frequencies can give reliable information on the time-dependent diffusion coefficient at very short times.

A spatial packet of arbitrary shape can be expanded in the following way [4.103]:

$$n(x) = \frac{1}{2\pi} \int_{-\infty}^{+\infty} \exp\left(-iqx + \sum_{k=1}^{\infty} \varkappa_k \frac{(iq)^k}{k!}\right) dq, \tag{4.23}$$

where $\varkappa_k$ are the semiinvariants of the distribution.

The time dependence of the position and shape of the packet enters into (4.23) via the semiinvariants. Now, from (4.23) one can obtain [4.103]

$$\frac{\partial n}{\partial t} - \sum_{r=1}^{\infty} \frac{(-1)^r}{r} \frac{d\varkappa_r}{dt} \frac{\partial^r n}{\partial x^r} = 0, \tag{4.24}$$

where the coefficients in front of $\partial^r n / \partial x^r$ have the following meanings:

$$\frac{d\varkappa_1}{dt} = \frac{d}{dt}\langle x\rangle \equiv v_d(t), \tag{4.25}$$

$v_d(t)$ being the time-dependent drift velocity of the packet;

$$\frac{1}{2}\frac{d\varkappa_2}{dt} = \frac{1}{2}\frac{d}{dt}\langle \Delta x^2\rangle \equiv D(t), \tag{4.26}$$

$D(t)$ being a time-dependent diffusion coefficient;

$$\frac{1}{3!}\frac{d\varkappa_3}{dt}\equiv C_3(t), \tag{4.27}$$

$C_3(t)$ being the coefficient for the packet asymmetry change; etc. In the long-time limit the packet becomes Gaussian and (4.24) reduces to the continuity equation

$$\frac{\partial n}{\partial t}+\frac{1}{e}\frac{\partial j}{\partial x}=0, \quad \text{where} \tag{4.28}$$

$$j=e\left(nv_{\mathrm{d}}-D_0\frac{\partial n}{\partial x}\right). \tag{4.29}$$

The most important characteristics of the packet are its drift velocity, $v_{\mathrm{d}}(t)$, and dispersion, $\langle \Delta x^2\rangle$. The time evolution of $\langle \Delta x^2\rangle$ is related to the time-dependent diffusion coefficient, $D(t)$, via (4.26) and this coefficient coincides with that present in Fick's law if the packet spreads out, remaining Gaussian. Nevertheless, in many cases the dispersion of a packet of a complicated form may depend linearly on time, even when Fick's law is not valid [4.104–106]. Figure 4.25 [4.106] shows an example of the time-evolution of an electron packet, propagating along the axis of a periodic heterostructure, which well illustrates this statement. In fact, though the spatial distribution is far from being Gaussian and thus Fick's law is not valid ($\partial n/\partial x$ tends to infinity on the heterojunctions), a linear time dependence of the dispersion is still achieved (Fig. 4.25a) and the packet spreads out similarly to a Gaussian one (Fig. 4.25b).

The coefficients D, C_3, etc. can be related to the electron velocity-coordinate correlation functions. In the case of $D(t)$, let us recall the two following relations, consistent with (4.26) [4.103]:

$$D(t)=\langle \Delta x_i(t)\,\Delta v_i(t)\rangle \quad \text{or} \tag{4.30}$$

$$D(t)=\langle \Delta x_i(0)\,\Delta v_i(t)\rangle +\int_0^t \langle \Delta v_i(0)\,\Delta v_i(t')\rangle\, dt', \tag{4.31}$$

where i enumerates the electrons, Δx_i is the deviation of an electron position from the center of the distribution, Δv_i is its velocity deviation from the average value, and $\langle\ \rangle$ means ensemble average.

The first term in the expression for $D(t)$ in (4.31) accounts for the initial electron distribution in real space. It is zero if the initial distribution is a Dirac function. Equation (4.31) can be rewritten as

$$\int_0^t \langle \Delta v_i(0)\,\Delta v_i(t')\rangle\, dt'=D^*(t)+\phi(0,t), \tag{4.32}$$

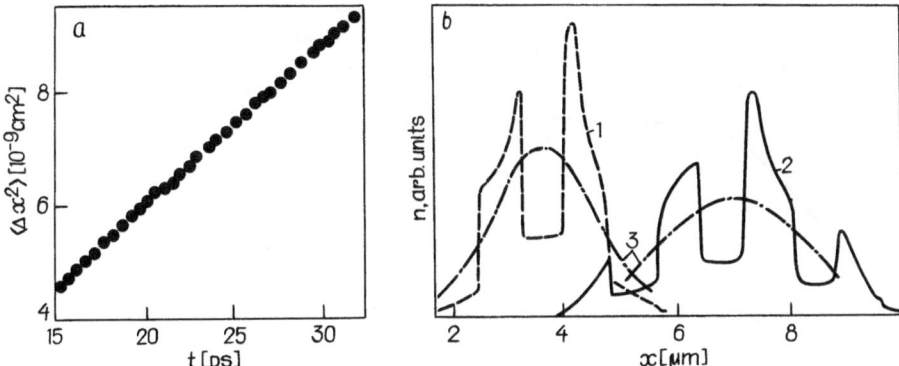

Fig. 4.25a, b. Time evolution of an electron packet (2000 electrons) propagating along the axis of a periodic heterostructure ($d = 0.8\,\mu m$) as simulated by the Monte Carlo calculations. The initial velocity distribution corresponds to the thermal equilibrium at $T_0 = 300\,K$ and the initial space distribution is assumed to be a Dirac function: **(a)** spatial dispersion versus time, and **(b)** spatial distribution at two different times: (1) 25 ps, (2) 50 ps and (3) Gaussian distributions

where $\phi(0, t)$ accounts for the initial velocity distribution and $D^*(t)$ represents the part of the time-dependent diffusion coefficient $D(t)$ which does not depend upon the initial conditions. When different electrons move uncorrelated, $D^*(t)$ assumes a simple form [4.105, 107]:

$$D^*(t) = \sum_j D_j(1 - e^{-t/\tau_j}), \quad \text{and} \tag{4.33}$$

$$\frac{dD^*(t)}{dt} = \langle \Delta v_i(0) \Delta v_i(t) \rangle, \tag{4.34}$$

where j enumerates independent fluctuation processes, characterized by their relaxation times τ_j; and D_j is the additive contribution of each relaxation process to the steady value of $D_0 = \sum_j D_j$. Note that $\phi(0, t) = 0$ if the initial velocity distribution satisfies the Boltzmann equation for the steady state.

Let us recall that microwave noise experiments usually give data on the spectral density of current fluctuations $S_j(\omega)$ under steady-state conditions. In order to relate $S_j(\omega)$ to the velocity-correlation function and determine the dependence of the diffusion coefficient $D^*(t)$, let us assume that the electrons move uncorrelated and the field is uniform (Chap. 2).

In this case, $S_j(\omega)$ is proportional to $S_v(\omega)$ and, according to (4.34), can be related to $D^*(t)$ in the following way [4.107]:

$$\frac{dD^*}{dt} = \frac{1}{4} \int_0^\infty \cos \omega t \, S_v(\omega) d\omega. \tag{4.35}$$

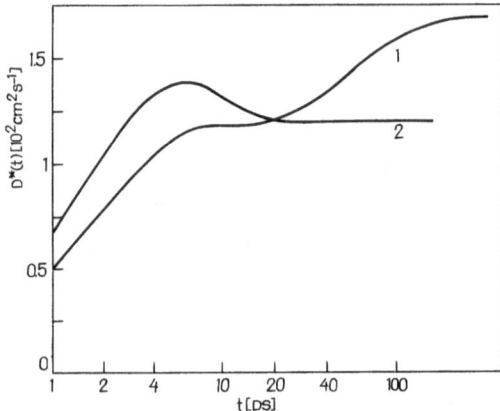

Fig. 4.26. Time dependence of $D^*(t)$ for n-Si at $T_0 = 78$ K; $E = 200$ V/cm reconstructed from the data reported in Fig. 4.22: (1) $E\|\langle 100\rangle$ and (2) $E\|\langle 111\rangle$

Thus, with the above arguments in mind, the steady-state noise experiments are found to give enough information to reconstruct only $D^*(t)$, that is, the initial-condition-independent term of the time-dependent diffusion coefficient $D(t)$.

Figure 4.26 shows the results for $D^*(t)$ as reconstructed from the data presented in Fig. 4.22. For $E\|\langle 100\rangle$ a long-relaxing positive contribution to $D^*(t)$ ($\tau \approx 50$ ps) is due to electron intervalley transfer involving the valleys which are nonequivalently oriented with respect to the electric field. For $E\|\langle 111\rangle$ all the valleys are equally oriented and an overshoot effect of $D^*(t)$ is evidenced.

References

4.1   G.B.Arthur, A.F.Gibson, J.W.Granville: J. Electronics **2**, 145–153 (1956)
4.2   A.F.Gibson, J.W.Granville, E.G.S.Paige: J. Phys. Chem. Sol. **19**, 198–217 (1961)
4.3   E.M.Conwell: *High Field Transport in Semiconductors* (Academic, New York, 1967)
4.4   K.Seeger: Phys. Rev. **114**, 476–481 (1959)
4.5   A.Vebra, J.Pozhela: Trudy Akademii Nauk Lit. SSR Ser. B, **2** (25), 99–105 (1961)
4.6   J.Zucker, V.J.Fowler, E.M.Conwell: J. Appl. Phys. **32**, 2606–2611 (1961)
4.7a T.N.Morgan: J. Phys. Chem. Sol. **8**, 245–249 (1959)
4.7b K.Seeger: *Semiconductor Physics.* An Introduction. 2nd ed. Springer Ser. Solid State Sci., Vol. 40 (Springer, Berlin, Heidelberg 1982)
4.8   V.Dienys, J.Pozhela: *Hot Electrons* (Mintis, Vilnius 1971)
4.9   G.H.Glover: J. Appl. Phys. **42**, 4025–4034 (1971)
4.10 E.T.Whittaker, G.N.Watson: *A Course of Modern Analysis* (Cambridge U. Press, Cambrigde, 1945) p. 229
4.11 J.B.Andersen, B.Majborn: IEEE Trans. MTT-**16**, 194–196 (1968)
4.12 L.D.Nielsen: IEEE Trans. MTT-**17**, 148–153 (1969)
4.13 L.V.Knishevskaja, M.N.Kotov, M.M.Jarmalis: Liet. Fiz. Rink. **20**, 47–53 (1980); [English transl.: Sov. Phys. – Collection **20**, 36–40 (1980)]
4.14 A.Dargys, J.Parsheliūnas: "Measurement of V-A Characteristics by High Field Microwave Techniques", Preprint, Inst. of Semicond. Phys., Acad. of Sc. of the Lithuanian SSR, Vilnius (1982)

4.15 A.B.Davydov, N.M.Cidilkovskii: Fiz. Tverd. Tela **8**, 120–123 (1966);
      [English transl.: Sov. Phys. – Solid State **8**, 92 (1966)]
4.16 G.H.Glover: J. Appl. Phys. **42**, 5590–5595 (1971)
4.17 L.D.Nielsen, T.Guldbrandsen: Proc. Intern. Symp. on Microwave Diagnostics of Semicond.,
      Porvoo, Finland, (July) 13–15, 1977, ed. by R.Paananen (The Swedish Acad. of Eng. Sc. in
      Finland, Helsinki 1977) pp. 409–420
4.18 H.J.Carlin, J.K.Pozhela: Proc. IEEE **53**, 1788–1790 (1965)
4.19 W.Schneider, K.Seeger: Appl. Phys. Lett. **8**, 133–135 (1966)
4.20 T.Banys, J.Pozhela, K.Repshas: Liet. Fiz. Rink. **6**, 415–425 (1966)
4.21 N.Braslau, P.S.Hauge: IEEE Trans. ED-17, 616–622 (1970)
4.22 J.Pozhela: *Plasma and Current Instabilities in Semiconductors*, Intern. Ser. Sci. Solid State,
      Vol. 18 (Pergamon, Oxford 1981)
4.23 A.Dargys, R.Sedrakyan, J.Pozhela: Phys. Status Solidi (a) **45**, 387–392 (1978)
4.24 G.A.Acket, J. de Groot: IEEE Trans. ED-14, 505–511 (1967)
4.25 S.G.Kalashnikov, V.E.Lyubchenko, N.E.Skvortsova: Fiz. Techn. Poluprovodn. **1**, 1445–1447
      (1967);
      [English transl.: Sov. Phys. – Semicond. **1**, 1206 (1967)]
4.26 C.Hamaguchi, T.Kono, Y.Inuishi: Phys. Lett. **24**, 500–501 (1967)
4.27 L.D.Nielsen: Phys. Lett. **38A**, 221–222 (1972)
4.28 A.Dargys, S.Zhilionis, A.Matulionis, J.Parsheliūnas, J.Pozhela, A.Poshkus, E.Shimulyte:
      Liet. Fiz. Rink. **17**, 493–500 (1977);
      [English transl.: Sov. Phys. – Collection **17**, 63–67 (1977)]
4.29 M.Kawashima, S.Kataoka: Japn. J. Appl. Phys. **18**, 1311–1316 (1979)
4.30 M.Inoue, K.Ashida, T.Sugino, J.Shirafuji, Y.Inuishi: Japn. J. Appl. Phys. **12**, 932 (1973)
4.31 A.F.Rudolph, A.Dargys: Phys. Status Solidi (a) **6**, K93–K95 (1971)
4.32 V.Dienys, A.Dargys: J. Physique **42**, Suppl. No. 10, C7, 33–49 (1981)
4.33 K.Hess, K.Seeger: Z. Phys. **218**, 431–436 (1969)
4.34 A.Dargys, T.Banys: Phys. Status Solidi (b)**52**, 699–706 (1972)
4.35 V.Dienys, Zh.Kancleris, Z.Martūnas: Fiz. Techn. Poluprovodn. **13**, 1706–1709 (1979);
      [English transl.: Sov. Phys. – Semicond. **13**, 994 (1979)]
4.36 V.Dienys, Zh.Kancleris, Z.Martūnas: Phys. Status Solidi (b) **101**, 145–152 (1980)
4.37 M.Abe, S.Kaneda: Japn. J. Appl. Phys. **11**, 1675–1683 (1972)
4.38 T.Kobayashi, Y.Nishida, K.Fujisawa: Japn. J. Appl. Phys. **13**, 645–650 (1974)
4.39 K.Hess, H.Kahlert: J. Phys. Chem. Sol. **32**, 2262–2265 (1971)
4.40 A.Philipp, F.Kuchar, K.Seeger: Phys. Status Solidi (b) **79**, 115–124 (1977)
4.41 M.T.Vlaardingerbroek, P.M.Boers, G.A.Acket: Phil. Res. Repts. **24**, 379–391 (1969)
4.42 K.Ashida, M.Inoue, J.Shirafuji: J. Phys. Soc. Jpn. **37**, 408–414 (1974)
4.43 G.H.Glover: J. Appl. Phys. **44**, 1295–1301 (1973)
4.44 G.Bauer, H.Kahlert: Phys. Rev. B5, 566–579 (1972)
4.45 H.Heinrich, K.Hess, W.Jantsch, W.Pfeiler: J. Phys. Chem. Sol. **33**, 425–431 (1972)
4.46 H.Kahlert, K.Hess, K.Seeger: Solid State Commun. **7**, 1149–1152 (1969)
4.47 T.M.Lifshits, A.Ya.Oleinikov, A.Ya.Shulman: Phys. Status Solidi **14**, 511–521 (1966)
4.48 E.Bonek: J. Appl. Phys. **43**, 5101–5109 (1972)
4.49 H.D.Rees: IBM J. Res. Dev. **13**, 537–542 (1969)
4.50 P.A.Lebwohl: J. Appl. Phys. **44**, 1744–1752 (1973)
4.51 W.Fawcett, H.Rees: Phys. Lett. **29A**, 578–579 (1969)
4.52 M.Abe, S.Yanagisawa, O.Wada, H.Takanashi: Appl. Phys. Lett. **25**, 674–675 (1974)
4.53 A.Matulionis, J.Pozhela, A.Reklaitis: Solid State Commun. **16**, 1133–1137 (1975)
4.54 J.Pozhela, A.Reklaitis: Fiz. Techn. Poluprovodn. **13**, 1127–1133 (1979);
      [English transl.: Sov. Phys. – Semicond. **13**, 660 (1979)]
4.55 S.Ashmontas, L.Subachius: Fiz. Techn. Poluprovodn. **13**, 1722–1727 (1979);
      [English transl.: Sov. Phys. – Semicond. **13**, 1002 (1979)]
4.56 T.Banys, A.Dargys, J.Pozhela: Phys. Status Solidi **36**, 755–760 (1969)
4.57 W.H.Haydl, S.Smith, R.Bosch: Appl. Phys. Lett. **37**, 556–557 (1980)

4.58   J.K.Pozhela, K.K.Repshas, V.I.Shilalnikas: Proc. Intern. Conf. on Phys. of Semicond., Exeter, England (1962) pp. 149–151
4.59   E.M.Conwell, J.Zucker: J. Appl. Phys. **36**, 2192–2196 (1965)
4.60   C.Hamaguchi, Y.Inuishi: J. Phys. Chem. Sol. **27**, 1511–1518 (1966)
4.61   S.Ashmontas, J.Pozhela, L.Subachius: Fiz. Techn. Poluprovodn. **11**, 357–364 (1977); [English transl.: Sov. Phys. – Semicond. **11**, 206 (1977)]
4.62   A.I.Veinger, L.G.Paritskii, E.A.Akopyan, G.Dadamirzaev: Fiz. Techn. Poluprovodn. **9**, 216–224 (1975); [English transl.: Sov. Phys. – Semicond. **9**, 144 (1975)]
4.63   S.P.Ashmontas, Yu.K.Pozhela, L.E.Subachius: Pisma v Zh. Eksp. Teor. Fiz. **33**, 580–583 (1981); [English transl.: Sov. Phys. – JETP Lett. **33**, 564–567 (1981)]
4.64   S.Ashmontas, A.Olekas: Fiz. Techn. Poluprovodn. **14**, 2196–2200 (1980);   · [English transl.: Sov. Phys. – Semicond. **14**, 1301 (1980)]
4.65   V.Bareikis, V.Viktoravichius, A.Galdikas, R.Miliushyte: Fiz. Techn. Poluprovodn. **12**, 156–160 (1978); [English transl.: Sov. Phys. – Semicond. **12**, 90 (1978)]
4.66   C. Canali, C.Jacoboni, G.Ottaviani, A.Alberigi-Quaranta: Appl. Phys. Lett. **27**, 278–280 (1975)
4.67   S.P.Ashmontas, A.P.Olekas: Fiz. Techn. Poluprovodn. **14**, 546–549 (1980); [English transl.: Sov. Phys. – Semicond. **14**, 321 (1980)]
4.68   Z.S.Gribnikov, V.L.Kochelap: Zh. Eksp. Teor. Fiz. **58**, 1046–1056 (1970); [English transl.: Sov. Phys. – JETP **58**, 562 (1970)]
4.69   A.N.Vystavkin, Sh.M.Kogan, T.M.Lifshits, P.P.Melnik: Radiotechnika i elektronika **8**, 994–1001 (1963)
4.70   F.G.Bass, Yu.G.Gurevich: *Hot Electrons and Electromagnetic Waves in Solid State and Gaseous Plasmas* (Nauka, Moscow 1975)
4.71   S.Ashmontas, J.Pozhela, K.Repshas: Phys. Status Solidi (b) **51**, 225–232 (1972)
4.72   S.Ashmontas, J.Pozhela, K.Repshas: Liet. Fiz. Rink. **10**, 897–901 (1970)
4.73   S.P.Kalvenas, Yu.K.Pozhela: Proc. 1st Biennal Cornell Conf. on Engineering Applications of Electronic Phenomena, Ithaca, NY, USA (1967) pp. 137–146
4.74   L.Leibler, J.Mycielsky, J.K.Furdyna: Phys. Rev. **B11**, 3037 (1975)
4.75   J.M.Pawlikowski: Phys. Status Solidi (a) **30**, K147 (1975)
4.76   A.Matulionis, J.Pozhela, E.Starikov: Liet. Fiz. Rink. **19**, 683–691 (1979); [English transl.: Sov. Phys. – Collection **19**, 46–52 (1979)]
4.77   S.Ashmontas, J.Pozhela, K.Repshas: Liet. Fiz. Rink. **11**, 243–245 (1971)
4.78   S.Ashmontas, J.Pozhela, K.Repshas, O.Vasilec: Proc. 12th Intern. Conf. on Phys. of Semicond., Stuttgart, ed. by M.H.Pilkuhn (Teubner, Stuttgart 1974) pp. 854–857
4.79   V.Bareikis, J.Pozhela, R.Shaltis: Liet. Fiz. Rink., **6**, 99–104 (1966)
4.80   J.P.Nougier, J.Comollonga, M.Rolland: J. Phys. E**7**, 287 (1974)
4.81   S.V.Gantsevich, V.L.Gurevich, R.Katilius: Rivista Nuovo Cimento **2**, 1–87 (1979)
4.82   V.Bareikis, J.Pozhela, I.Matulioniene: Proc. 9th Intern. Conf. on Phys. of Semicond., Moscow, USSR, Vol. 2 (Nauka, Leningrad 1968) pp. 760–765
4.83   J.P.Nougier, M.Rolland: Phys. Rev. **138**, 5728 (1973)
4.84   V.Bareikis, R.Barkauskas, A.Galdikas, R.Katilius: Fiz. Techn. Poluprovodn. **14**, 1760–1767 (1980); [English transl.: Sov. Phys. – Semicond. **14**, 1046 (1980)]
4.85   V.Bareikis, A.Galdikas, R.Miliushyte, J.Pozhela, V.Viktoravichius: J. Physique **42**, Suppl. No 10, C7, 215–220 (1981)
4.86   R.E.Robson: Phys. Rev. Lett. **31**, 825 (1973)
4.87   R.Katilius, R.Miliushyté: Zh. Eksp. Teor. Fiz. **79**, 631–640 (1980); [English transl.: Sov. Phys. – JETP **52**, 320 (1980)]
4.88   P.J.Price: IBM J. Res. Dev. 3, 191–193 (1959)
4.89   I.B.Levinson, A.J.Matulis: Zh. Eksp. Teor. Fiz. **54**, 1466 (1968); [English transl.: Sov. Phys. – JETP **27**, 786 (1968)]

4.90 V.Bareikis, A.Galdikas, R.Miliushyte, V.Viktoravichius: Proc. 6th Intern. Conf. on Noise in Physical Systems, ed. by P.H.E.Meijer, R.D.Mountain, R.J.Soulen (National Bureau of Standards, Washington 1981) p. 406
4.91 V.Bareikis, V.Viktoravichius, A.Galdikas, R.Miliushyte and Yu.Pozhela: Fiz. Techn. Poluprovodn. **16**, 1816–1819 (1982); [English transl.: Sov. Phys. – Semicond. **16**, 1165 (1982)]
4.92 P.J.Price: J. Appl. Phys. **31**, 949 (1960)
4.93 P.M.Tomchuk, A.A.Chumak: Fiz. Tverd. Tela **14**, 2347–2351 (1972); [English transl.: Sov. Phys. – Solid State **14**, 2031 (1972)]
4.94 L.G.Hart: Canad. J. Phys. **49**, 1469 (1971)
4.95 V.Bareikis, V.Viktoravichius, A.Galdikas, R.Miliushyte: Fiz. Techn. Poluprovodn. **12**, 151–156 (1978); [English transl.: Sov. Phys. – Semicond. **12**, 85 (1978)]
4.96 J.P.Nougier, M.Rolland, D.Gasquet, M.Savelli: Proc. 4th Intern. Conf. on Physical Aspects of Noise in Solid State Devices, Noordwijkerhout, The Netherlands (1975) pp. 19–22
4.97 J.W.Holm-Kennedy, K.S.Champlin: J. Appl. Phys. **43**, 1889 (1972)
4.98 V.Bareikis, V.Viktoravichius, A.Galdikas: Fiz. Techn. Poluprovodn. **16**, 1868–1870 (1982); [English transl.: Sov. Phys. – Semicond. **16**, 1202 (1982)]
4.99 R.Brunetti, C.Jacoboni, F.Nava, L.Reggiani, G.Bosman, R.J.J.Zijlstra: J. Appl. Phys. **52**, 6713–6722 (1981)
4.100 J.Pozhela, A.Reklaitis: Solid State Electron. **23**, 927–933 (1980)
4.101 V.Bareikis, V.Viktoravichius, A.Galdikas, R.Miliushyté: Fiz. Techn. Poluprovodn. **14**, 1427–1429 (1980); [English transl.: Sov. Phys. – Semicond. **14**, 847 (1980)]
4.102 V.Bareikis, J.Pozhela, I.Vaitkevichiūte: Liet. Fiz. Rink. **6**, 437–440 (1966)
4.103 A.Matulionis, J.Pozhela, E.Starikov: Fiz. Techn. Poluprovodn. **16**, 601–606 (1982); [English transl.: Sov. Phys. – Semicond. **16**, 388 (1982)]
4.104 A.Matulionis: Liet. Fiz. Rink. **20**, 65–76 (1980); [English transl.: Sov. Phys. – Collection **20**, 48–57 (1980)]
4.105 V.Bareikis, A.Matulionis, J.Pozhela, S.Ashmontas, A.Reklaitis, A.Galdikas, R.Miliushyte, E.Starikovas: In *Hot Electron Diffusion*, ed. by J.Pozhela (Mokslas, Vilnius 1981)
4.106 A.Matulionis, J.Pozhela, E.Starikov: Phys. Status Solidi (a) **68**, K149–K152 (1981)
4.107 A.Matulionis, J.Pozhela, E.Starikov: Liet. Fiz. Rink. **21**, 45–57 (1981)

5. Multivalued Distributions of Hot Electrons Between Equivalent Valleys

Marion Asche

With 20 Figures

Let us consider a many-valley semiconductor (Si as a typical example) subject to an external electric field. When the intervalley transition rate increases sharply with carrier heating, transverse fluctuations of the electric field, instead of decaying spontaneously, may lead to a repopulation of the valleys, even if these are equally oriented with respect to the applied field. As a consequence, a steady state which in the absence of fluctuations is symmetric around the field direction may, owing to the fluctuations, decay into new steady states which lose this symmetry (states of broken symmetry). Under this condition, transverse to the direction of the current flow, a sample exhibits regions with different transverse fields, and in turn negative differential conductivity (NDC) may occur on account of the anisotropy in the conductivity. The position of the wall between regions with different transverse fields becomes very sensitive to external disturbances and can be switched by these influences. For the case of electrons in Si we shall show that the results obtained from different experiments and Monte Carlo calculations agree very well and fully support the evidence of this spontaneous symmetry breaking.

In Sect. 5.1 the creation of the transverse fields is analyzed for a two-valley model and the appearance of NDC is explained accordingly. Section 5.2 reports the results of an experimental proof for some predicted effects, as obtained for n-Si at low temperatures. A theoretical analysis of the conditions which allow a multivalued electron distribution (MED) to be realized is given in Sect. 5.3. Numerical results for the transverse fields and the drift velocity as functions of the heating field are presented in Sect. 5.4. In Sect. 5.5 the necessary condition for MED, that is, a sharp change of the intervalley transfer rate with carrier heating, is investigated. Section 5.6 reports further experimental investigations which were performed for other orientations of the current density different from that of Sect. 5.1. Finally, Sect. 5.7 provides a brief retrospect of the efforts made to prove the existence of MED, in view of its future possibilities.

5.1 Multivalued Electron Distribution (MED) as Spontaneous Symmetry Breaking

In systems operating under strongly nonlinear conditions, fluctuations can lead to spontaneous symmetry breaking (for a general review of this argument see [5.1]). Such a condition can be accomplished at high electric fields in the case of

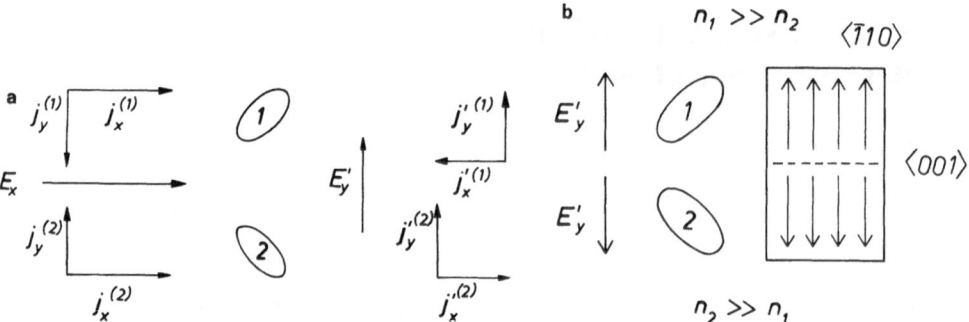

Fig. 5.1a, b. Electric fields and currents for a simple model semiconductor consisting of two equivalent valleys (**a**). Cross section in the middle of a sample with transverse fields (arrows) and carrier redistributions in the two layers (**b**)

many-valley semiconductors. In fact, for appropriate crystallographic orientations, the uniform distribution of a high steady-state field may become unstable when the carrier transition rate between equivalent valleys sharply increases with carrier energy, and thus with increasing field strength.

For illustrative purposes let us consider a brick-shaped sample with the two transverse dimensions much smaller than the logitudinal one, and assume a semiconductor band model consisting of two equivalent minima with ellipsoidal equienergetic surfaces (Fig. 5.1a). If we apply an electric field E_x symmetrically oriented with respect to both valleys, the components of the current density of each valley in the field direction $j_x^{(1)}$ and $j_x^{(2)}$ are equal. However, because of the anisotropic conductivity in each valley associated with the nonspherical energy surfaces, the components of the current density of each valley transverse to the field direction $j_y^{(1)}$ and $j_y^{(2)}$ are equal but oppositely oriented. The current density in each valley is therefore turned from the direction of the applied field into the direction of smaller effective mass according to $j^{(\alpha)} = en^{(\alpha)} \mu^{(\alpha)} E$. Here e is the electron charge: $n^{(\alpha)}$ and $\mu^{(\alpha)}$ denote the carrier density and mobility in valley α, respectively. The components of the mobility tensor $\mu^{(\alpha)}$ are given by $\mu_i^{(\alpha)} = e\langle\tau_i\rangle/m_i$, with $\langle\tau_i\rangle$ the averaged ith component of the momentum relaxation time and m_i^{-1} the ith component of the inverse effective-mass tensor.

Similarly, a field fluctuation E_y' will cause parallel directed components $j_y'^{(1)}$ and $j_y'^{(2)}$ and oppositely directed components $j_x'^{(1)}$ and $j_x'^{(2)}$.

Now, to first order there are two terms which contribute to the energy fluctuation, namely, $j_y^{(\alpha)} E_y'$ and $j_y'^{(\alpha)} E_y$, originated in the fluctuating field and current, respectively. As can be easily argued from Fig. 5.1, both these contributions are negative for Valley 1 and therefore the carriers of this valley will be cooled. On the other hand, both contributions are positive for Valley 2; therefore the carriers of this valley will be heated. The resulting difference in the valley energies permits an electron redistribution between the two valleys. Consequently, when an intervalley scattering process with the assistance of a phonon becomes possible, the hotter valley will finally be depleted. Therefore all current contributions of Valley 1 will be increased at the expense of Valley 2, and the conductivity of the two-valley semiconductor, which in the absence of fluctuations was symmetric around the field direction, will now become asymmetric.

5.1.1 States of Broken Symmetry – Layers of Transverse Fields

If $j_y^{(1)} < (j_y^{(2)} + j_y'^{(1)} + j_y'^{(2)})$, that is, the only component opposite to E_y' is smaller than the sum of the others directed parallel to E_y', the fluctuation will be damped. When, on the other hand, $j_y^{(1)} > (j_y^{(2)} + j_y'^{(1)} + j_y'^{(2)})$, the fluctuation will be enhanced and, in turn, a finite transverse field built up.

In a similar way a fluctuation of the opposite sign will cause a depletion of the population in Valley 1. In this case, a finite transverse field will be built up if the component $j_y^{(2)}$, oppositely oriented to E_y', is greater than the sum of the others directed parallel to E_y'.

Summarizing, it can be stated that if the value of the difference between the currents $j_y^{(\alpha)}$ of Valleys 1 and 2 is greater than the corresponding value of the sum of the fluctuation currents $j_y'^{(\alpha)}$, a symmetric steady state with equal electron densities of both valleys is unstable. The spontaneous symmetry breaking of the electron distribution between equivalent valleys leads to new asymmetric steady states of the semiconductor. These states are characterized by the presence of electric fields which are oppositely directed and perpendicular to the total current density j_x[1]. Since the two possible asymmetric states, respectively with higher population of Valley 1 or 2, are degenerate, the boundary conditions determine how the sample decays into regions of either of the two states. For the case reported in Fig. 5.1 theoretical considerations [5.3, 4] predict that, in absence of intervalley transfer on the surfaces with normals in y direction, the electrons populate Valley 1 on one surface and Valley 2 on the opposite one. Consequently the sample decays into layers, each with the transverse field of that sign (compare Fig. 5.1a) which belongs to the predominant population of the valley preferred by the boundary condition of the adjacent surface (Fig. 5.1b). By symmetry requirements the wall between the layers (interlayer wall) is located halfway between the surfaces if the crystal is perfectly homogeneous and the field is oriented exactly symmetrically with respect to the considered valleys.

The presence of a small additional transverse electric field[2] in the y direction will assist the development of the field fluctuation of the same sign and damp the

---

1 If the heating field is not directed along a high-symmetry axis of the crystal, the directions of the current density and field strength do not coincide, as was predicted in [5.2a] and experimentally shown for the first time in [5.2b]. In bounded samples this effect leads to transverse field components (the Sasaki field). By analogy, the appearance of transverse fields for symmetry directions was, in the beginning, called "multivalued Sasaki effect".

2 Of course, when this transverse field is higher than the fluctuation field, the latter will no longer play any role.

development of that opposite to its orientation. Therefore it will shift the interlayer wall in such a way that the layer with the field in the same direction will grow. The additional field may be caused by an inhomogeneity in the crystal, a contact, a magnetic field, a mechanical stress, etc. The shift of the interlayer wall towards one of the surfaces gives rise to a high potential difference between equivalent points 1 and 2 on the pair of opposite surfaces, as indicated in Fig. 5.2 (Sect. 5.2.1).

5.1.2 Negative Differential Conductivity (NDC) Caused by Transverse Fields

For the case of the two-valley semiconductor, the symmetry breaking may lead to a reduction of the total current density j_x along the direction of the applied field with increasing field strength. In fact, for E'_y oriented as in Fig. 5.1 with growing repopulation between the valleys, the increase of $j_x'^{(1)}$ does not compensate the change of $j_x'^{(2)}$, but both add so that the total current density j_x will be a decreasing function of field, thus leading to an N-shaped NDC region in the current-voltage characteristics.

As is well known, a state with NDC is unstable and, in this case, it will cause a domainization of the field along the x direction of the sample. For values of the applied field strength above a critical value E_{low} a static high-field domain with field strength E_{high} is built up near one of the electrodes or near an inhomogeneity and will extend through the sample with rising applied field strength E_x.

Contrary to the usual NDC, which appears in presence of a longitudinal applied field only [5.5–8], in the present case NDC is due to the presence of built-in transverse fields.

5.2 Experimental Evidence of MED

For the investigations under consideration, n-Si is chosen because it has a simple many-valley band structure and fulfills the requirements with respect to intervalley scattering processes.

The Si conduction band consists of three pairs of equivalent minima (valleys) in the $\langle 100 \rangle$ directions. Each valley is characterized by a longitudinal and transverse effective mass, $m_{lo} = 0.92\,m_0$ and $m_{tr} = 0.19\,m_0$, m_0 being the free-electron mass. If a sample is cut from a slide oriented normal to $\langle \bar{1}10 \rangle$, with j_x along the $\langle 110 \rangle$ direction the valleys in the $\langle 100 \rangle$ and those in the $\langle 010 \rangle$ directions play the roles of Valleys 1 and 2 of Fig. 5.1. At the same time, the valleys in the $\langle 001 \rangle$ direction do not interfere with the phenomena expected from the simplified model, since in the $(\bar{1}10)$ plane considered their effective masses do not depend on the field direction. As a matter of fact by increasing the energy level of the two $\langle 001 \rangle$ valleys by the application of an uniaxial pressure it can be proven that their influence is not essential indeed [5.9].

Now a strong nonlinearity of the system, which is the necessary condition for symmetry breaking [5.1], can be realized by phonon-assisted intervalley scattering of hot electrons, since the scattering rate increases exponentially with carrier energy at low temperatures. This condition, because of a lower intervalley scattering rate at ionized impurities, is more favorably met in n-Si than in n-Ge.

5.2.1 Evidence for Layered Structures

An experimental proof of the stratification of the transverse fields due to MED for a field applied along a $\langle 110 \rangle$ direction can be obtained by measuring the potential differences between adjacent Probes 1–3–2 or 1–4–2, respectively (Fig. 5.2, insert). These measurements exhibit a high positive value of the voltage between Probes 1 and 3 and a high negative value between Probes 3 and 2. Accordingly, between Probes 1 and 2 a small value remains only, which agrees with the partial compensation of the opposite fields (compare Fig. 5.1b). As an example, for one sample at $E_x = 105$ V/cm, with U_{ij} denoting the potential difference between Probes i and j, and d_{ij} their distance, the values are $\Delta U_{13}/d_{13} = 47.6$ V/cm, $\Delta U_{32}/d_{32} = -44.6$ V/cm, and $\Delta U_{12}/d_{12} = 3$ V/cm. Quantitatively, the same picture is obtained with Probe 4 replacing Probe 3. In this way it can be shown that, above a critical value of the applied field strength, the small voltage which appears between Probes 1 and 2 on the $(\bar{1}10)$ surfaces ($\times \times \times$ in Fig. 5.2) can be ascribed to the difference of the two opposite transverse fields.

A magnetic field $\boldsymbol{B}$ in the $\langle 001 \rangle$ direction creates a Hall field parallel to the $\langle \bar{1}10 \rangle$ direction and, as mentioned in Sect. 5.1.1, this field shifts the interlayer wall in such a way that the layer with the transverse field in the direction of the Hall field will grow [5.3, 4, 10, 11]. Figure 5.2 shows the transverse voltage difference between Probes 1 and 2 normalized to the width of the sample as a function of the heating field applied in the absence of $\boldsymbol{B}$ and for $B = \mp 0.066$ T. For values of E_x

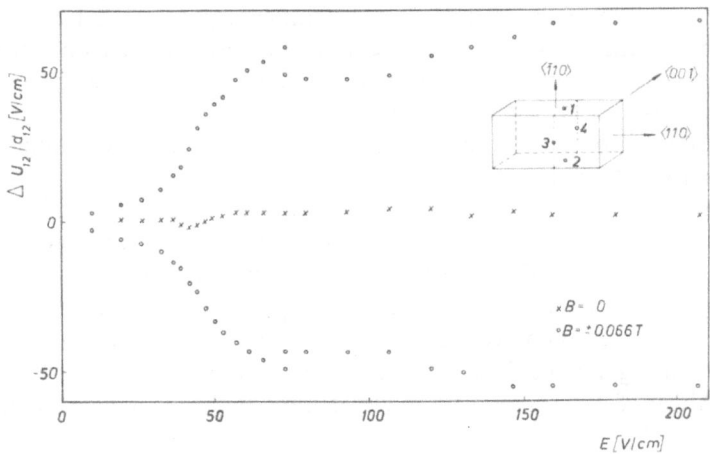

Fig. 5.2. Transverse fields versus applied fields with and without magnetic field for n-Si at 27 K. *Insert*: crystallographic orientation of the sample and position of the probes

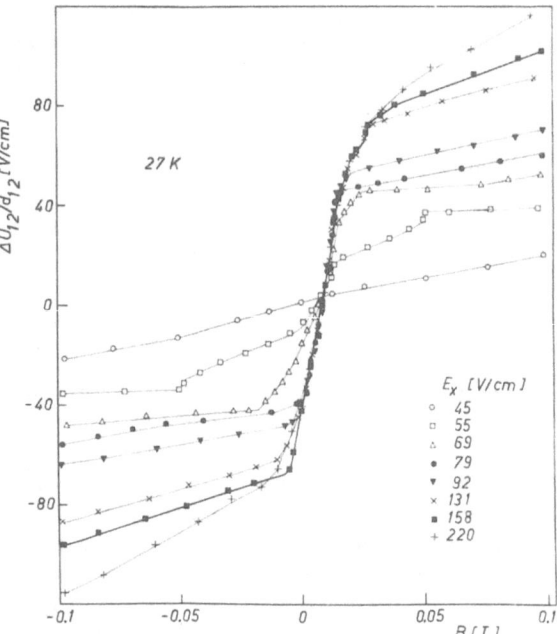

Fig. 5.3. Transverse fields versus magnetic field for n-Si at 27 K. Values of the applied field are indicated in the figure

lower than 40 V/cm a usual Hall field is observed. Above 40 V/cm [3], the value of the transverse field exhibits an anomalous steep increase while the longitudinal static high-field domain (Sect. 5.2.2) is found to extend further into the sample, with a value which remains nearly constant for an applied electric field above E_{high}. The behavior is almost symmetric with respect to a change in the polarity of the magnetic field. If in the absence of the magnetic field the interlayer wall is not located in the middle of the sample, but a significant potential difference between the probes on the pair of $(\bar{1}10)$ surfaces is already present, then only one type of the magnetic field polarity switches the wall remarkably, i.e., the polarity for which the wall is shifted to the opposite surface. In this case, too, the transverse potential difference remains equal, but opposite in sign according to the polarity of **B** [Ref. 5.11, Fig. 1].

A detailed analysis of $\Delta U_{12}/d_{12}$ as a function of **B** is shown in Fig. 5.3. Here for fields applied below 40 V/cm the common Hall effect is observed. Above a critical value of the applied field E_x, 45 V/cm in the figure at weak B $\Delta U_{12}/d_{12}$ exhibits a steep rise. At applied field values above E_{high} (about 79 V/cm for this sample) and in the same region of weak B, the slope of all curves remains almost constant with increasing applied field, that is, as long as MED is realized (thick line in Fig. 5.3). Always for fields above E_{high}, by increasing the magnetic field an

---

3 In presence of longitudinal domains of high- and low-field strength, E_x corresponds to the applied potential divided by the length of the sample.

abrupt transition towards a weaker dependence of the transverse field is observed; this corresponds to the fact that the interlayer wall has already been shifted to one of the surfaces. For $E_x \gtrsim 200$ V/cm this abrupt transition is smoothed out, indicating that there is no spontaneous symmetry breaking of the equivalent valleys left, but only an anomalously high carrier redistribution between them. This means that the Hall field is strongly enhanced by the intervalley redistribution of the electrons, which in turn is due to the different influence the Hall field has on carrier heating in both valleys [5.12]. While the increase at weak magnetic fields is jumplike due to the layered structure, the increase with **B** becomes smoother for the homogeneous case in the absence of MED. Of course, for high **B**, when in the case of MED the interlayer wall has already been shifted to the surface, the redistribution is similar for both cases.

5.2.2 Current Saturation and Longitudinal Domains

Another consequence of MED is the reduction of the current density along the applied field, mentioned in Sect. 5.1.2, which can lead to an NDC region and in turn to a domainization of the sample. This phenomenon was observed in [5.11, 13] as a saturation of the current density; which occurs in a certain region of values of the field strengths applied along a $\langle 110 \rangle$ direction as shown in Fig. 5.4 (Curve a). The critical field strength E_{low}, for which a high-field domain appears at one contact, is clearly exhibited.

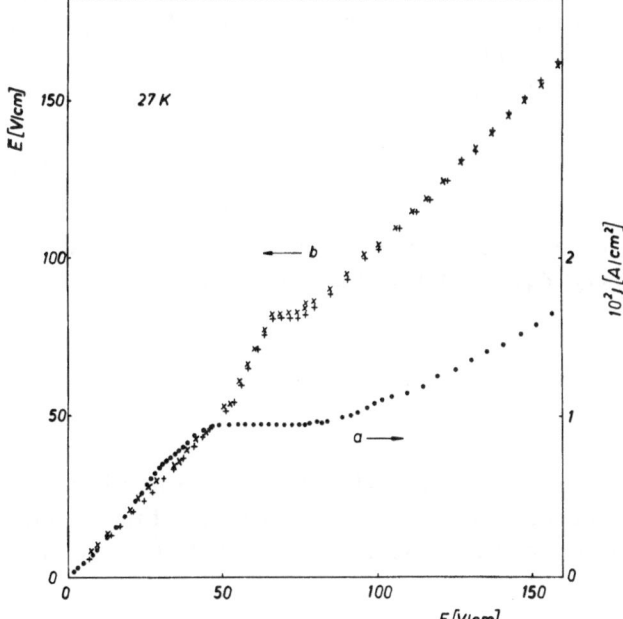

Fig. 5.4. Current density (a) and longitudinal fields (b) at the Probes 3 ($\times$) and 4 ($+$) (*insert* of Fig. 5.2) versus applied field

The domainization of the longitudinal field is to be seen by the behavior of the averaged field strength $\bar{E}$ between the current contact at low potential and probes at the (001) surfaces (Curves b in Fig. 5.4). It should be noted that the probes are approximately situated halfway between the ($\bar{1}$10) surfaces to minimize the influences from the transverse fields [5.11]. When the current begins to saturate, the two curves (b) show a superlinear rise of the field strength at these probes. This corresponds to the buildup of a static high-field domain at the current contact [5.14]. Then the two curves (b) exhibit a plateau. This means that the high-field domain has already reached the position of the probes and extends further into the direction of the other current contact. When the high-field domain reaches this contact, the field strength in the whole sample rises by further increasing the applied field. Also the inverse effect, which corresponds to a domain nucleated near the current contact at the higher potential, is often observed.

5.3 Theoretical Analysis of MED (Homogeneous Case)

Following [5.3, 15] an analytical theory can be developed by introducing the following simplifying assumptions:

(1) In each valley α $(\alpha = 1, 2, 3)$ the carrier distribution function $f^{(\alpha)}(k)$ only weakly deviates from the isotropic part $f_0^{(\alpha)}$ $(\mathscr{E})$.

(2) Transitions between perpendicular valleys (f-type scattering), characterized by an out scattering time $\tau^{(\alpha)}$, have negligible effects for the energy and momentum balance and only determine the valley population, thus leading to an electron density in the valley given by

$$n^{(\alpha)} = n\tau^{(\alpha)} \left/ \sum_{\beta=1}^{3} \tau^{(\beta)} \right. . \tag{5.1}$$

(3) A momentum relaxation time can be introduced, the anisotropy of which is independent of carrier energy $\mathscr{E}$, thus allowing the mobility of the valley α to be written as

$$\mu^{(\alpha)} = a^{(\alpha)} \mu(E^{(\alpha)})$$
$$a_{ij}^{(\alpha)} = g_{ik}^{(\alpha)} g_{jk}^{(\alpha)} M_{kk} , \tag{5.2}$$

where sum over repeated indices is implied. Here $[g^{(\alpha)}]$ is the matrix transforming[4] the principal axes of each valley to the reference frame of the sample; $M = 3\mu/(2\mu_{\mathrm{tr}} + \mu_{\mathrm{lo}})$ is a dimensionless diagonal tensor with $\mu_{11} = \mu_{\mathrm{lo}}$ and $\mu_{22} = \mu_{33} = \mu_{\mathrm{tr}}$.

---

4 It consists of the transformation $[B^{(\alpha)}]$ from the axes of the valley to the main crystallographic axes and $[A]$ from them to the reference frame of the sample.

Now $\mu(E^{(\alpha)})$ and $\tau^{(\alpha)} = \tau(E^{(\alpha)})$ depend only on the effective field strength $E^{(\alpha)}$ given by $E^{(\alpha)^2} = Ea^{(\alpha)}E$; the latter ones are reduced to the usual expressions in the case of an isotropic momentum relaxation time [5.16].

For a homogeneous brick-shaped sample with the two transverse dimensions d_y and d_z smaller than the longitudinal one, the definition of the current density

$$j = en \sum_\alpha \tau(E^{(\alpha)}) \mu(E^{(\alpha)}) a^{(\alpha)} E \bigg/ \sum_\beta \tau(E^{(\beta)}) \tag{5.3a}$$

and the conditions $j_y = j_z = 0$ yield

$$j_x = en \sum_\alpha \phi^{(\alpha)} a_{xi}^{(\alpha)} E_i \bigg/ \sum_\beta \tau(E^{(\beta)}). \tag{5.3b}$$

With $\phi^{(\alpha)}$ abbreviating $\mu(E^{(\alpha)})\tau(E^{(\alpha)})$, for $\vartheta_y = E_y/E_x$ and $\vartheta_z = E_z/E_x$ the following expressions are obtained:

$$\vartheta_y = -\frac{\sum_\alpha \phi^{(\alpha)} a_{yx}^{(\alpha)} \sum_\beta \phi^{(\beta)} a_{zz}^{(\beta)} - \sum_\alpha \phi^{(\alpha)} a_{zx}^{(\alpha)} \sum_\beta \phi^{(\beta)} a_{yz}^{(\beta)}}{\sum_\alpha \phi^{(\alpha)} a_{yy}^{(\alpha)} \sum_\beta \phi^{(\beta)} a_{zz}^{(\beta)} - \sum_\alpha \phi^{(\alpha)} a_{zy}^{(\alpha)} \sum_\beta \phi^{(\beta)} a_{yz}^{(\beta)}}$$

$$\vartheta_z = \frac{\sum_\alpha \phi^{(\alpha)} a_{yx}^{(\alpha)} \sum_\beta \phi^{(\beta)} a_{zy}^{(\beta)} - \sum_\alpha \phi^{(\alpha)} a_{zx}^{(\alpha)} \sum_\beta \phi^{(\beta)} a_{yy}^{(\beta)}}{\sum_\alpha \phi^{(\alpha)} a_{yy}^{(\alpha)} \sum_\beta \phi^{(\beta)} a_{zz}^{(\beta)} - \sum_\alpha \phi^{(\alpha)} a_{zy}^{(\alpha)} \sum_\beta \phi^{(\beta)} a_{yz}^{(\beta)}}. \tag{5.4}$$

5.3.1 General Discussion: The Case of Si with Current Density in a ($\bar{1}$10) Plane

Since the analytical expressions for a general orientation become too complicated to allow clear insight into the expected effects and because available experimental data for Si refer to the case of a current density in a ($\bar{1}$10) plane, the following analysis will be limited to this case.

By indicating with ψ the angle between the $\langle 110 \rangle$ direction and the current density vector (Fig. 5.5), for the nonvanishing terms of (5.2) one obtains

$$a_{xx}^{(1)} = a_{xx}^{(2)} = 1 + a_1(1 - 3\cos 2\psi)/4,$$

$$a_{xx}^{(3)} = 1 - a_1(1 - 3\cos 2\psi)/2,$$

$$a_{xy}^{(1)} = -a_{xy}^{(2)} = 3a_1(\cos\psi)/2, \quad a_{xz}^{(1)} = a_{xz}^{(2)} = 3a_1(\sin 2\psi)/4,$$

$$a_{xz}^{(3)} = -3a_1(\sin 2\psi)/2, \quad a_{yy}^{(1)} = a_{yy}^{(2)} = 1 - a_1/2,$$

$$a_{yy}^{(3)} = 1 + a_1, \quad a_{yz}^{(1)} = -a_{yz}^{(2)} = -3a_1(\sin\psi)/2,$$

$$a_{zz}^{(1)} = a_{zz}^{(2)} = 1 + a_1(1 + 3\cos 2\psi)/4,$$

$$a_{zz}^{(3)} = 1 - a_1(1 + 3\cos 2\psi)/2$$

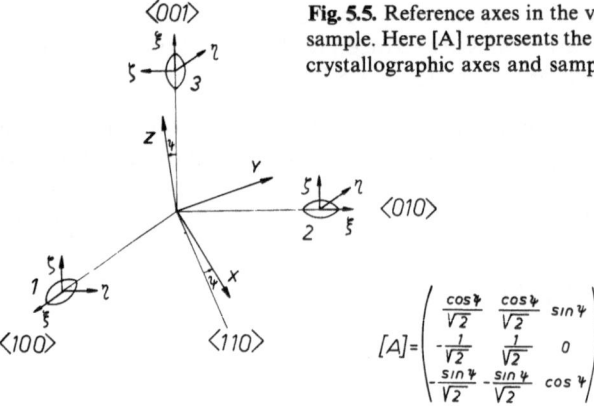

Fig. 5.5. Reference axes in the valleys and reference frame of the sample. Here [A] represents the matrix for the transition between crystallographic axes and sample reference frame

$$[A] = \begin{pmatrix} \frac{\cos\psi}{\sqrt{2}} & \frac{\cos\psi}{\sqrt{2}} & \sin\psi \\ -\frac{1}{\sqrt{2}} & \frac{1}{\sqrt{2}} & 0 \\ -\frac{\sin\psi}{\sqrt{2}} & -\frac{\sin\psi}{\sqrt{2}} & \cos\psi \end{pmatrix}$$

with $a_1 = (\mu_{tr} - \mu_{lo})/(2\mu_{tr} + \mu_{lo})$. Introducing

$$\Phi_0 = (\phi^{(1)} - \phi^{(2)})/\sum_\alpha \phi^{(\alpha)}$$

$$\Phi = (\phi^{(1)} + \phi^{(2)} - 2\phi^{(3)})/\sum_\alpha \phi^{(\alpha)} \tag{5.5}$$

the total current density is

$$j_x = en \sum_\alpha \phi^{(\alpha)} [1 + \Phi a_1(1 - 3\cos 2\psi)/4 + \vartheta_y \Phi_0 3 a_1 (\cos\psi)/2$$

$$+ \vartheta_z \Phi 3 a_1 (\sin 2\psi)/4] E_x / \sum_\alpha \tau^{(\alpha)}. \tag{5.6}$$

The ratios of the transverse fields to the applied field are

$$\vartheta_y = -\frac{\Phi_0 3 a_1 \{\cos\psi [4 + \Phi a_1(1 + 3\cos 2\psi)] + \Phi 3 a_1 \sin 2\psi \sin\psi\}}{2\{(1 - \Phi a_1/2)[4 + \Phi a_1(1 + 3\cos 2\psi)] - (\Phi_0 3 a_1 \sin\psi)^2\}},$$

$$\vartheta_z = -\frac{(\Phi_0 3 a_1)^2 \cos\psi \sin\psi + \Phi 3 a_1 (1 - \Phi a_1/2) \sin 2\psi}{(1 - \Phi a_1/2)[4 + \Phi a_1(1 + 3\cos 2\psi)] - (\Phi_0 3 a_1 \sin\psi)^2}, \tag{5.7}$$

and the arguments $E^{(\alpha)}$ of the $\phi^{(\alpha)}$ are given by

$$Ea^{(1)}E = \{1 + a_1(1 - 3\cos 2\psi)/4 + \vartheta_y 3 a_1 \cos\psi + \vartheta_z 3 a_1 (\sin 2\psi)/2$$

$$+ \vartheta_y^2 (2 - a_1)/2 - \vartheta_y \vartheta_z 3 a_1 \sin\psi + \vartheta_z^2 [1 + a_1(1 + 3(\cos 2\psi))/4]\} E_x^2,$$

$$Ea^{(3)}E = \{1 - a_1(1 - 3\cos 2\psi)/2 - \vartheta_z 3 a_1 \sin 2\psi + \vartheta_y^2 (1 + a_1)$$

$$+ \vartheta_z^2 [1 - a_1(1 + 3\cos 2\psi)/2]\} E_x^2, \tag{5.8}$$

where for $Ea^{(2)}E$ only terms proportional to ϑ_y change their signs in comparison to $Ea^{(1)}E$.

Inserting (5.8) into (5.7) these equations may be written in the form

$$\vartheta_y = R(\vartheta_y, \vartheta_z),$$
$$\vartheta_z = Q(\vartheta_y, \vartheta_z). \tag{5.9}$$

Eq. (5.9) have to be discussed with respect to the possibility of MED. They are strongly nonlinear, of transcendent type, and their right-hand sides are limited; furthermore, they always have the common solutions corresponding to the Sasaki effect [5.2].

From (5.7) it is to be seen that for $\psi = \pi/2$ (the $\langle 001 \rangle$ direction) only the solution $\vartheta_y = \vartheta_z = 0$ exists; thus no MED can appear with a field applied along this direction.

For $\psi = 0$ (the $\langle 110 \rangle$ direction) the z component again only has the solution $\vartheta_z = 0$, but ϑ_y reads

$$\vartheta_y = -\frac{3a_1(\phi^{(1)} - \phi^{(2)})}{(2-a_1)(\phi^{(1)} + \phi^{(2)}) + 2(1+a_1)\phi^{(3)}}. \tag{5.10}$$

Besides the trivial solution $\vartheta_y = 0$, which corresponds to $\phi^{(1)} = \phi^{(2)}$ – that is, equal population of Valleys 1 and 2 – there may be nontrivial solutions with nonsymmetric population of the Valleys 1 and 2, this because for $\vartheta_y \neq 0$ the effective fields differ for both valleys, as can be seen from (5.8).

For $\tan \psi = 1/\sqrt{2}$ (the $\langle 111 \rangle$ direction) the trivial solutions of (4.7) $\vartheta_y = \vartheta_z = 0$ mean equal population of all valleys, see (5.5). Further on, nontrivial solutions with nonsymmetric population of the valleys may exist.

For other orientations besides the common nonvanishing solutions, non-trivial ones can exist, too.

The nontrivial solutions for ϑ_y or ϑ_z lead to a conductivity which decreases with increasing applied field. Furthermore for ϑ_y and ϑ_z increasing strongly with E_x, even NDC can be obtained (detailed conditions were reported in [5.15]).

To carry out the analysis of the nontrivial solutions for each E_x the left- and right-hand sides of (5.9) can be drawn as functions of ϑ_y and ϑ_z, and their intersections determined.

5.3.2 Current Density Along a $\langle 110 \rangle$ Direction

An example of the above-mentioned procedures is illustrated in Fig. 5.6 for the case of a current density along a $\langle 110 \rangle$ direction. Under this configuration only one of the transverse field components in (5.9) does not vanish. The critical-field strengths for a nontrivial solution are given when the curves, according to the left- and right-hand sides of (5.10), are touching; this means that their slopes

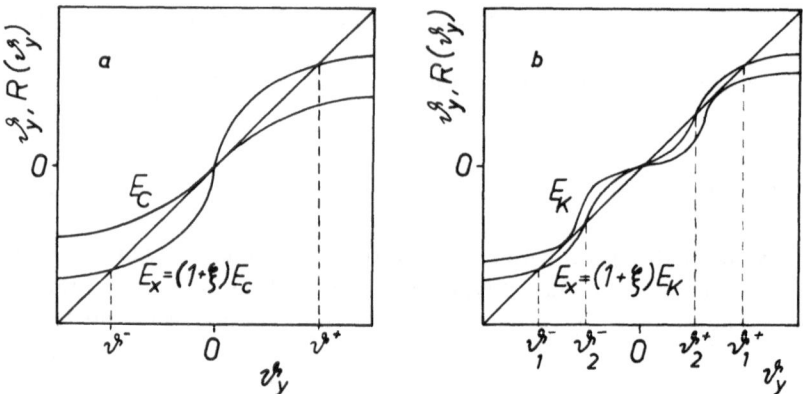

Fig. 5.6a, b. Both sides of (5.10) as functions of ϑ_y near the critical electric fields of E_C type (a), and of E_K type (b)

have to be equal:

$$1 = \frac{dR(\vartheta_y)}{d\vartheta_y}. \tag{5.11}$$

If this occurs in the vicinity of $\vartheta_y = 0$ (Fig. 5.6a), one obtains

$$\phi^{(1)} = \phi^{(2)} \quad \text{and} \quad \frac{d\phi^{(1)}}{d\vartheta_y} = -\frac{d\phi^{(2)}}{d\vartheta_y} = \frac{3a_1 E_x}{\sqrt{2(2-a_1)}} \frac{d\phi(E^{(1)})}{dE^{(1)}}, \tag{5.12}$$

and the use of (5.11) with $E^{(1)} = \sqrt{2(2-a_1)}\, E_x/2 = E_C$ leads to

$$-\frac{9a_1^2}{(2-a_1)^2} \frac{E_C}{\phi(E_C)} \frac{d\phi(E_C)}{dE_C} = 1 + \frac{(1+a_1)}{(2-a_1)} \frac{\phi(\sqrt{2(1+a_1)/(2-a_1)}\, E_C)}{\phi(E_C)}$$

or

$$-\left(\frac{\mu_{tr} - \mu_{lo}}{\mu_{tr} + \mu_{lo}}\right)^2 \frac{d\ln\phi(E_C)}{d\ln E_C} = 1 + \frac{\mu_{tr}\phi(\sqrt{2\mu_{tr}/(\mu_{tr}+\mu_{lo})}\, E_C)}{(\mu_{tr}+\mu_{lo})\phi(E_C)}. \tag{5.13}$$

The condition for the existence of a pair of nontrivial solutions $\vartheta^{\pm}$ can be fulfilled if $\phi^{(\alpha)} = \mu(E^{(\alpha)})\tau(E^{(\alpha)})$ exhibits a pronounced decrease with increasing fields. Since $\mu(E^{(\alpha)})$ is a slowly varying function, the required steep change of $\phi^{(\alpha)}$ with E_x depends critically on the intervalley scattering time $\tau(E^{(\alpha)})$.

When more solutions exist we have to decide whether they are stable or unstable. These investigations were performed in [5.15], and here only the results are used.

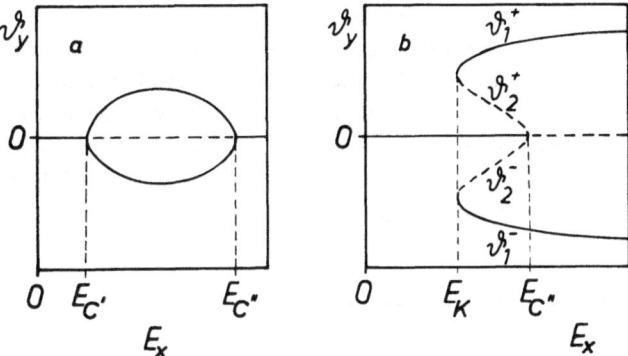

Fig. 5.7a, b. Possible values of the ratio of transverse to applied electric fields versus fields applied along a ⟨110⟩ direction

To discuss the behavior of ϑ_y in the neighborhood of the critical point ($\vartheta_y = 0$, $E_x = E_C$) we insert $E_x = E_C(1 + \xi)$ into (5.8) and develop in a power series $\phi^{(\alpha)}$ with respect to ϑ_y and ξ. To the first nonvanishing order in ϑ_y and ξ, this leads to a relation of the type [5.15, 17]

$$\vartheta_y[B(E_C)\xi + \vartheta_y^2 A(E_C)] = 0. \tag{5.14}$$

When $A(E_C)$ and $B(E_C)$ have different signs, a pair of nontrivial solutions $\vartheta^\pm = \pm\sqrt{B(E_C)\xi/A(E_C)}$ is obtained for positive ξ, and the critical-field strength is termed $E_{C'}$. When they have the same sign, $\vartheta^\pm$ is obtained for negative ξ, and the critical-field strength is designated as $E_{C''}$. Figure 5.7a illustrates the dependence of $\vartheta^\pm$ on E_x, if $E_{C'}$ and $E_{C''}$ are the minimum and maximum field strength for MED, respectively. The broken lines indicate the nonstable solutions and the solid lines the stable ones, respectively [5.15].

If, however, $E_{C''}$ (i.e., $\xi < 0$) is realized for the lower limit of the field region with MED, in the vicinity of $\vartheta_y = 0$ the function $R(\vartheta_y)$ becomes more complicated. Then a critical-field strength E_K (Fig. 5.6b) exists for which at once two pairs of additional solutions, besides the trivial one, appear at $\vartheta_y \neq 0$ (in contrast with a critical field of E_C type, with one pair of nontrivial solutions in the vicinity of $\vartheta_y = 0$). This case is shown in Fig. 5.7b, the higher pair of values for the intersection of ϑ_y and $R(\vartheta_y)$ corresponds to stable solutions $\vartheta_1^\pm$, while the lower pair $\vartheta_2^\pm$ corresponds to unstable solutions. If we obtain two pairs of solutions for $E_x > E_K$, then, by further increasing the heating field, $\vartheta_2^\pm$ shifts towards $\vartheta_y = 0$ (as may be seen from Fig. 5.6b) and only one pair of nontrivial solutions will remain for fields above $E_{C''}$. A similar situation can appear for the high-field limit of the region with nontrivial solutions ϑ_y, but mixed types of Fig. 5.7a and b, may be found for the upper and lower limit of the critical-field region.

The situation above described remains valid for values of ψ which correspond to directions close to ⟨110⟩, since the electron density in the group of

Valleys 1 and 2 essentially does not change when compared to that of the Valleys 3^5. At increasing values of ψ, the influence of Valleys 3 on MED increases (remember $\phi^{(1)} = \phi^{(2)} = \phi^{(3)}$ for tan $\psi = 1/\sqrt{2}$ in the absence of MED) and instead of the MED between Valleys 1 and 2, now a redistribution of the carriers among all three pairs of valleys has to be considered. For this case to occur, the limiting value of ψ depends on the values of the microscopic parameters used, and it can be determined by a quantitative theory only.

5.3.3 Current Density Along a ⟨111⟩ Direction

When the current density is along or close to a ⟨111⟩ direction, a simple analytical theory is not available for the general case. However, it is possible to treat the problem when either $\vartheta_y = 0$ or $\vartheta_z = 0$, that is, by shunting one of the transverse field components. The condition $\vartheta_z = 0$ leads to solutions for ϑ_y, which, when compared with the case of the ⟨110⟩ direction, differ only quantitatively. On the other hand, the condition $\vartheta_y = 0$ offers the possibility of investigating the conditions of MED on account of ϑ_z, that is, when a redistribution of carriers between the Valleys 3 and the group of Valleys 1 and 2 takes place. This case leads to new types of solutions. In particular, for the total current density along the ⟨111⟩ direction (5.7) yields

$$\vartheta_z = -\frac{\sqrt{2}\,a_1(\phi^{(1)} - \phi^{(3)})}{(2 + a_1)\phi^{(1)} + (1 - a_1)\phi^{(3)}},\tag{5.15}$$

with the arguments of $\phi^{(\alpha)}$ equal to

$$E^{(1,2)} = \sqrt{1 + 2^{1/2}a_1\vartheta_z + (1 + a_1/2)\vartheta_z^2}\,E_x,$$
$$E^{(3)} \ = \sqrt{1 - 2^{3/2}a_1\vartheta_z + (1 - a_1)\vartheta_z^2}\,E_x.\tag{5.16}$$

In complete analogy to the discussion for the case of the ⟨110⟩ direction, as a criterion for the appearance of MED in the vicinity of $\vartheta_z = 0$, one obtains

$$1 = -a_1^2\,\frac{d\phi^{(\alpha)}}{dE_S}\,\frac{E_S}{\phi^{(\alpha)}} = -a_1^2\,\frac{d\ln\phi^{(\alpha)}}{d\ln E_S},\tag{5.17}$$

with $E_S = E^{(1,2)} = E^{(3)}$. Very close to the point $(\vartheta_z = 0, E_x = E_S)$, the solution of (5.15) has the form [5.15]

$$\vartheta_z[\vartheta_z - C(E_S)\xi] = 0,\tag{5.18}$$

---

5 For the same reason we do not expect MED near a ⟨001⟩ direction, although from (5.4) MED could be possible.

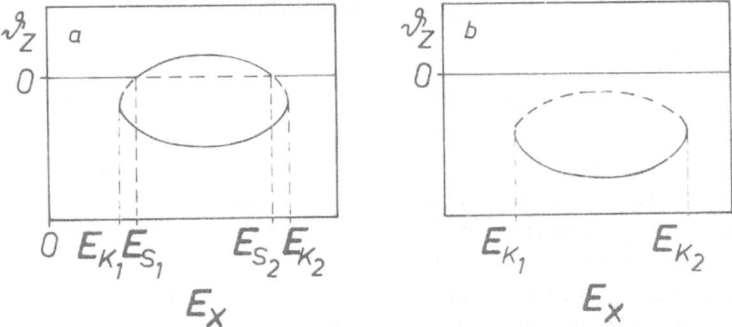

Fig. 5.8a, b. Possible values of the ratio of the transverse electric field component E_z to the applied field E_x versus fields applied along a $\langle 111 \rangle$ direction

which, in contrast with (5.14), exhibits the same number of solutions for both sides of $E_S(\xi \gtrless 0)$, as shown in Fig. 5.8a. The critical-field strengths E_K for the lower and upper limits of the region with MED are again given by the intersection of the left- and right-hand side of (5.15) for $\vartheta_z = 0$. While the upper branch, limited between E_{S_1} and E_{S_2}, represents the stable solution related to a predominant population of the group of the Valleys 1 and 2, the lower branch, limited between E_{K_1} and E_{K_2}, exhibits the stable solution $\vartheta_z < 0$, which is related to a predominant population of the Valleys 3.

Instead of the situation shown in Fig. 5.8a, an isolated closed loop $\vartheta_z(E_x)$ can be obtained (Fig. 5.8b). This means that owing to a predominant population of the Valley 3, a stable as well as an unstable nontrivial solution will appear[6], while the trivial solution remains stable for all fields. Depending on the parameters involved, type a or b of Fig. 5.8 will be realized.

Since the peculiarities corresponding to the nontrivial solutions for ϑ_y or ϑ_z show up in the current-voltage behavior, one can even obtain isolated loops theoretically for $j_x(E_x)$.

5.4 Quantitative Calculations for the Case of Si

While by analytical calculations general conclusions may be drawn for high-symmetry directions only, numerical calculations will not suffer this restriction. Thus, in the following, some numerical calculations obtained for the case of Si will be briefly reviewed.

---

6 When one does not choose the special brick-shaped sample geometry with $\vartheta_y = 0$, the Valleys 3 are not distinguished from the others and the problem has a threefold symmetry; in this case, any one of the valleys can become predominantly populated.

5.4.1 Monte Carlo Calculations of Intervalley Scattering Time and Effective Mobility

To evaluate $\vartheta(E_x)$ and $j_x(E_x)$ we must first calculate $\mu(E^{(\alpha)})$ and $\tau(E^{(\alpha)})$. This can be performed by the Monte Carlo method and the results obtained in this way can be used for all orientations if the effective-field method is applied[7].

Intravalley as well as intervalley scattering mechanisms have been taken into account. For intravalley scattering, the long-wavelength acoustic phonons and ionized impurities have been considered and found to be of great importance. For intervalley scattering, we recall that there are processes between the pair of valleys on the same axis (g type) and between valleys on perpendicular axes (f type). The calculations [5.15, 17–19] were performed using the values from [5.20] as reported in Table 5.1. However, to simplify calculations for f processes the interactions with the LA and TO phonons were combined (by choosing values of the phonon energy and coupling constant representative for them both). Furthermore, the interactions with the TA phonon, which owing to its low energy can play a significant role at low temperatures even if its deformation potential constant is weak, was accounted for by a variable coupling strength given by $\eta \times 0.15 \times 10^8$ eV/cm with η as an adjustable parameter. In addition, intervalley scattering at ionized impurities is included by an energy-independent relaxation time τ_0 (intervalley impurity scattering time).

Table 5.1. Parameters of intervalley phonons in Si [5.20]

Type	Branch	$\hbar\omega_j$[meV]	T_j[K]	D_{jk}[10^8 eV/cm]
g	LO ($\Delta_{2'}$)	62	720	7.5
f	TO (S_1)	59	684	2
f	LA (S_1)	47	545	4.3
f	TA (S_4)	18	210	0.15
g	TA (Δ_5)	12	140	0.65

Figure 5.9 shows the scattering rate τ^{-1} for the f-intervalley phonons as a function of the effective field at 27 K. As expected, the scattering at TA phonons predominates in the region of low fields, while processes from LA and TO phonons become prevailing in the high-field region. The influence of g-type processes on the f-intervalley electron transfer is shown in the insert for the case of coupling between electrons and g-TA phonons (again, the continous and dashed lines refer to TA and LA plus TO phonons, respectively).

---

7 The method of the effective fields, however, requires us to substitute the anisotropic scattering probabilities with some effective probabilities, which only depend on the angle between the initial and final momenta of the carrier. This approximation is usually found to be satisfactory.

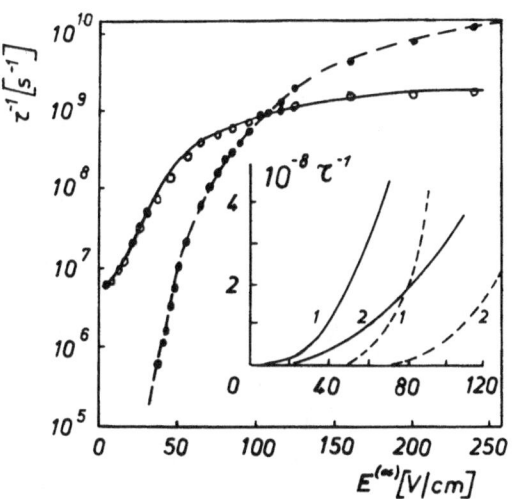

Fig. 5.9. Reciprocal scattering time of f-type intervalley processes versus effective field for TA phonons (○○○) and LA plus TO phonons (●●●) *Insert:* influence of g scattering with TA phonons on τ^{-1}: (*1*) $D_{TA}^{(g)} = 6.5 \times 10^7\,\mathrm{eV/cm}$; (*2*) $D_{TA}^{(g)} = 1.3 \times 10^8\,\mathrm{eV/cm}$

As a check of the model used, we have found that the mobility obtained from the results of Monte Carlo calculations [5.15, 17, 18] agrees with the experimental values extrapolated from time-of-flight measurements (Chap. 3).

5.4.2 Numerical Results for the Transverse Fields

With the results of Sect. 5.4.1 we are now in the position to calculate $\vartheta_y(E_x)$ and $v_x(E_x)$ [5.15, 17–19]. The results for the $\langle 110 \rangle$ case are presented in Figs. 5.10a, and b for different values of η and τ_0^{-1}. When the results are compared with the experimental data of Figs. 5.2, 4, we conclude that small values of η have to be chosen depending on the value of the additional intervalley impurity scattering. This is because experimental results do not exhibit a pronounced "camel back" behavior for ϑ_y, and only a single region, with a current density which decreases at increasing E_x, is observed. An analysis of the effect which ionized impurities produce on the intervalley scattering shows that ionized impurities determine the lower limit of E_x for MED and their effect has to be included in order to achieve agreement with experimental data. For instance, the choice of values $\eta \simeq 0.1$ and $\tau_0^{-1} \simeq 10^7\,\mathrm{s}^{-1}$ allows us to describe the observed effects satisfactorily. The values so obtained for $D_{TA}^{(f)}$ and τ_0^{-1} are reasonable with respect to the known properties of n-Si.

The numerical results for $\vartheta_z(E_x)$, for the case when the current density is in a direction very close to the $\langle 111 \rangle$ direction, are reported in Fig. 5.11 for different values of $\delta\psi$, where $\psi = \psi_0 + \delta\psi$ with $\tan\psi_0 = 1/\sqrt{2}$. The calculations have been performed with $\eta = 0.1$ and $\tau_0^{-1} = 2 \times 10^6\,\mathrm{s}^{-1}$. The curve in the lower part of the figure refers to the nontrivial solution for $\delta\psi = 0$ (i.e., for the $\langle 111 \rangle$ direction), and is a closed loop in accordance with Fig. 5.8b. For values of $\delta\psi$ which correspond to angular deviations less than $1°$ in both directions (Curves 2 and 3

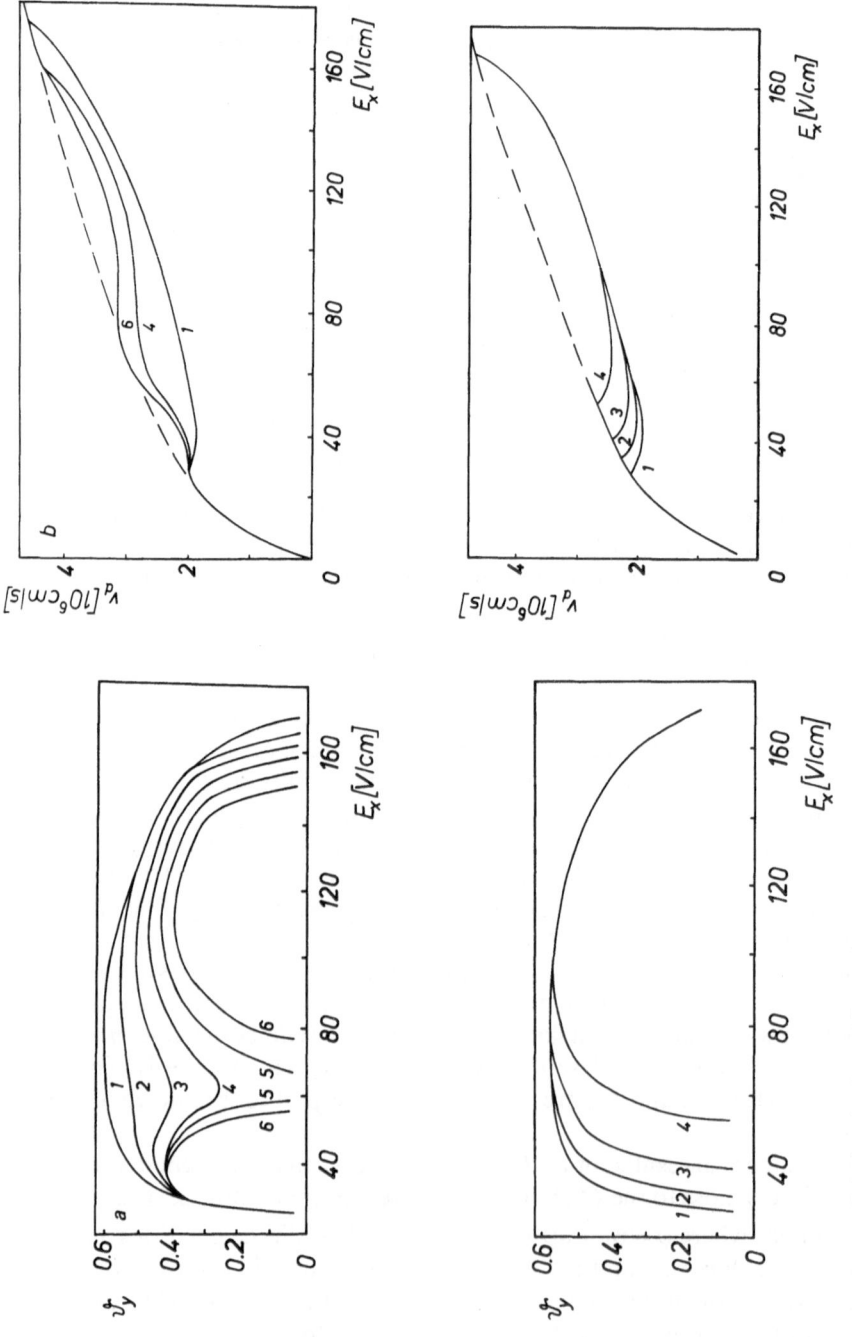

Fig. 5.10a, b Calculated values of $\vartheta_y = E_y/E_x$ (**a**) and drift velocity (**b**) versus E_x applied along the $\langle 110 \rangle$ direction. *Upper figure* refers to $\tau_0^{-1} = \infty$ and $\eta = 0.1$ (*1*), 0.2 (*2*), 0.3 (*3*), 0.6 (*4*), 0.8 (*5*), 1.0 (*6*). *Lower figure* refers to $\eta = 0.1$ and $\tau_0^{-1} = 5 \times 10^5$ (*1*), 2×10^6 (*2*), 8×10^6 (*3*), 3.2×10^7 (*4*)

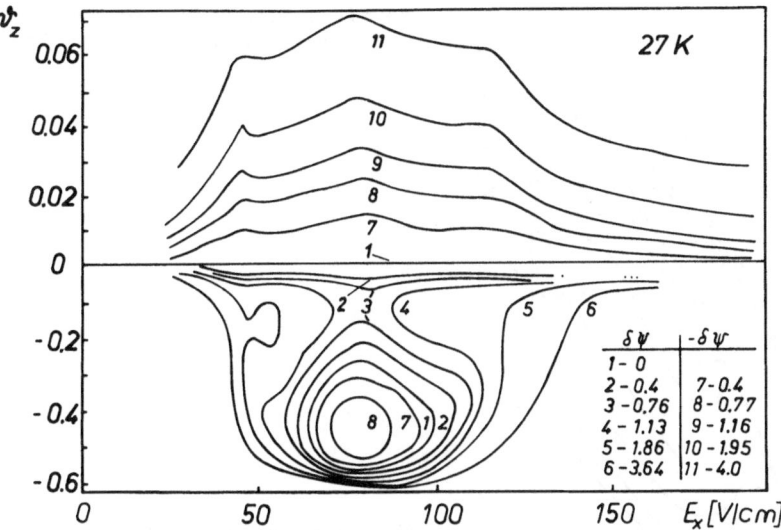

Fig. 5.11. Calculated values of $\vartheta_z = E_z/E_x$ versus E_x applied near the $\langle 111 \rangle$ direction for $\psi = \psi_0 + \delta\psi$ with $\psi_0 = 35.25°$

for ψ departing from ψ_0 towards the $\langle 001 \rangle$ direction and Curves 7 and 8 for ψ departing from ψ_0 towards the $\langle 110 \rangle$ direction) closed loops are still found, but, in addition solutions with opposite signs become available (as reported in the upper part of Fig. 5.11). These additional solutions are due to the usual Sasaki effect according to the carrier redistribution into the Valleys 3 or into the group of Valleys 1 and 2, respectively. For deviations greater than 1°, closed loops disappear for $\delta\psi < 0$, i.e., when j_x departs from the $\langle 111 \rangle$ towards the $\langle 110 \rangle$ direction, and they finally tend to assume a shape which is specular with the $\delta\psi > 0$ case but characterized by an abnormally large value of $|\vartheta_z|$ (compare Curves 6 and 11 in Fig. 5.11). The existence of more than one value of ϑ_z for a fixed E_x is limited to a narrow range of values of $\delta\psi$. Already for $\delta\psi = 3,64°$, as represented by Curve 6, we find only a region with anomalously great, but everywhere monovalued, Sasaki fields.

Varying the parameters η and τ_0^{-1} brings the result that the region of E_x, characterized by the existence of nontrivial solutions, smoothly decreases with increasing intervalley impurity scattering (especially on the low-field side). The above region of E_x is also diminished at increasing intervalley scattering with TA phonons, but with a more specific dependence on the different f type phonons.

5.4.3 Numerical Results for the Drift Velocity

The numerical values for the drift velocities as functions of the applied field [5.15, 19] are reported in Fig. 5.12 for various directions. The calculations were performed by using $\eta = 0.1$ and $\tau_0^{-1} = 10^8$ s^{-1} for all orientations except the

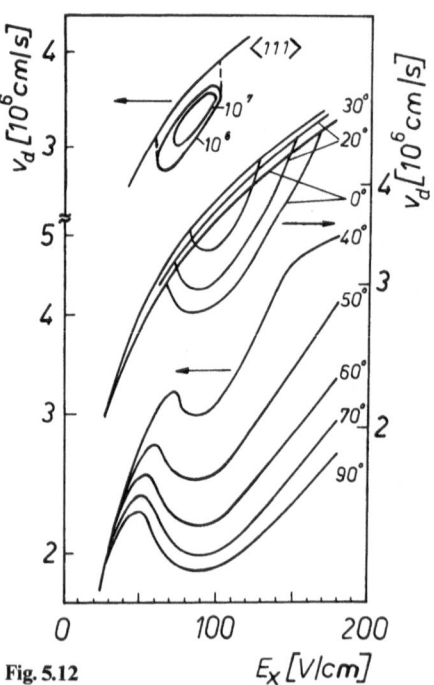

Fig. 5.12. Calculated drift velocities versus field for the indicated orientations at 27 K

Fig. 5.13. Regions for the existence of different types of NDC and MED (see text) as functions of the current-density orientation in the plane $(\bar{1}10)$ and of the impurity intervalley scattering time. Calculations are performed for 27 K with $\eta = 0.1$

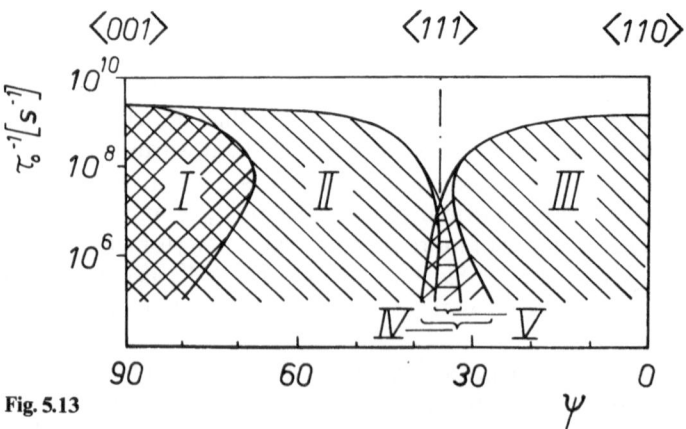

Fig. 5.13

$\langle 111 \rangle$ direction, for which the values of intervalley impurity scattering rate are indicated in s^{-1} units.

For orientations lying between $\langle 110 \rangle$ and $\langle 111 \rangle$ directions – Region III in Fig. 5.13 – a state with equally populated Valleys 1 and 2 is not stable and the effect of MED is exhibited by the lower branches of the curves, while the upper ones refer to the case of shunted ϑ_y.

For orientations lying between $\langle 111 \rangle$ and $\langle 001 \rangle$ directions, NDC is observed because of the predominant population of Valleys 3. Partly this is connected with the term $\Phi a_1 (1 - 3 \cos 2\psi)/4$ in (5.6), which describes the change of the effective masses along the direction of the applied field due to the intervalley transfer (Region I in Fig. 5.13). Such an effect was known long ago for n-GaAs [5.6, 7], as well as for n-Ge along a $\langle 111 \rangle$ direction [5.8] and for n-Si along a $\langle 001 \rangle$ direction (see [5.14], for instance). On the other hand, in (5.6) there are terms with ϑ_z, which dominate if j_x is declining remarkably from the direction of the symmetry axis of one valley (Region II in Fig. 5.13).

For orientations very close to the $\langle 111 \rangle$ direction, theory predicts that MED should lead to velocity-field curves in the form of isolated loops (Fig. 5.12) with correspondent closed loops for the transverse field, as already mentioned in Sect. 5.3.3. Whether or not these loops contribute to the current-voltage characteristics depends on the boundary conditions [5.15, 17]. In the absence of intervalley relaxation on the surface and for an inhomogeneous field distribution, solutions can exist for which the loops lead to a N-type NDC (broken lines indicated in Fig. 5.12). In Fig. 5.13 the Region IV is ascribed to loops due to the MED between the Valleys 1 and 2, and the Region V is ascribed to loops due to the MED between the Valleys 3 and either of the Valleys 1 or 2, respectively. Very close to a $\langle 111 \rangle$ direction, when $\eta = 0.1$, NDC disappears already for values of τ_0^{-1} between 10^7 and 10^8 s^{-1}, i.e., for these orientations MED requires a crystal purity higher than for any other orientation, as shown in Fig. 5.13.

Similar investigations for a (100) plane can be found in [5.15, 19].

5.5 Low Temperature as Condition for MED

The considered symmetry breaking requires a steep increase of the intervalley scattering rate at increasing field, as can be seen from the logarithmic derivative in (5.13, 17). This condition is achieved in practice when the carrier temperature is lower than the equivalent temperature of the phonon involved in the transition. In fact, as can be seen from Figs. 5.9 and 5.14, at increasing electric field strength the dependence of $\tau^{-1}(E^{(\alpha)})$ on electric field becomes smoother and, in turn, only the trivial solution will exist.

With rising lattice temperature the effect of carrier heating on the repopulation rate becomes less pronounced, as shown in Fig. 5.14. Accordingly, in comparison with the results at 27 K (Sect. 5.3.2), the numerical calculations [5.15, 19] for the $\langle 110 \rangle$ direction show a remarkable weakening of the MED effect at temperatures above 40 K, and predict that the MED effect vanishes above 50 K.

Experimental results of the current density versus applied electric fields [5.21, 22] are shown in Fig. 5.15. Here the arrows indicate the values of the threshold field for the current saturation. It is clearly shown that this value systematically increases with increasing temperature, especially above 40 K, and that the effect of current saturation vanishes for $T_0 > 50$ K. These results are in agreement with the numerical calculations performed with $\eta = 0.1$ and $\tau_0^{-1} \gtrsim 10^7$ s^{-1}.

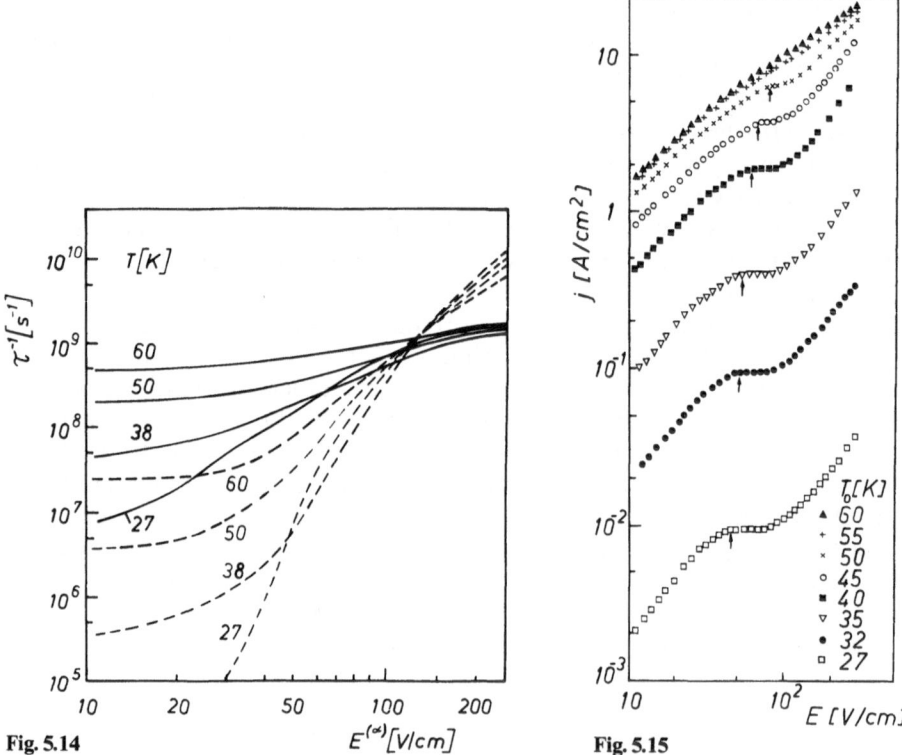

Fig. 5.14

Fig. 5.15

Fig. 5.14. f-type intervalley scattering rates versus effective-field strength for TA phonons (——) and LA plus TO phonons (----) for the indicated temperatures

Fig. 5.15. Current densities versus field applied along the $\langle 110 \rangle$ direction at the indicated temperatures

Concerning the effect of the lattice temperature on the transverse fields, it is found from experiments that the presence of a magnetic field parallel to a $\langle 001 \rangle$ direction (Sect. 5.2.1) permits one to analyze the involved dependences in a better way because spurious effects, which may influence the position of the interlayer wall, can be excluded. Therefore the potential difference ΔU_{12} measured between probes on opposite $(\bar{1}10)$ surfaces will be ascribed to the transverse field strength in a more reliable way. Values of $\Delta U_{12}/d_{12}$ (see insert of Fig. 5.2) are reported in Fig. 5.16 [5.21] as a function of $\boldsymbol{B}$ for $E_x = 106$ V/cm (for this value of the applied field there is no longitudinal domainization due to NDC) at the indicated lattice temperatures. One can see that for $T_0 \leq 50$ K the steep rise in the region of weak $\boldsymbol{B}$ coincides for all curves with MED. Furthermore, the value of the transverse field measured at high $\boldsymbol{B}$ is found to decrease for T_0 above 40 K.

For $T_0 > 50$ K the change of slope, from the steep rise at weak $\boldsymbol{B}$ to the smooth rise at strong $\boldsymbol{B}$, becomes more broadened than at lower temperatures.

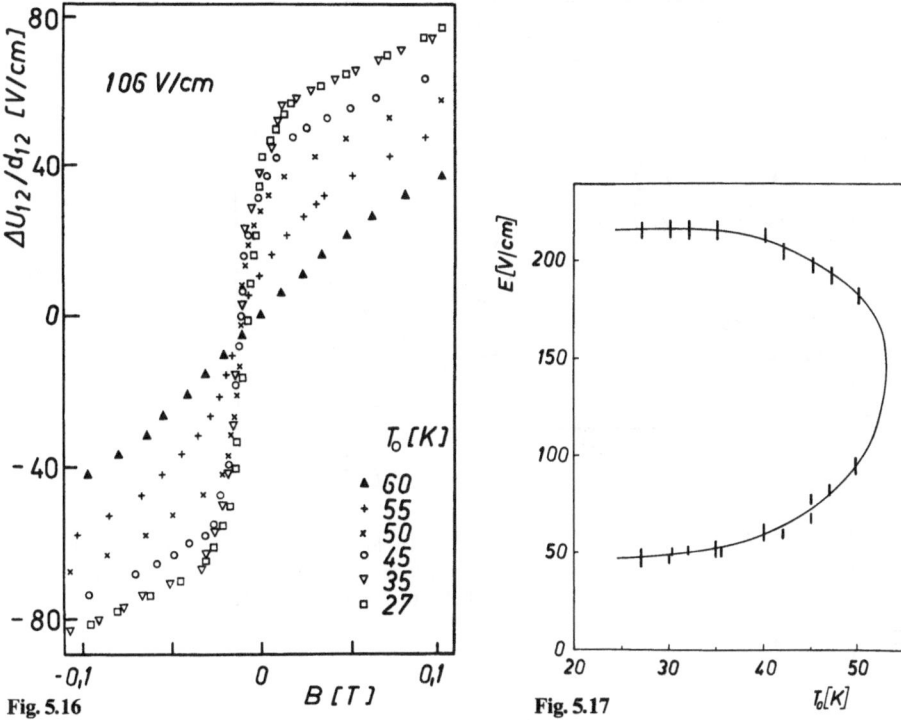

Fig. 5.16. Transverse electric fields versus magnetic field as measured at the indicated temperatures

Fig. 5.17. Region of applied fields and lattice temperatures for the existence of MED as obtained from experiments with E_x along a $\langle 110 \rangle$ direction

Furthermore, above 55 K the initial steep rise, due to the switching of the wall, disappears and only an anomalously strong Hall effect is observed [5.12]. Consequently, *the existence or absence of this abrupt change in slope, as exhibited by the transverse field versus magnetic field, can be taken as a criterion for the existence or absence of* MED. When this criterion is applied to n-Si for a current density along a $\langle 110 \rangle$ direction, the values reported in the upper part of Fig. 5.17 are obtained for the existence of MED [5.22].

5.6 Experimental Results in Si for a Current Density Nonsymmetrically Oriented

5.6.1 Multivalued Transverse Fields for Current Densities Between $\langle 110 \rangle$ and $\langle 111 \rangle$ Directions

Due to the carrier redistribution between the Valleys 1 and 2, multivalued transverse fields are possible for orientations of the current density which range

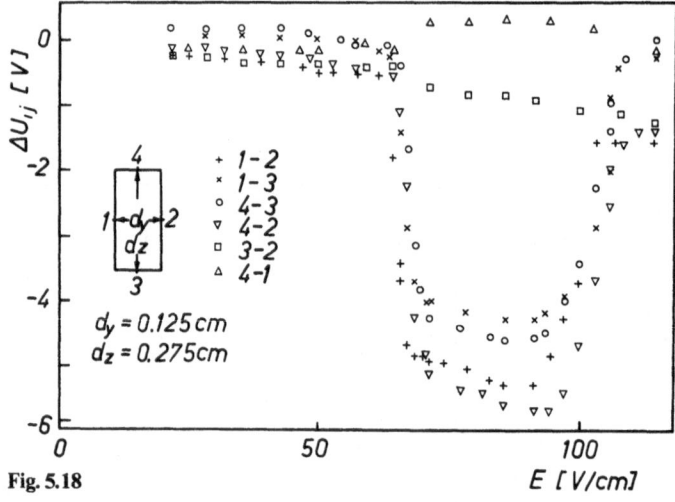

Fig. 5.18. Potential differences, measured between the probes indicated in the insert, versus applied field for $\psi = 33.5°$ at 27 K

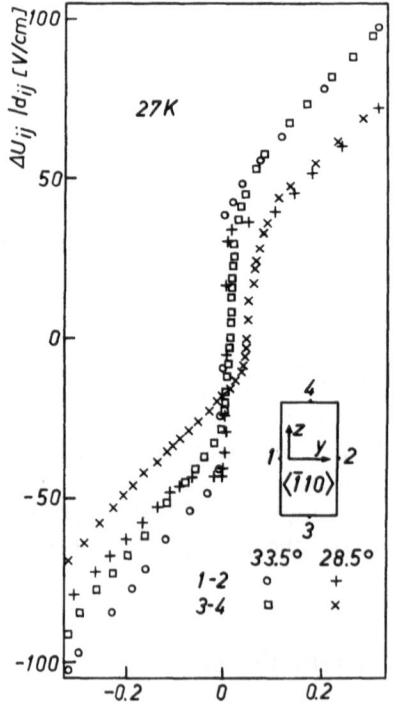

Fig. 5.19. Transverse electric fields E_y and E_z versus magnetic field as measured at the indicated orientations of E_x (ψ counted from $\langle 110 \rangle$ towards $\langle 111 \rangle$ direction)

from $\langle 110 \rangle$ up to nearly $\langle 111 \rangle$ directions. In contrast with the $\langle 110 \rangle$ direction, in the vicinity of the $\langle 111 \rangle$ direction an electron repopulation into either Valleys 1 or 2 leads to transverse fields with components parallel or opposite to E_y, respectively, and E_z (but equal in sign in both cases).

Figure 5.18 reports the values of the potential differences, measured between the probes indicated in the insert, as functions of the heating field for an orientation of the current density departing by $\psi = 33.5°$ from the $\langle 110 \rangle$ direction.

With the exception of ΔU_{32} and ΔU_{41}[8], the potential differences appear and vanish sharply at fixed values of the applied field. Furtheron it is noticed that the region of heating fields where high transverse fields appear is significantly smaller than for the $\psi = 0$ case. These observations indicate the presence of loops, theoretically predicted for both the transverse field and the current density close to the $\langle 111 \rangle$ direction (see Region IV in Fig. 5.13), in contrast with the small Sasaki fields expected in the absence of MED.

Again the existence of MED can be clearly evidenced by applying a magnetic field perpendicular to the current density as well as to the investigated transverse field component E_y, and then, analyzing with the criterion reported at the end of Sect. 5.5, how the potential difference between probes on opposite $(\bar{1}10)$ surfaces is influenced. As shown in Fig. 5.19, an abrupt change of the potential difference ΔU_{12} is observed at increasing values of $\boldsymbol{B}$ parallel to $\hat{z}$ for both the sample orientations. This exhibits how the wall between the regions with predominant population of either Valley 1 or 2 is shifted between the $\langle \bar{1}10 \rangle$ surfaces. At the same time, a measurement of ΔU_{34} in dependence on B_z shows no change. This is expected due to the mere interchange of the roles of Valleys 1 and 2 for this configuration.

In agreement with theory (see Region V of Fig. 5.13) experiments show that the MED between Valleys 1 plus 2 and Valleys 3 is possible for orientations of the current density very close to the $\langle 111 \rangle$ direction in very pure samples. Otherwise, the same effect can be achieved by applying $\boldsymbol{B}$ along the $\langle 1\bar{1}0 \rangle$ direction, since in this case the total electric field vector is turned towards the $\langle 001 \rangle$ direction. This time the potential difference ΔU_{34} is very sensitive to a change of the value of $\boldsymbol{B}$ around a critical value, corresponding to the sudden repopulation from Valleys 1 or 2, respectively, into the Valley 3. This value is higher for greater deviations of the current vector from $\langle 111 \rangle$ towards $\langle 110 \rangle$ directions; see Fig. 5.19 for 28.5°. At this critical value of B_y, ΔU_{12} jumps from its value due to the transverse fields E_y in the case of the population of one of the Valleys 1 or 2 to zero.

5.6.2 Current-Voltage Characteristics in the $(\bar{1}10)$ Plane

According to the theoretical considerations of Sect. 5.4.3, at low lattice temperatures and for not too high scattering rates on ionized impurities, NDC in the current-voltage characteristics should be observed for almost every orientation of the current density in the $(\bar{1}10)$ plane. This expectation had been verified

---

8 Both the pairs of Probes 3 and 2 as well as 4 and 1 lie almost on equipotential lines, because, for the sample considered above, the interlayer wall is not located in the middle of the sample, but near the surface.

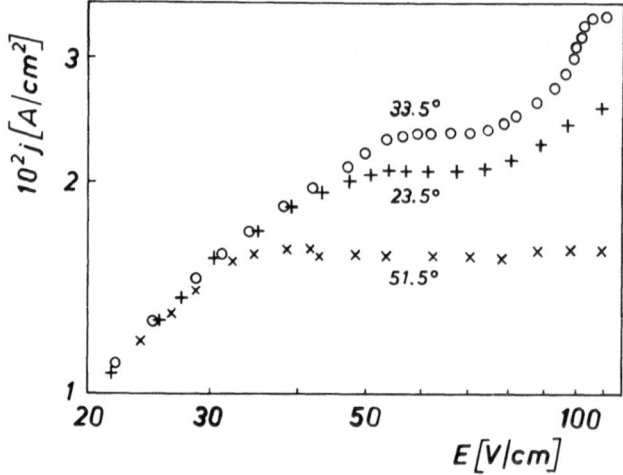

Fig. 5.20. Current density as a function of applied electric field for the indicated orientations at 27 K. Angles are measured from $\langle 110 \rangle$ direction in the $(\bar{1}10)$ plane

first for the case of orientations between $\langle 001 \rangle$ and $\langle 111 \rangle$ directions. In this case the combined effects of the change of effective mass of the total carrier ensemble in the current direction as due to electron redistribution between the valleys and of the transverse fields occur (Region II in Fig. 5.13). On the other side, for the case of orientations between $\langle 111 \rangle$ and $\langle 110 \rangle$ directions, when MED is realized (Regions III and IV in Fig. 5.13), the existence of NDC was verified, too. Typical experimental results are reported in Fig. 5.20. Here, as expected, for both cases a saturation of the current density is observed. This occurs because the NDC leads to the appearance of static high-field domains, as proved experimentally by measuring the potential drop along the sample. Only for the $\langle 111 \rangle$ direction the saturation effect has not been detected. Probably the reason is that even at the small acceptor concentration of 5×10^{12} cm^{-3}, the value of τ_0^{-1} is not low enough for NDC to occur in this direction (Fig. 5.13). Otherwise, the experimental results compare well with numerical data when values of τ_0^{-1} in the range from 10^7 to 10^8 s^{-1} are used [5.19]. The different origins of NDC for orientations between $\langle 001 \rangle$ and $\langle 111 \rangle$ directions and between $\langle 111 \rangle$ and $\langle 110 \rangle$ directions can be evidenced by the influence of magnetic fields.

5.7 Conclusion: Historical Survey and Future Investigations

The MED among equivalent valleys treated in the present chapter was at the origin of a triple-valued dependence of the Sasaki field, theoretically predicted in 1962 by *Reik* and *Risken* [5.23] in n-Ge when the current density was oriented near a $\langle 110 \rangle$ direction for the limiting case of negligible intervalley scattering.

However, these authors paid no attention to this effect, since it vanished for an appropriate choice of the scattering rate which provided better agreement between experimental and numerical results.

The problem of the possible existence of multivalued transverse fields was theoretically considered in 1970 by *Gribnikov* et al. [5.3], and the first experimental indication with respect to this subject was reported in 1972 by *Astrov* and *Kastalskii* [5.29]. In the mean time, most of the interest was devoted to the similar problem of the transverse negative resistance [5.24–28]. These authors, recognizing the necessity of a strong intervalley repopulation rate, required in [5.3], performed measurements in Ge at low temperatures. By choosing the heating field along a $\langle 100 \rangle$ direction, they investigated the electric response in the $\langle 110 \rangle$ direction for both the closed and open circuit conditions. In a certain region of applied fields, they observed the appearance of transverse potential differences, which they correlated with MED on account of fluctuations. The experimental data, however, do not allow one to draw such conclusions unambiguously [5.15].

Early attempts to measure transverse instabilities in n-Si were made in 1972 by *Gram* et al. [5.30]. However, their efforts were unsuccessful because the lattice temperature they chose was not low enough to satisfy the required steep increase of the intervalley repopulation rate with heating field strength.

As reported in this chapter, the existence of MED has been definitely confirmed in recent years. Therefore, in perspective its consequences with respect to other phenomena should open fruitful areas of investigation. These pertain to the optical properties with the interaction of light with free carriers as well as microwave transduction phenomena probably accompanied by resonances.

Acknowledgement. I want to thank Prof. Dr. O. G. Sarbey for valuable discussions and Dr. H. W. Streitwolf and Dr. L. Reggiani for a critical reading of the manuscript.

References

5.1   H.Haken: *Synergetics, An Introduction*, 3rd. ed., Springer Ser. Syn., Vol. 1 (Springer, Berlin, Heidelberg 1983)
5.2a M.Shibuya: Phys. Rev. **99**, 1189–1195 (1955);
5.2b W.Sasaki, M.Shibuya: J. Phys. Soc. Jpn. **11**, 1202–1203 (1956)
5.3   Z.S.Gribnikov, V.A.Kochelap, V.V.Mitin: Zh. Eksp. Teor. Fiz. **59**, 1828–1845 (1970)
5.4   Z.S.Gribnikov, V.V.Mitin: Zh. Eksp. Teor. Fiz. Pisma **14**, 272–276 (1971), Phys. Status Solidi (b) **68**, 153–164 (1975)
5.5   B.K.Ridley, T.B.Watkins: Proc. Phys. Soc. Lond. **78**, 293–304 (1961)
5.6   C.Hilsum: Proc. IRE **50**, 185–189 (1962)
5.7   J.B.Gunn: J. Phys. Soc. Jpn. **21**, 505–508 (1966)
5.8   A.A.Kastalskii, S.M.Ryvkin: Zh. Eksp. Teor. Fiz. **7**, 446–450 (1968)
5.9   M.Asche, H.Kostial: Solid State Commun. **39**, 457–460 (1981)
5.10 Z.S.Gribnikov, V.V.Mitin: Fiz. Tekh. Poluprov. **9**, 276–281 (1975)
5.11 M.Asche, H.Kostial, O.G.Sarbey: J. Phys. **C13**, L645–649 (1980)
5.12 M.Kriechbaum, H.Heinrich, J.Waida: J. Phys. Chem. Sol. **33**, 829–838 (1972)
5.13 M.Asche, H.Kostial: Phys. Status Solidi (b) **93**, K89–92 (1979)

5.14 L.F.Kurtenok, E.A.Movchan, O.G.Sarbey, V.V.Mitin, M.Asche: Phys. Status Solidi (a) **48**, 323–328 (1978)

5.15 M.Asche, Z.S.Gribnikov, V.V.Mitin, O.G.Sarbey: *Gorjachie electroni v mnogodolinnikh poluprovodnikakh* (Naukova dumka, Kiev 1982) Chap. 3

5.16 H.F.Budd: Phys. Rev. **131**, 1520–1524 (1963)

5.17 Z.S.Gribnikov, V.M.Ivashchenko, V.V.Mitin, O.G.Sarbey: *Chislennii raschet mnogosnachnikh raspredelenii elektronov po dolinam* (preprint No **8**, Inst. Fiz. AN, Kiev 1981)

5.18 M.Asche, Z.S.Gribnikov, V.M.Ivashchenko, H.Kostial, V.V.Mitin, O.G.Sarbey: Zh. Eksp. Teor. Fiz. **81**, 1347–1361 (1981)
[English transl.: Sov. Phys. – JETP **54**, 715–722 (1982)]

5.19 M.Asche, Z.S.Gribnikov, V.M.Ivashchenko, H.Kostial, V.V.Mitin: Phys. Status Solidi (b) **114**, 429–438 (1982)

5.20 M.H.Jørgensen: Phys. Rev. **B18**, 5657–5666 (1978)

5.21 M.Asche, H.Kostial, O.G.Sarbey: J. Physique **42**, C7-323–328 (1981)

5.22 H.Kostial, L.F.Kurtenok: Phys. Status Solidi (b) **109**, K109–113 (1982)

5.23 H.A.Reik, H.Risken: Phys. Rev. **126**, 1737–1746 (1962)

5.24 E.Erlbach: Phys. Rev. **132**, 1976–1979 (1963)

5.25 M.Shyam, H.Kroemer: Appl. Phys. Lett. **12**, 283–285 (1968)

5.26 C.Hammar: Phys. Rev. **B4**, 2560–2566 (1971)

5.27 N.N.Grigorev; I.M.Dykman, P.M.Tomchuk: Fiz. Tverd. Tela **8**, 1083–1089 (1974)

5.28 T.K.Gaylord, T.A.Robson: Phys. Lett. **38A**, 493–494 (1972)

5.29 Ju.A.Astrov, A.A.Kastalskii: Fiz. Tekh. Poluprov. **6**, 323–328 (1972)

5.30 N.O.Gram, M.N.Jørgensen, N.I.Meyer: Proc. 11th Intern. Conf. Phys. Semicond., Warsaw, Poland, 1972 (PWN, Warsaw 1973) pp. 622–629.

6. Streaming Motion of Carriers in Crossed Electric and Magnetic Fields

Susumu Komiyama, Tatsumi Kurosawa, and Taizo Masumi

With 16 Figures

In general, the carrier distribution function in momentum space is of particular importance in the investigation of high-field transport phenomena. The widely accepted terminology "hot electron" implicitly assumes that the distribution function is of quasi-Maxwellian type, characterized by an electron temperature higher than that of the thermal bath. This model corresponds to a quite isotropic distribution function and can be justified if scattering events can be regarded as quasi-elastic processes (e. g., acoustic-phonon and/or impurity scatterings). This means that the energy lost by a carrier per collision is only a minute fraction of the energy before the collision.

On the other hand, a quite different type of high-field transport phenomena characterized by a highly anisotropic distribution function can be attained in pure semiconductors at low temperatures. This happens when optical-phonon emission predominates over all other scattering mechanisms. In fact, by applying a strong electric field it is possible to accelerate a carrier at rest directly up to an energy equal to that of the optical phonon $\hbar\omega_{op}$. Provided that the interaction with optical phonons is adequately strong, the carrier having reached the energy $\mathscr{E} = \hbar\omega_{op}$ will emit almost immediately an optical phonon by dissipating nearly all of its kinetic energy. The carrier, thereby scattered near to the ground state $\mathscr{E} \cong 0$, will be accelerated again to $\mathscr{E} = \hbar\omega_{op}$ and will repeat thereafter the same process until this is interrupted by some other collision event. Such a repeated motion is often referred to as streaming motion [6.1–3], and it makes the distribution function far from a Maxwellian type.

By applying a magnetic field perpendicular to the electric field, it is even possible to bring about more remarkable distribution functions (e. g., a carrier accumulation in a limited area in momentum space) which eventually lead to a population inversion [6.4]. Furthermore, in the case of intense microwave-electric fields, the application of magnetic fields introduces a very specific motion of carriers (referred to as streaming cyclotron motion) and makes the distribution function even more peculiar, including a carrier bunching in momentum space [6.5].

A considerable number of experimental investigations on these phenomena have recently been reported. Magnetotransport studies on Ag-halides [6.6], CdS [6.7], p-Ge [6.8], and p-Si [6.9] have provided evidence for streaming motion and for carrier accumulation. Far-infrared emission from accumulated carriers has been observed in p-Ge [6.10–12] and, in turn, has encouraged further work seeking for far-infrared amplification [6.13]. Microwave-conductivity experiments on Ag-halides have supported the picture of streaming cyclotron motion in alternating electric fields [6.14, 15].

The purpose of this chapter is to describe new aspects of these phenomena by reviewing the recent experimental achievements and to discuss several further properties which are expected from a theoretical ground. After elucidating the importance of optical-phonon scattering in semiconductors (Sect. 6.1), we first describe in Sect. 6.2 streaming motion in the most simple situation, the case of dc electric fields without magnetic field. In Sect. 6.3 we study effects of applying transverse magnetic field. In Sect. 6.4 we deal with the phenomena in intense microwave fields. Finally, in Sect. 6.5, we briefly review theoretical predictions, which have not yet been observed experimentally.

6.1 Substantial Role of Optical-Phonon Scattering

Streaming motion can be expected under the condition that the optical-phonon scattering predominates over the acoustic-phonon and/or impurity scatterings. As will be illustrated in the following, this situation is realized in pure materials at low temperatures.

In ionic crystals, charge carriers interact strongly with optical phonons via polar optical coupling (Chap. 2). In many covalent semiconductors, also, the interaction via the optical-deformation potential is considerably strong (Chap. 2). Accordingly, for a variety of materials, the probability that a carrier emits an optical phonon is very high if its energy exceeds the optical-phonon energy $\hbar\omega_{op}$. On the other hand, the scattering probabilitites due to other

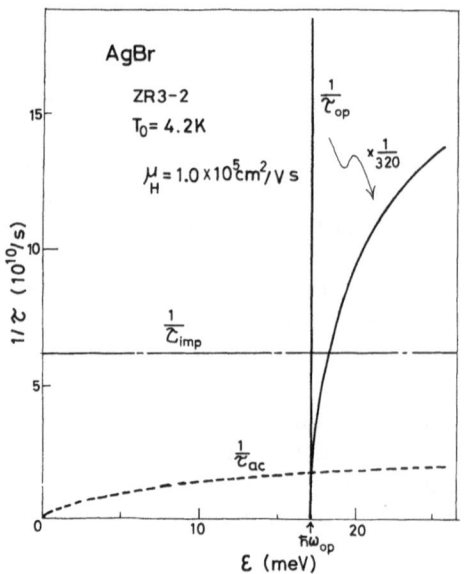

Fig. 6.1. Scattering rates versus electron energy in a crystal of AgBr at 4.2 K: $1/\tau_{imp}$ refers to (neutral) impurities; $1/\tau_{ac}$ to acoustic phonons; and $1/\tau_{op}$ to (longitudinal) optical phonons [6.6]. The low-field mobility of electrons is 1×10^5 cm^2/Vs

mechanisms are relatively low if a pure crystal at low temperatures is considered. Consequently, the total scattering probability of a carrier is markedly different according to whether the carrier energy is below or above $\hbar\omega_{op}$.

A typical example of such a situation is shown in Fig. 6.1. Here the scattering rate of electrons is shown as a function of electron energy for the case of a zone-refined crystal of AgBr at $T_0 = 4.2 K$ [6.6]. The solid line, drawn according to the conventional perturbation treatment with the *Fröhlich* coupling constant $\alpha_p \simeq 1.6$, evidences the high values of the scattering rate for spontaneous emission of optical phonons, $1/\tau_{op}$, in the energy range $\mathscr{E} > \hbar\omega_{op}$. On the other hand, in the range $\mathscr{E} < \hbar\omega_{op}$ the scattering rates due to neutral impurities and acoustic phonons are markedly low. Other scattering mechanisms such as absorption of optical phonons and intercarrier scattering are negligible because of the low lattice temperature and low carrier density (e. g., typically $\leq 10^9$ cm^{-3}). Similar features of the scattering probabilities are also found in a number of semiconductors with low doping levels.

Table 6.1. Quantities characteristic of carrier streaming motion for several materials. The electron is assumed except for germanium and silicon. The superscripts li and h for p-Ge and p-Si denote the light and the heavy holes. Also, $B_{\zeta=1}$ is the magnetic field at which $\zeta = 1$. Values for p-Si provide only a rough measure because of the presence of a significant nonparabolicity and warping of the band

	m/m_0	$\hbar\omega_{op}$ [meV]	A [10^{12} s^{-1}]	$e/(mA)$ [cm^2/Vs]	V_{op} [10^7 cm/s]	T_{op}^0 [ps] at 1 kV/cm	$B_{\zeta=1}$ [T] at 1 kV/cm
AgCl	0.43	23	112	36	1.37	3.2	0.74
AgBr	0.287	17	94	64	1.45	2.2	0.70
CdS	0.17	38.2	32	320	2.81	3.0	0.36
GaAs	0.0665	36.7	6.9	3800	4.43	1.6	0.23
p-Ge	0.35^h	37	2.5^{h,li}	2000^h	1.94^h	3.6^h	0.52^h
	0.043^{li}			16000^{li}	5.53^{li}	1.3^{li}	0.18^{li}
p-Si	$\sim 0.5^h$	63	3.8^h	910^h	$\sim 2^h$	$\sim 5.3^h$	$\sim 0.5^h$
	$\sim 0.16^{li}$				$\sim 3.7^{li}$	$\sim 2.5^{li}$	$\sim 0.27^{li}$

To compare the strength of the optical-phonon scattering among several materials, a characteristic frequency A is listed in Table 6.1. In this way the rate of phonon emission $1/\tau_{op}$ is obtained from A through the simplified relation

$$\frac{1}{\tau_{op}(\mathscr{E})} = A(\mathscr{E}/\hbar\omega_{op} - 1)^{1/2}, \tag{6.1}$$

which holds for the nonpolar coupling case as well as, when $\mathscr{E}/(\hbar\omega_{op}) \simeq 1$, for the polar case [6.17]. An effective mobility, $e/(mA)$, where e is the charge and m the effective mass of the carrier, is also listed in the same table. It enables us to measure the relative strength of the optical phonon to other scattering sources through the ratio $\mu/[e/(\text{mA})]$, where μ is the low-field mobility in the given material. Thus, recalling that low temperatures are considered, values of $\mu/[e/(\text{mA})]$ much greater than unity are indicative of strong optical-phonon coupling.

6.2 Picture of Simple Streaming Motion

Let us first consider a carrier, at rest at time $t=0$, which is completely free from any scattering mechanism except for optical-phonon emission. With an electric field E applied, the carrier is accelerated along the field direction with a rate $\dot{v}=eE/m$ (Fig. 6.2). The kinetic energy of the carrier reaches the optical-phonon energy at time $t=T_{\text{op}}^0$ given by

$$(eE/m)T_{\text{op}}^0 = V_{\text{op}}, \tag{6.2}$$

where

$$(1/2)mV_{\text{op}}^2 = \hbar\omega_{\text{op}}. \tag{6.3}$$

If E is not too high and the optical-phonon interaction is adequately strong, the carrier will almost immediately emit an optical phonon and return to about its ground state $\mathscr{E} \simeq 0$. Thereafter, by repeatedly emitting phonons at time interval T_{op}^0, the carrier will perform a streaming motion. The carrier distribution function thereby will take a needlelike shape, as sketched in the right half of Fig. 6.2. Under this condition the drift velocity of the carrier v_{d}^{s} is, clearly,

$$v_{\text{d}}^{\text{s}}=\tfrac{1}{2} V_{\text{op}}. \tag{6.4}$$

The Hall mobility, as determined by the streaming motion, is given by [6.6]

$$\mu_{\text{H}}=\tfrac{1}{3}(e/m)T_{\text{op}}^0. \tag{6.5}$$

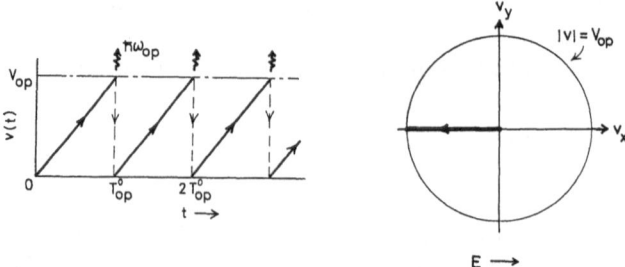

Fig. 6.2. Ideal streaming motion of a carrier; the velocity versus time (*left*) and the velocity distribution (*right*)

In actual experimental situations, two conditions are necessary for realizing the streaming motion. The first is that a carrier could reach V_{op} without suffering scatterings in the passive region ($\mathscr{E} \leq \hbar \omega_{op}$), that is, the following condition should be satisfied:

$$T_{op}^0 < \tau \equiv (\bar{\tau}_{imp}^{-1} + \bar{\tau}_{ac}^{-1})^{-1}, \tag{6.6}$$

where $\bar{\tau}_{imp}$ and $\bar{\tau}_{ac}$ are the impurity and acoustic-phonon scattering times averaged over $\mathscr{E} \leq \hbar \omega_{op}$. This condition sets a lower limit, E_{min}, to the electric field required for the streaming motion. The second is that the carrier should not penetrate significantly into the active region ($\mathscr{E} > \hbar \omega_{op}$). By allowing a penetration up to $1.2 V_{op}$ in average, the following condition[1] is obtained by integrating (6.1) over t:

$$T_{op}^0 > 10.3/A. \tag{6.7}$$

This condition sets an upper limit E_{max} to the electric field required for the streaming motion.

According to relations (6.6, 7), the value of E_{min}, for a crystal with a low-field mobility $\mu = e\tau/m$, may be obtained from the relation $\mu E_{min} = V_{op}$, and the value of E_{max} from $(e/m)(10.3/A)E_{max} = V_{op}$. Values of V_{op} and T_{op}^0 are listed for several materials in Table 6.1. For example, the region for streaming motion is $40 \sim 1 \times 10^3 \text{V/cm}$ for p-Ge, $200 \sim 1 \times 10^4 \text{V/cm}$ for CdS (electron), and $100 \sim 3 \times 10^4 \text{V/cm}$ for AgCl (electron) in typically pure samples [6.18]. Thus a streaming motion situation can be expected to occur in a number of readily accessible materials.

However, the needlelike distribution function of streaming carriers, as shown in Fig. 6.2, should not be taken too naively in actual physical conditions. In the passive region, impurities or acoustic phonons will occasionally scatter the carriers out of the trajectory for streaming motion, and thus tend to produce a diffuse distribution of carriers over the whole sphere $\mathscr{E} < \hbar \omega_{op}$. Furthermore, a slight (but finite) penetration of streaming carriers into the active region, which is inevitably introduced by the application of strong electric fields, will spread the needlelike distribution. For a typical example the reader can refer to the results reported in Fig. 6.5a, which are obtained by a Monte Carlo simulation on a realistic model.

Analytical approaches can also provide reasonable carrier distribution functions on the basis of the Boltzmann equation [6.19]. However, direct experimental observation of the momentum distribution function, such as to allow comparison with the theories, has not yet been reported in the literature.

---

1 As compared to [Ref. 6.18, Eq. (3.6)], this relation (6.7) is more stringent and gives smaller values for E_{max} by a factor of about 3.

Experimentally, a streaming motion situation was first indicated by *Pinson* and *Bray* [6.2], by studying on p-Ge at 77 K the infrared absorption caused by the direct transition of hot holes from the heavy to the split-off band. The energy distribution function of hot (heavy) holes so obtained exhibited a decay above the optical-phonon energy steeper than that expected from a Maxwellian distribution function. Furthermore, an observed dependence of the absorption coefficient on the polarization of the incident light indicated a strong anisotropy with the field orientation in the distribution function which, in turn, directly suggested the streaming character of hot carriers in p-Ge. Similar measurements have been repeated more recently by *Vorob'ev* et al. [6.21] to study the distribution function of light holes [6.20] as well as of heavy holes. Since the pioneer work of *Pinson* and *Bray*, several theoretical investigations about streaming motion have been reported [6.3, 19, 22]. Nevertheless, the phenomenon did not receive due attention from experimentalists, for over a decade, until recent magnetotransport measurements on purer materials at lower temperatures made possible a deeper understanding of the phenomenon [6.6–9].

A simple indication of streaming motion, which comes from transport measurements, is provided by the saturation of the carrier drift velocity at $V_{op}/2$ in the electric field range $E_{max} > E > E_{min}$. This has been observed at helium temperatures for the case of heavy holes in Ge [6.8, 16] and Si [6.9, 16], and for photoexcited electrons in AgBr, AgCl [6.6], and CdS [6.7].

Alternative and even more convincing evidence comes from Hall-effect measurements. Figure 6.3 shows the electron Hall mobility μ_H as a function of electric fields in several crystals of AgCl and AgBr [6.6]. The measurements are carried out on photoelectrons (with a density $\sim 10^7$ cm^{-3}) created by a pulsed light at band-gap energy by using a blocking electrode technique. The theoretical

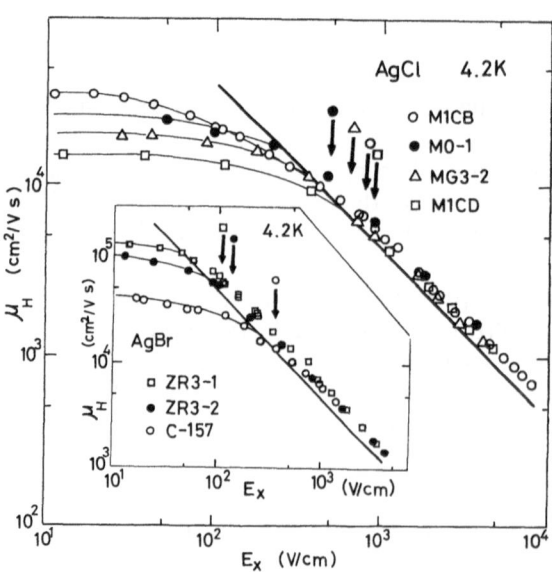

Fig. 6.3. Hall mobility versus electric field for electrons in several crystals of AgCl. Theoretical values for ideal streaming motion (6.5) are represented by (——). The value of E_{min} derived from (6.6) is marked by the arrow for each crystal. The inset shows similar data for AgBr

results, as obtained from (6.5) by using the values of m and $\hbar\omega_{op}$ given in Table 6.1, are indicated by the solid line. The value of E_{min}, above which the occurrence of streaming motion is theoretically expected, is marked by an arrow for each sample. When E increases above E_{min} the measured values of the Hall mobility come close to the theoretical values given by (6.5) for each sample. Similar agreement is also found by performing measurements on photoelectrons in CdS [6.7] and on heavy holes in p-Ge [6.8]. An indication of streaming motion is obtained also in p-Si. In this material the phenomena are, however, complicated because of a significant band warping. One remarkable effect is the occurrence of a negative Hall effect [6.9]. In p-Ge, such an anomalous Hall effect is not observed because the band warping is much less pronounced.

6.3 Streaming Motion in Magnetic Fields

6.3.1 Accumulation of Carriers in Momentum Space

We shall consider the effect of applying a magnetic field $\boldsymbol{B} = (0, 0, B)$ to an ensemble of carriers ideally streaming in an electric field $\boldsymbol{E} = (E, 0, 0)$. To begin with, let us assume that the free motion of the carriers is described by the classical equation of motion,

$$m \frac{d}{dt} \boldsymbol{v} = e \, (\boldsymbol{E} + \boldsymbol{v} \times \boldsymbol{B}). \tag{6.8}$$

The trajectory in velocity space of the streaming motion, which at $B = 0$ is represented by a straight line (Fig. 6.2), is now curved by the Lorentz force $e(\boldsymbol{v} \times \boldsymbol{B})$. For a fixed E, the curvature will increase with increasing B. Noting that the free motion of a carrier described by (6.8) is a cyclotron oscillation with the center point in velocity space at $C = (0, -E/B, v_z)$, we can realize that the curved trajectory for streaming motion is represented by an arc on the plane $v_z = 0$ with the center point at C. Therefore, the trajectory is determined by the ratio E/B.

For convenience, let us define a normalized field ζ,

$$\zeta \equiv V_{op}/(E/B), \tag{6.9}$$

which specifies the position of Point C relative to the circle $|\boldsymbol{v}| = V_{op}$. According to the range of variability of ζ, $(0 \leq \zeta < \infty)$, the streaming motion expected at different strengths of E and B can be classified into three stages, as schematically shown in Fig. 6.4 [6.6, 22]. The expected features of the carrier kinetics at each stage from (a) to (c) shall be described in the following.

a) When $\zeta < 1$, the Point C is located outside the circle $|\boldsymbol{v}| = V_{op}$. The only effect of applying B is to curve the trajectory toward the negative v_y direction. In real space the current carried by the streaming carriers is accordingly deflected,

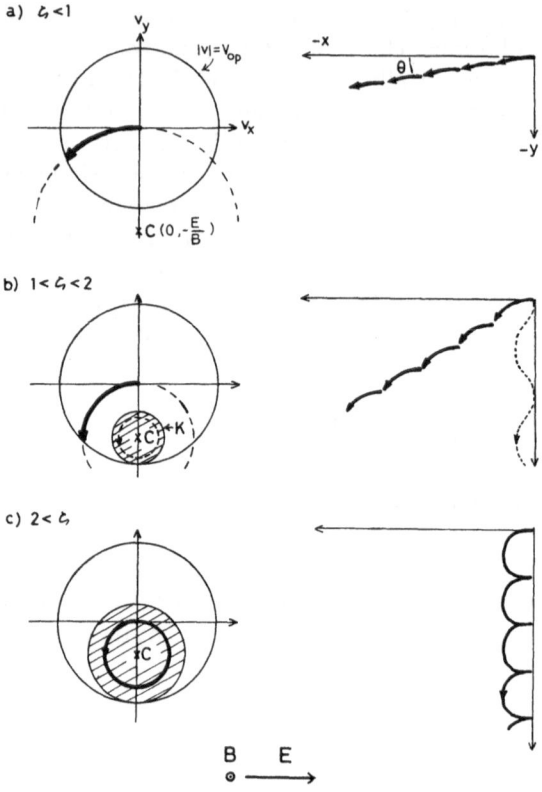

a) $\zeta < 1$

b) $1 < \zeta < 2$

c) $2 < \zeta$

B E
○ ——→

Fig. 6.4a–c. Different stages of streaming motion for $\boldsymbol{B} = (0, 0, B)$ and $\boldsymbol{E} = (E, 0, 0)$, classified by the normalized field $\zeta \equiv V_{op}/(E/B)$. Electron trajectories in velocity space ($v_z = 0$) are illustrated in the *left half*, and those in real space in the *right half*

from the direction of the electric field, toward the negative y direction by a Hall angle θ. As the trajectory for streaming motion is uniquely determined by ζ, the Hall angle is expected to be a function of ζ. A straightforward analysis of (6.8) for the case of ideal streaming motion [6.6] gives

$$\tan\theta = 2\zeta^{-2} \arccos\left(1 - \tfrac{1}{2}\zeta^2\right) - (4\zeta^{-2} - 1)^{1/2}. \tag{6.10}$$

At this stage, a variation of E and B which keeps ζ constant does not affect the phenomena except that it changes the repetition rate of the cyclic streaming motion. The time interval of this repetition is given by [6.6]

$$T_{op}(\zeta) = T_{op}^0 \zeta^{-1} \arccos\left(1 - \tfrac{1}{2}\zeta^2\right), \tag{6.11}$$

where T_{op}^0 is defined by (6.2).

b) When ζ exceeds 1, Point C enters the circle $|v| = V_{op}$. While no significant alteration arises in streaming motion, except for an increase in the curvature of the trajectory, a round area in which the trajectories of the cyclotron motion do not cross the circle $|v| = V_{op}$ newly appears around Point C within the circle.

This area, indicated by shading and labeled as K in Fig. 6.4b, develops along the v_z direction to yield a spindle-shaped region tangent to the sphere $|v| = V_{op}$ (as shown below in Fig. 6.7). Accordingly, a carrier in this area, if present, performs a sustained cyclotron motion, as shown by the dashed circle, and drifts along the y direction in real space without being interrupted by the optical-phonon emission. As assumed at the beginning of this section, if the carriers are ideally streaming so that no broadening is involved in the distribution function, Region K is empty and its presence would be physically of no significance; the phenomena expected at this stage would be a trivial analogue of those at Stage (a).

In actual physical conditions, however, it may occasionally happen that a streaming carrier penetrates into the active region $|v| > V_{op}$ deeply enough to jump into Region K after an optical-phonon emission. Although the probability of such an event occurring is low, an appreciable distribution of carriers may be collected in Region K. This is because streaming carriers repeat the process of phonon emission very frequently (with frequency $1/T_{op}$), whereas, once having jumped into Region K, carriers are able to stay there for a long time. In fact, once collected, these carriers are removed out of Region K only by impurity or acoustic-phonon scatterings. This effect was first demonstrated under realistic physical conditions by *Maeda* and *Kurosawa* [6.4] through a Monte Carlo simulation on p-Ge. The degree of carrier accumulation, which will be determined by the balance between the rates of carriers coming into and going out of Region K, has been calculated for realistic physical conditions by *Kurosawa* [6.17]. A plausible carrier distribution on the $v_z = 0$ plane, as calculated by a Monte Carlo simulation for the case of electrons in AgBr, is shown in Fig. 6.5 together with the result for $B = 0$ [6.17]. In addition to the optical-phonon emission, the calculation includes the impurity and the acoustic-

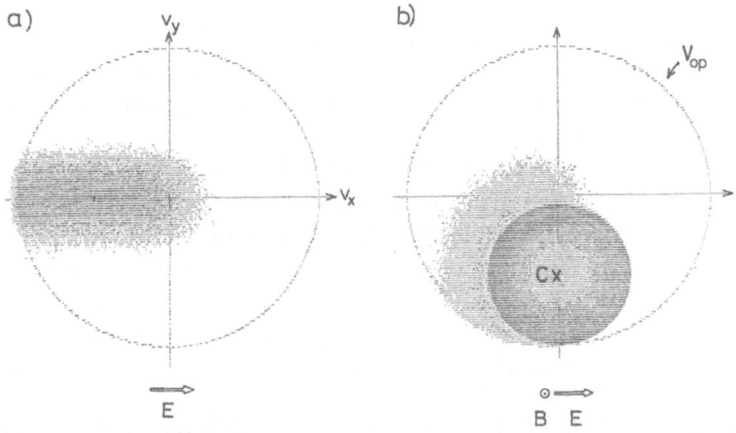

Fig. 6.5a, b. Electron distribution in the $v_z = 0$ plane, as calculated by a Monte Carlo simulation on AgBr [6.17]: **(a)** $E = 1200\ \text{V/cm}$ and $B = 0$; **(b)** $E = 3600\ \text{V/cm}$ and $B = 4.6\ \text{T}$ ($\zeta = 1.83$)

phonon scatterings which are fitted to the Hall-effect data on a specimen with mobility $1.5 \times 10^5 \ cm^2/Vs$. The broadening in the distribution function is primarily caused by a slight penetration of streaming carriers into the active region, scarcely discernible in the figure. A similar distribution function including carrier accumulation has also been obtained analytically by *Brosens* and *Devreese* [6.23] through a Boltzmann equation approach.

The accumulation of carriers into Region K will lead to population inversion, if the accumulation process is strong enough [6.4]. This type of population inversion is of particular interest since it occurs in the continuum of the energy band, in contrast to the usual population inversion realized between discrete energy levels in many existing laser systems.

c) When ζ is greater than 2, the trajectory for streaming motion is closed within the circle $|v| = V_{op}$. This implies that a carrier starting at $v = 0$ never acquires enough energy to emit an optical phonon, and streaming motion becomes impossible. If carriers were free from any other scattering mechanism but the optical-phonon emission, the average motion would be reduced to the simple drift motion along y direction with velocity $v = (0, -E/B, 0)$. In real cases, however, residual events of impurity and acoustic-phonon scatterings play a significant role in the carrier kinetics, and retain the drift motion v_x of carriers in the x direction to a finite small level.

Though the entire discussion here has relied upon the classical picture of the carrier motion, a quantum-mechanical approach leads to similar results [6.24]. One effect to be distinguished is that the electronic states in Region K are quantized into Landau levels. The quantization is confirmed by the observation of cyclotron emission associated with carrier transitions between adjacent Landau levels [6.11].

6.3.2 Galvanomagnetic Measurements

A number of relevant effects supporting the above picture of hot-carrier kinetics can be observed through magnetotransport measurements. The most direct indication of streaming motion and accumulation of carriers is provided by the measurement of the Hall angle θ.

Figure 6.6 shows $\tan\theta$ as a function of ζ for electrons in several crystals of silver halides [6.6]. In the measurements, the time-integrated photo-currents in the x and the y-directions, Q_x and Q_y as induced by a pulse of light, are measured as a function of B_z for given values of E_x. As the measurements are carried out at low levels of photoexcitation by using a fast pulse technique, the Hall electric field E_y does not develop. Hence the Hall angle is defined by $\tan\theta = Q_y/Q_x$. The solid line in the figure indicates the theoretical values calculated according to (6.10).

In the range $\zeta < 1$ all data points, corressponding to different samples and different values of E_x, fall close to the theoretical line. This fact definitely supports the picture of curved streaming motion, as shown in Fig. 6.4a.

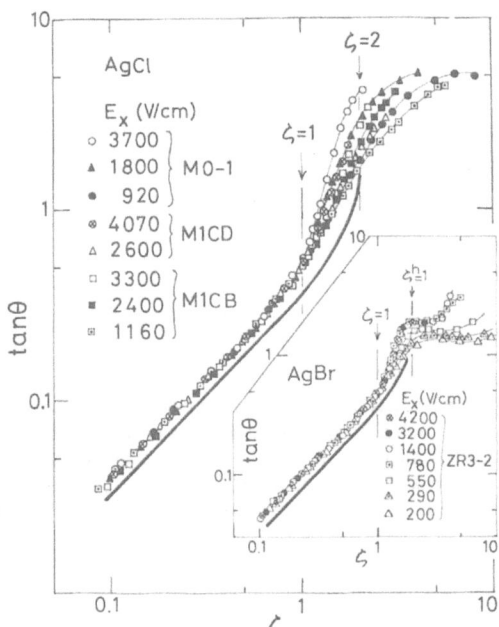

Fig. 6.6. Tangent of the Hall angle, $\tan\theta \equiv Q_y/Q_x$, versus ζ for AgCl and AgBr (*in the inset*) for the given values of E_x. (——) indicates the theoretical values as obtained from (6.10) in the case of ideal streaming motion

In the range $\zeta > 1$, a distinct upward departure of the data from the theoretical line is observed. This can be ascribed to a carrier accumulation into Region K. Since these accumulated carriers drift primarily in the y direction in real space, the total current due to both the streaming and the accumulated carriers yields a Hall angle larger than that expected only for streaming carriers. Therefore, the marked scatter of the data points in the range $\zeta > 1$ is interpreted as a consequence of the difference in degree of accumulation (at a fixed ζ), which depends on the sample purity and the strength of electric field.

The ratio of the number of accumulated carriers, n_k, to that of streaming carriers, n_s, can be estimated from the magnitude of the discrepancy between the measured and the theoretical values of $\tan\theta$ [6.6]. As an example, for the high-purity AgCl crystal (M0 − 1), n_k/n_s attains a value of about 0.3 at $E_x = 3700\,\text{V/cm}$ and $\zeta = 1.4$; under this condition the relative volume of the spindle Region K is 0.048 of the whole sphere $|v| \leq V_{op}$. Although there are difficulties in providing a quantitative account for the observed values of n_k/n_x [6.18, 24], its dependence on E_x and sample purity has been qualitatively explained through a simple rate-equation approach [6.4, 6].

In the range $\zeta > 2$ the experimental results can be reasonably interpreted on the basis of the model previously illustrated in Fig. 6.4 and labeled as stage (c).

The characteristic features, described above for the $\tan\theta$ measurements, equally apply for both AgCl and AgBr. However, one aspect that distinguishes AgBr from AgCl is the accumulation of positive holes coexisting with that of electrons. This accumulation of holes in AgBr causes a bend-over region in the

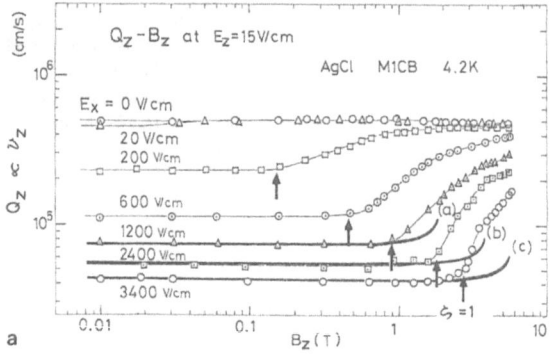
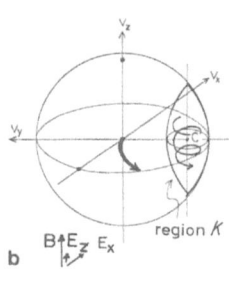

Fig. 6.7. (a) Response current Q_z in a crystal of AgCl as a function of B_z for a given $E_z = 15$ V/cm and for the indicated values of E_x. The arrow marks for each E_x the magnetic field position at which $\zeta = 1$. (———) indicate the theoretical values obtained from (6.11) for an ideal streaming motion. **(b)** Trajectory in velocity space for streaming motion and the spindle-shaped Region K. The rapid increase of Q_z in the presence of small E_z in the range $\zeta > 1$ (Fig. 6.7a) is explained by the helical motion of accumulated carriers

curves of $\tan\theta$ versus ζ in the ζ range above $\zeta^h = 1$, where ζ^h is defined by (6.9) for holes. The finite contribution of holes in AgBr is contrasted to the situation of AgCl, where holes are known to be self-trapped [6.25].

Other evidence for streaming motion and accumulation of carriers in silver halides – evidence which may be even more conclusive – is provided by the results reported in Fig. 6.7a. The current response Q_z to a weak electric field E_z, additionally applied along the magnetic field direction, is here plotted against B_z for several fixed values of E_x. The response current Q_z is proportional to the carrier mean free time between collisions. The solid thick line in the figure indicates the drift velocities, $v_z = (e/m) \, [T_{op}(\zeta)/2]E_z$, which are theoretically expected for ideally streaming carriers with the time of free flight $T_{op} (\zeta)$ given by (6.11). The Lines (a), (b) and (c) are for $E_x = 1200$, 2400, and 3400V/cm, respectively. The agreement of the theoretical lines with the relevant data is almost perfect in the range of low B, i.e., $\zeta < 1$. It should be remarked that here the comparison is of absolute character, no adjustable parameter for the fitting being required [6.6]. Therefore, the agreement must be considered more than satisfactory and taken as proof of the physical soundness of the picture of streaming motion (Fig. 6.4a).

In the range of higher B, i.e., $\zeta > 1$, the experimental values of Q_z start increasing rapidly, and deviate largely from those predicted by the simple streaming motion scheme. This manifests the advent of a carrier accumulation in Region K, where the mean free time between scatterings is much longer than that of the streaming carriers (Fig. 6.7b).

Hall-effect data in p-Ge indicating carrier accumulation are shown in Fig. 6.8 [6.8]. In this case the currents are transmitted through ohmic contacts to establish the Hall field E_y in the specimen, and the Hall angle is defined by

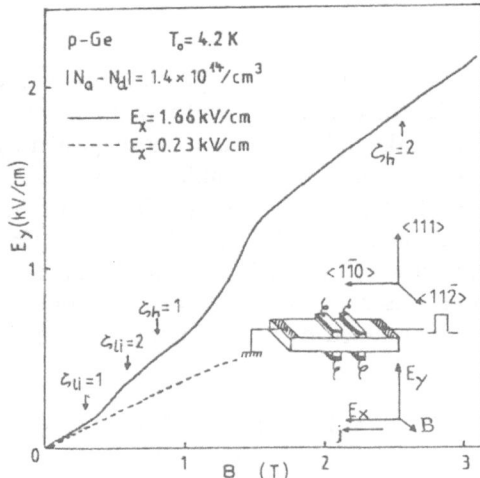

Fig. 6.8. Hall electric field E_y versus B in p-Ge: E_x is almost independent of B, and E_y is nearly proportional to $\tan\theta \equiv E_y/E_x$. The magnetic field positions at which $\zeta_{li,h} = 1$ or 2 are marked by arrows. Here $\zeta_{li,h}$ denotes ζ for light and heavy holes, respectively

$\tan\theta = E_y/E_x$, where E_x is nearly independent of B. To enable the detection of E_x and E_y at low temperatures a capacitive probe technique [6.26] is introduced, as sketched in the figure: For the low value of E_x (0.23kV/cm), E_y (the dashed line) is a smoothly increasing function of B. For the high value of E_x (1.66kV/cm), E_y (the solid line) exhibits a steplike increase above $\zeta_{li} = 1$ and $\zeta_h = 1$, indicating the accumulations of light and heavy holes, respectively. Here $\zeta_{li,h}$ are defined by (6.9) for the respective holes with $E = (E_x^2 + E_y^2)^{1/2}$.

Noticeable in p-Ge is the light hole accumulation; in the range $\zeta_{li} > 1 > \zeta_h$, the population of light holes increases remarkably [2] (e.g., by a factor $2 \sim 3$) since the accumulation can take place only for light holes [6.8, 18].

Another indication of the carrier accumulation in p-Ge can be also found from the behavior of the current density j as measured in a sweep of B at fixed E_x [6.8].[3]

In addition to AgCl, AgBr, and p-Ge, photocurrent investigations on CdS [6.7] and Si [6.9] also provide reliable indications of the streaming motion and the accumulation of carriers. In particular, the case of Si shows up as an interesting subject because of its pronounced band warping.

---

2 Note that heavy holes contribute significantly to the accumulation of light holes because the optical-deformation scattering does not distinguish the interband transition from the intraband one.

3 Prior to these works, a rapid decrease of the current density with increasing B above $\zeta_{li,h} > 1$ had been observed on short samples of p—Ge, in which E_y is effectively short-circuited [6.27]. Although the interpretation of the authors did not concern the scheme discussed here, this observation also provides another indication of the accumulation phenomena.

6.3.3 Far-Infrared Emission

An important consequence of carrier accumulation in p-type materials is the emission of far-infrared radiation. Under the condition $\zeta_{li} > 1 > \zeta_h$, in a p-type material accumulation takes place only in the light-hole band, as suggested by the results for p-Ge reported in Fig. 6.8. The light-hole accumulation gives rise to a marked increase in the emission intensity of the radiation associated with the transition of holes from the light to the heavy band (Fig. 6.9a), as was first pointed out theoretically by *Andronov* et al. [6.13]. Emission of far-infrared radiation via this mechanism has been observed experimentally for p-Ge in [6.10–12]. Figure 6.10 shows the results obtained by using a photoconductive

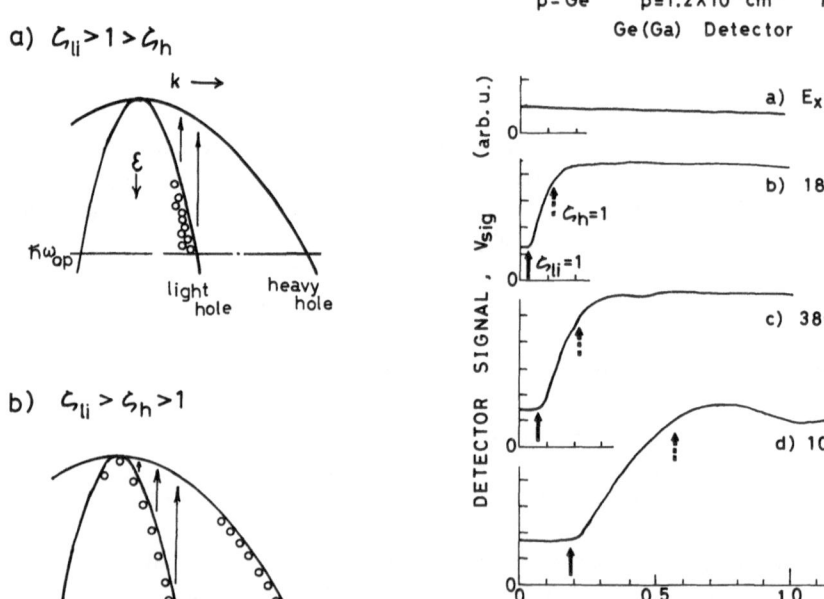

Fig. 6.9a, b. Radiative transition of accumulated light holes to the heavy-hole band for the two cases **(a)** and **(b)**

Fig. 6.10. Far-infrared radiation from p-Ge, detected with a Ge/Ga detector, as a function of B at different values of E_x [6.11]

detector made by Ge doped with Ga at a concentration $\sim 1 \times 10^{15}$ cm^{-3} [6.11]. It reports the detector signals as a function of the magnetic field B at several fixed levels of E_x. Since the detector has an excellent sensitivity over a wide spectral range of radiation above a threshold photon energy of about ~ 9meV, the signals approximately represent the intensities of the radiation integrated over energy. (The emission band is expected to lie below 33meV.) At high values of E_x, the intensity of the radiation begins to increase abruptly at $\zeta_{li} = 1$, owing to the setting in of the light-hole accumulation.[4] The increase of the radiation intensity levels off in the region of higher magnetic fields $\zeta_h > 1$, where the accumulation of heavy holes suppresses that of light holes (Fig. 6.9b). The interpretation is assured by the spectroscopic analysis of the emitted radiation [6.11].

An interesting question arising from the emission studies is the possibility of far-infrared amplification [6.13]. However, whether or not an amplification might occur in actual physical conditions seems to be an open problem. This is because the matrix elements for radiative transition between the light- and the heavy-hole bands in p-Ge are not large [6.28]. Hence, in order that a gain of radiation may overcome a probable loss in the crystal, the presence of population inversion between the two bands is not enough, but the concentration of light holes has to be considerably high in absolute terms. This implies that a *strong accumulation* should be attained in a crystal with a *sufficiently high (total) carrier concentration*. In general, these two requirements are conflicting[5], and the experimental efforts should be addressed to compromise them. As an example, there is an attempt to observe laser action by applying very high electric fields ($\sim 6 kV/cm$) to a specimen with a relatively high carrier concentration ($\sim 10^{15}$ cm^{-3}) [6.30].

6.4 Streaming Motion in Microwave Fields

6.4.1 Streaming Cyclotron Motion

In this section we shall deal with carrier kinetics in strong microwave fields linearly polarized $E_\omega = (E_\omega \cos\omega t, 0, 0)$ in the presence of transverse magnetic fields $B = (0, 0, B)$. The physical setup is identical to that for cyclotron resonance experiments, except that it includes very strong microwave fields. An important aspect of the cyclotron resonance phenomena in intense microwave fields is that a change in the effective mass of hot carriers can be analyzed by studying the shift of the resonance peak position; this cannot be investigated in the dc field condition. The knowledge of the effective mass, together with the information about the scattering time, as obtained from the width of the resonance lines, enables us to achieve a deeper understanding of the hot-carrier phenomena.

Provided the field strength is strong enough, another significant aspect is that carriers can be repeatedly accelerated from rest up to the optical-phonon energy within half a period of the microwave field. In such a physical condition – which we call the regime of *streaming cyclotron motion* – the phenomena become completely different from the usual cyclotron resonance. To clarify the physical

---

4 Radiation in the region of lower B, $\zeta_{li} < 1$, is attributed to the radiative transition of streaming light holes to the heavy-hole band.

5 However, *Vorob'ev* et al. [6.29] have suggested a population inversion for a crystal with a relatively high carrier concentration ($N_{imp} = 1 \times 10^{15}$ cm^{-3}) on the basis of studies of far-infrared absorption by light holes. Thus the problem may be controversial.

a)

b)

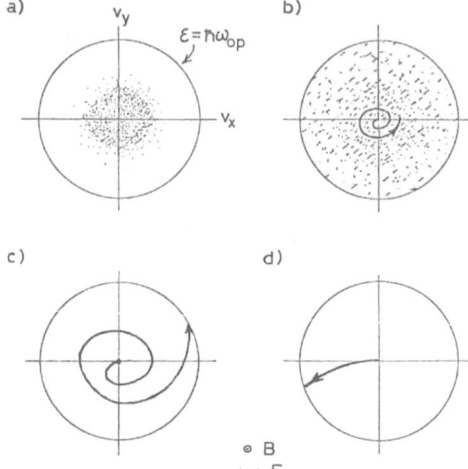

c)

d)

Fig. 6.11a–d. Sketch of the expected variation in hot-carrier kinetics under the cyclotron resonance condition at increasing E_ω; (a) ordinary hot-electron regime, (b) prestreaming regime, (c) streaming cyclotron resonance, and (d) streaming cyclotron motion. Typical trajectories in velocity space are illustrated in cases (b–d) by continuous curves. Shading points in (a) and (b) give a rough idea of the carrier distribution. In contrast to the situations (a) and (b), the carrier distribution becomes strongly time dependent in (c) and (d)

implications of the above statements and to simplify the description of the experiments, in the following we shall illustrate qualitatively the change that occurs in the cyclotron resonance phenomena at increasing the field strength E_ω. Figure 6.11 summarizes schematically the main expectations.

a) At low values of E_ω, the high-energy tail of the carrier distribution function does not reach $\mathscr{E} = \hbar\omega_{op}$. Optical-phonon emission can therefore be neglected. Under this condition, scattering is ruled by impurities and/or acoustic phonons. The energy distribution function is determined by the balance between the energy gained from the field and that lost through the acoustic-phonon emission. The average energy of carriers will increase with increasing E_ω.

b) At intermediate values of E_ω the high-energy tail of the carrier distribution will reach $\mathscr{E} = \hbar\omega_{op}$. Accordingly, emission of optical phonons sets in and suppresses the increase of carrier average energy, while the momentum relaxation is still dominated by impurity and/or acoustic-phonon scatterings.

c) At high values of E_ω, it becomes possible for carriers to be resonantly accelerated up to $\mathscr{E} = \hbar\omega_{op}$ before being scattered by impurities or acoustic phonons. Both the momentum loss and the energy loss are dominated by the optical-phonon emission. This type of streaming motion, which we shall call *streaming cyclotron resonance*, appears when E_ω is high enough to satisfy the following condition, analogous to (6.6):

$$T_{op}^\omega \equiv 2mV_{op}/(eE_\omega) < \tau. \tag{6.12}$$

Here, T_{op}^ω is the traveling time for a carrier starting from $v=0$ to achieve $v = V_{op}$. It is worth noting that under cyclotron resonance conditions the acceleration induced by the field $E_\omega \cos \omega t$ is equivalent to that by a dc field with intensity $E = E_\omega/2$. It should be noted that the *streaming cyclotron resonance* is essentially a resonance phenomenon and, as such, should be distinguished from the *streaming cyclotron motion* illustrated below.

   d) At the highest values of E_ω, such that the condition

$$T^\omega_{op} < \pi/\omega \tag{6.13}$$

is satisfied, carriers can be accelerated to $\hbar\omega_{op}$ within half a period of the microwave field, and the acceleration process is possible even when $\omega \neq \omega_c$; ω_c being the cyclotron angular frequency. We shall call such a condition *streaming cyclotron motion*. The physical situation differs completely from that of the cyclotron resonance phenomena because the acceleration process of carriers from $\mathscr{E} = 0$ to $\mathscr{E} = \hbar\omega_{op}$ is essentially not a resonance acceleration. It is theoretically predicted that systematic patterns are eventually introduced into the carrier distribution function, as will be described later.

6.4.2 Experimental Results and Interpretation

All the four stages of hot-carrier kinetics described above have been met in performing cyclotron resonance experiments on photocarriers in AgBr. Here the intensity of the microwave fields at 35GHz is extended up to 3kV/cm [6.14, 15]. In the measurements, to attain strong microwave fields, a pulsed magnetron is incorporated in a reflection-type spectrometer with a sample cavity of TE_{112} mode [6.24]. As shown in Fig. 6.12, the absorption line is found to vary significantly with increasing E_ω. Although the hole contribution is also recognized in the absorption line, in the following we shall consider only the absorption due to electrons. At increasing E_ω the absorption line gets remarkably broad and the position of the maximum absorption point B_p shifts to higher magnetic fields. The peak position B_p and the half-line width $\Delta B/2$ are plotted as a function of E_ω in Figs. 6.13 and 14, respectively. Here $\Delta B/2$ is defined as the field difference between the position of the half-maximum point on the higher magnetic field side and the resonance position in the limit of low E_ω. For convenience of description we divide the whole range of E_ω into the four regions: (A) $E_\omega \leq 30$V/cm, (B) $30 \leq E_\omega \leq 100$V/cm, (C) $100 \leq E_\omega \leq 330$V/cm, and (D) 330V/cm $\leq E_\omega$. Each stage in the range from (A) to (D) in the experiments corresponds, respectively, to the hot-carrier situation from (a) to (d) illustrated in Fig. 6.11.

   In Range (A), the peak position shifts slowly to higher values with increasing E_ω (Fig. 6.13). This is caused by the heating of carriers, which gives rise to an increase of the effective mass due to the polaron effect [6.31–33]. As E_ω increases within Range (A), the carrier distribution function spreads over higher energy region, where the band nonparabolicity becomes increasingly pronounced. This should give rise to an inhomogeneous broadening of the absorption line [(A) in Fig. 6.14].

   In Range (B), the shift of B_p tends to saturate at increasing E_ω (Fig. 6.13, Range B). This suggests that a further increase in the average energy of carriers is

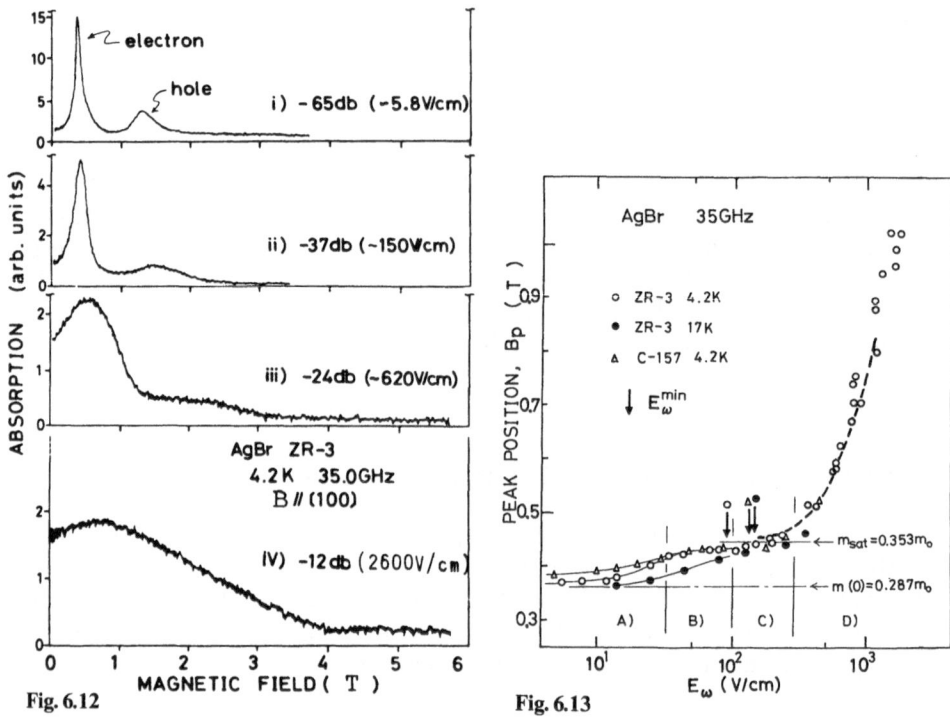

Fig. 6.12. Variation with E_ω of the cyclotron resonance line due to photocarriers in AgBr at 4.2 K [6.14]

Fig. 6.13. Peak position B_p in the cyclotron resonance line as a function of E_ω [6.14,15]. (---) indicates results of a Monte Carlo calculation [6.5]

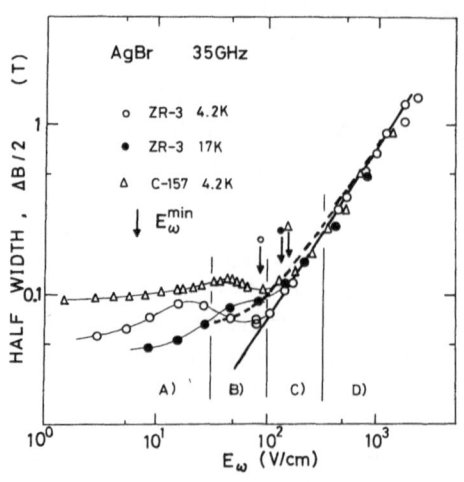

Fig. 6.14. Half width $\Delta B/2$ of the resonance line as a function of E_ω [6.14,15]. (——) indicates the values given by (6.14), under streaming cyclotron resonance. (---) is a result of Monte Carlo calculation [6.5]

suppressed by the onset of optical-phonon emission [Fig. 6.11(b)]. In this range of E_ω, the line width stops increasing, and even decreases with E_ω (Fig. 6.14, Range B). This effect may be related to the transitional character of stage (b), where the carrier system transfers from the nonstreaming regime [Stage (a)] to the streaming cyclotron resonance regime [Stage (c)].

In Range (C), the minimum field strength $E_\omega^{\min}$ at which streaming cyclotron resonance is expected to set in can be obtained from (6.12), and the value is marked by an arrow for each experimental condition in Figs. 6.13, 14. In Fig. 6.14 we can note that the line width begins to increase steeply with increasing E_ω above $E_\omega^{\min}$. That this lines broadening is caused by the repeated emission of optical phonons in the regime of streaming cyclotron resonance is demonstrated by the excellent agreement found between the experimental values of $\Delta B/2$ and the calculations (solid line) performed using the theoretical expression, given by

$$\Delta B/2 = 2B/(\omega T_{op}^\omega). \tag{6.14}$$

Equation (6.14) is obtained by substituting the collision time τ with $T_{op}^\omega/2$ in the ordinary relation for the cyclotron resonance line width $2B/(\Delta B) = \omega\tau$. From Fig. 6.13 we can also note that the peak position B_p remains almost constant in this range of E_ω. From the picture of streaming cyclotron resonance, the values of B_p in this range are thought to correspond to "an effective mass" of the streaming carriers; that is to say, the observed cyclotron mass value $m_{sat} = eB_p/\hbar\omega = 0.353m_0$, which is about 23% larger than the cold polaron mass $m(0) = 0.287m_0$, should be regarded as a mass value averaged over the cyclotron spiral trajectory from $\mathscr{E} = 0$ to $\mathscr{E} = \hbar\omega_{op}$.

In the Range (D), E_ω satisfies (6.13), which gives $E_\omega > 330\,\text{V/cm}$ with $\omega = 2\pi \times 35\,\text{GHz}$. The feature of the phenomena observed in this range is characterized by a steep increase of the peak position towards higher values of the magnetic field (Fig. 6.13, Range D) and a simultaneous marked increase of the line width (Fig. 6.14, Range D). The steep increase of B_p is related to a peculiarity of the streaming cyclotron motion, in which the aspect of resonance vanishes. In fact, the peak position in the absorption line has no longer a direct correspondence with the carrier effective mass. The increase of B_p can be qualitatively explained by a more efficient acceleration of carriers at higher magnetic fields $\omega_c > \omega$ [6.15]. The magnetic field position at which the absorption practically vanishes (as indicated by an arrow in Fig. 6.12) coincides with the calculated field at which the streaming motion disappears [6.14], namely,

$$\zeta_\omega \equiv V_{op}/(E_\omega/B) \simeq 2.$$

A Monte Carlo calculation has been carried out for the case of AgBr under the condition described above [6.5, 15]. The results well reproduce both the steep increase of B_p and the marked increase in $\Delta B/2$ as shown by the dashed line in Figs. 6.13, 14, thus supporting the picture of streaming cyclotron motion.

6.4.3 Phenomena Predicted by Theory

As described in the previous section, the main features of the experimental results are well reproduced by the calculation. Furthermore, the theoretical analysis reveals several remarkable aspects of the streaming cyclotron motion which have not been explored experimentally.

A significant example is the prediction of a bunching of carriers in momentum space [6.5]. According to the calculation on streaming cyclotron motion, the moments when they arrive at $\mathscr{E} = \hbar\omega_{op}$ usually converge to particular phase values of the microwave field, which are uniquely determined by E_ω/ω and ω_c/ω irrespective of initial conditions. This implies that the carriers are bunched onto a particular *point* also in momentum space. The bunching is quite rapid under appropriate conditions; for instance, when $\omega_c/\omega = 0.4$–0.8 only half a period π/ω suffices for a practical bunching to be completed. In actual physical conditions, a finite spread of carrier distribution is introduced around the bunched point by the effect of residual scatterings in the passive region and, more significantly, by the finite penetration of carriers into the active region. Figure 6.15 reports the calculation of *Kurosawa* [6.5], and shows a sequential change of the bunched-carrier distribution during half a period of the microwave field under the conditions $E_\omega = 1200\,\text{V/cm}$ and $\omega_c/\omega = 0.8$. The dissipative current J_x carried by the bunched carriers is shown as a function of time by the solid line in the inset, together with the microwave field $E_x = E_\omega \cos \omega t$ shown by the dotted line. The arrow indicates the moment at which the distribution is displayed. The systematic spikes in J_x correspond to the arrivals of the bunched carriers at $\mathscr{E} = \hbar\omega_{op}$.

When the magnetic field is sufficiently strong to satisfy the condition

$$\zeta_\omega \equiv V_{op}/(E_\omega/B) > 1,$$

a carrier accumulation into a limited region of momentum space is expected, similar to the case of dc electric field. In particular, if ω_c/ω is an odd integer (≥ 3), a coherent cyclotron motion is induced [6.5]. The region where the carriers

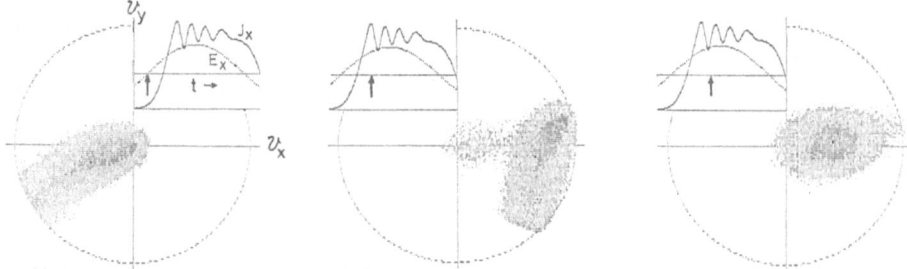

Fig. 6.15. Sequential change in the distribution function of bunched electrons on the $v_z = 0$ plane, according to a simulative calculation performed for the case of AgBr at 4.2 K and $E_\omega = 1200\,\text{V/cm}$ with $\omega_c/\omega = 0.8$ and $\omega = 2\pi \times 35\,\text{GHz}$ [6.5]

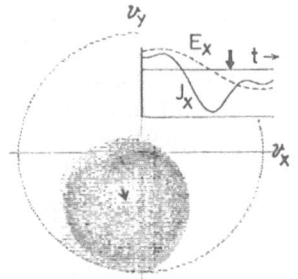

Fig. 6.16. Example of the distribution function at $\omega_c/\omega = 3$, according to a simulative calculation with the same parameters as for Fig. 6.15 [6.5]

accumulate travels elliptically within the sphere $\mathscr{E} \leq \hbar\omega_{op}$ with angular frequency ω, while the accumulated carriers rotate with cyclotron frequency ω_c within the region. As a consequence of the superposition of these two different frequencies, the accumulation region is deformed and attains an asymmetric shape when ω_c/ω is an odd integer. Accordingly, as shown by an arrow in Fig. 6.16, a net dipole moment rotating with ω_c appears to generate the higher harmonic at ω_c.

6.5 Future Perspective

In this chapter, by presenting experimental evidence along with the results of Monte Carlo simulations, we have illustrated the validity of simple pictures of the hot-carrier phenomena related to streaming motion and carrier accumulation in momentum space.

The unique aspect of streaming motion and accumulation effects is connected with the realization of very specific distribution functions which can be directly manipulated by varying the external fields. A number of remarkable effects are theoretically predicted from these specific characteristics, and in turn the theoretical predictions give a strong motivation to experiments. The far-infrared amplification in dc fields (Sect. 6.3.3), the carrier bunching effect, and the higher harmonic generation in microwave field (Sect. 6.4.3) are significant examples. However, other fascinating possibilities, as briefly described in the following, still remain.

Kurosawa [5.34, 35] and *Al'ber* et al. [6.36, 37] have predicted induced cyclotron emission (negative cyclotron resonance) for accumulated carriers in dc electric fields in conjunction with the possibility of a population inversion over Landau levels within Region K. Moreover, it is predicted that microwave fields, additionally applied to an intense dc electric field, can induce coherence in the cyclotron motions of accumulated carriers, which possibly brings about a superradiance [6.17]. In order to explore these effects, it is essential to carry out cyclotron resonance experiments by applying weak microwave fields on the carriers accumulated under the influence of intense dc electric and magnetic fields. Experimental investigations of this type, however, have not yet been undertaken except for preliminary attemps on p-Ge by *Gavrilenko* et al. [6.38]. It is also expected that streaming motion and accumulation of carriers possibly lead to negative differential dc conductivities (NDC) through several different mechanisms [6.3, 4, 17, 39, 40]. Among them, the so-called negative mass amplification and generation (NEMAG), which is specifically expected for streaming carriers in a warped band [6.3], may be of relevant interest. The negative Hall effect, already observed experimentally on p-Si [6.9], provides a good indication for the possibility of NEMAG. Yet another effect expected is the resonance of the streaming carriers with an alternating electric field of frequency corresponding to the repetition rate of optical-phonon emissions [6.41, 42]. Comprehensive investigations of all these phenomena have not yet been undertaken, and are left to future experiments.

References

6.1 W.Shockley: Bell Syst. Tech. J. **30**, 990 (1951)
6.2 W.E.Pinson, R.Bray: Phys. Rev. A **136**, 1449 (1964); Phys. Rev. Lett. **11**, 268 (1963)
6.3 T.Kurosawa: J. Phys. Soc. Jpn. **21**, Suppl., 424 (1966), T.Kurosawa, H.Maeda: J. Phys. Soc. Jpn., **31**, 668 (1971)
6.4 H.Maeda, T.Kurosawa: Proc. 11th Intern. Conf. Phys. Semicond., Warsaw (Elsevier, New York 1972) p. 602
6.5 T.Kurosawa: J. Phys. Soc. Jpn., **49**, Suppl. A, 345 (1980)
6.6 S.Komiyama, K.Kajita, T.Masumi: Phys. Rev. **B20**, 5192 (1979)
6.7 Y.Iye, K.Kajita: Solid State Commun., **17**, 957 (1975)
6.8 S.Komiyama, R.Spies: Phys. Rev. **B23**, 6839 (1981); J. Physique, **42**, Suppl. 10, C7-387 (1981)
6.9 K.Kajita: Solid State Commun., **31**, 573 (1979); Physica **117, 118**B, 223 (1983)
6.10 Yu.L.Ivanov: Pis'ma Zh. Eksp. Teor. Fiz., **34**, 539 (1981);
     [English transl.: Sov. Phys. – JETP Lett. **34**, 515 (1981)]
6.11 S.Komiyama: Phys. Rev. Lett., **48**, 271 (1982)
6.12 V.I.Gavrilenko, V.N.Murzin, S.A.Stoklitskii, V.P.Chebotayev: Pis'ma Zh. Eksp. Teor. Fiz. **35**, 81 (1982);
     [English transl.: Sov. Phys. – JETP Lett. **35**, 97 (1982)]
6.13 A.A.Andronov, V.A.Kozlov, L.S.Mazov, V.N.Shastin: Sov. Phys. – JETP Lett. **30**, 551 (1979)
6.14 S.Komiyama, T.Masumi: Solid State Commun. **26**, 381 (1978); J. Magn. Mater., **11**, 59 (1979)
6.15 S.Komiyama, T.Masumi, T.Kurosawa: Proc. 14th Intern. Conf. Phys. Semicond., Edinburgh, (The Institute of Physics, Bristol 1978) p. 335
6.16 C.Jacoboni, L.Reggiani: Adv. Phys. **28**, 493 (1979)
6.17 T.Kurosawa: J. Physique **42**, Suppl. 10, C7-377 (1981)
6.18 S.Komiyama: Adv. Phys. **31**, 255 (1982)
6.19 J.T.Devreese, R.Evrard: Phys. Status Solidi (b), **78**, 85 (1976);
     J.T.Devreese, F.Brosens, R.Evrard, E.Katheuser: J. Phys. Soc. Japn. **49**, Suppl. A, 341 (1980)
6.20 L.E.Vorob'ev, Yu.K.Pozhela, A.S.Reklaytis, E.S.Smirnitskaya, V.I.Stafeev, A.B.Fedortsov: Fiz. Tekh. Poluprovodn, **12**, 1585 (1978);
     [English transl.: Sov. Phys. – Semicond., **12**, 935 (1978)]
6.21 L.E.Vorob'ev, Yu.K.Pozhela, A.S.Reklaytis, E.S.Smirnitskaya, V.I.Stafeev, A.B.Fedortsov: Fiz. Tekh. Poluprovodn, **12**, 742 (1978);
     [English transl.: Sov. Phys. – Semicond., **12**, 433 and 754 (1978)]
6.22 I.I.Vosilyus, I.B.Levinson: Zh. Exp. Teor. Fiz., **52**, 1013 (1967);
     [English transl.: Sov. Phys. – JETP **25**, 672 (1967)]
6.23 F.Brosens, J.T.Devreese: Solid State Commun. **44**, 597 (1982)
6.24 S.Komiyama: In *Polarons and Excitons in Polar Semiconductors and Ionic Crystals*, ed. by J.T.Devreese and F.Peeters (Plenum, New York, 1984) p. 41
6.25 M.Höhne, M.Stasiw: Phys. Status Solidi, **28**, (b) 247 (1968)

6.26 S.Komiyama: Appl. Phys. **25**, 303 (1981)
6.27 L.K.Gvardzhaladze, Yu.L.Ivanov: Fiz. Tekh. Poluprov., **7**, 1328 (1973);
      [English transl.: Sov. Phys. – Semicond. **7**, 890 (1974)]
6.28 E.O.Kane: J. Phys. Chem. Sol. **1**, 82 (1956)
6.29 L.E.Vorob'ev, F.I.Osokin, V.I.Stafeev, V.N.Tulupenko: Pis'ma Zh. Eksp. Teor. Fiz., **34**, 125
      (1981);
      [English transl.: Sov. Phys. – JETP Lett. **34**, 118 (1981)]
6.30 L.E.Vorob'ev, F.I.Osokin, V.I.Stafeev, V.N.Tulupenko: Pis'ma Zh. Eksp. Teor. Fiz. **35**, 360
      (1982);
      [English transl.: Sov. Phys. – JETP Lett. **35**, 440 (1982)] J.K.Pozhela, E.V.Starikov,
      P.N.Shiktorov, L.E.Vorobjev, F.I.Osokin, V.E.Stafeev, V.N.Tulupenko: Physica **117, 118**B,
      226 (1983)
6.31 D.M.Larsen: Phys. Rev. **144**, 697 (1966)
6.32 J.W.Hodby: Solid State Commun. **7**, 811 (1969)
6.33 H.Tamura, T.Masumi: J. Phys. Soc. Jpn. **30**, 1763 (1971)
6.34 T.Kurosawa: Solid State Phys. **11**, 217 (1976) in Japanese
6.35 T.Kurosawa: Solid State Commun. **24**, 357 (1977)
6.36 Ya.I.Al'ber, A.A.Andronov, V.A.Valov, V.A.Kozlov, I.P.Ryazantseva: Solid State Commun.
      9, 955 (1976)
6.37 Ya.I.Al'ber, A.A.Andronov, V.A.Valov, V.A.Kozlov, A.M.Lerner, I.P.Ryazantseva: Sov.
      Phys. JETP **45**, 539 (1977)
6.38 V.I.Gavrilenko, E.P.Dodin, Z.F.Krasil'nik, Yu.N.Nozadrin, M.D.Chernobrovtseva: Pis'ma
      Zh. Eksp. Teor. Fiz. **35**, 432 (1982);
      [English transl.: Sov. Phys. – JETP Lett. **35**, 535 (1982)]
6.39 A.A.Andronov, V.A.Valov, V.A.Kozlov, L.S.Mazov: Solid State Commun. **36**, 603 (1980)
6.40 V.L.Gurevich, D.A.Parshin: Solid State Commun. **37**, 511 (1981)
6.41 A.A.Andronov, V.A.Kozlov: Sov. Phys. JETP Lett. **17**, 87 (1973)
6.42 A.Chenis, A.Matulis: J. Phys. Soc. Jpn. **49**, Suppl. A, 337 (1980)

7. Hot Electrons in Semiconductor Heterostructures and Superlattices

Karl Hess and Gerald J. Iafrate

With 18 Figures

Parallel to the revolutionary development of large-scale integrated silicon technology, much progress has been made in the basic understanding of superlattices and lattice-matched III-V compound heterostructure layers in general [7.1]. Two new epitaxial technologies have emerged in recent years – molecular-beam epitaxy (MBE) [7.2] and metalorganic chemical vapor deposition (MOCVD) [7.3] – which have opened a variety of new possibilities. These include superlattice transport as discussed by *Esaki* et al. [7.4–6], the fabrication of materials exhibiting extremely high mobilities [7.7], quantum-well heterostructure lasers [7.8, 9], planar doped barrier structures [7.10], and real-space transfer switching [7.11]. More recent research has also concentrated on strained-lattice mismatch superlattices [7.12] and on type II superlattices [7.6].

The theoretical understanding of many heterostructure properties has kept pace with experimental developments. Much work has been done on the band-edge discontinuity problem [7.13–15], deviations from ideal interface structures [7.16, 17], superlattice disordering [7.18], two-dimensional excitons, effective-mass levels of shallow impurities [7.19], and many other important material properties. The mobility enhancement is understood to some extent [7.20–22] and hot-electron and optical properties have been studied in some detail [7.1, 11].

Semiconductor devices involving heterolayers have displayed encouraging results and many groups are pursuing investigations of the high electron mobility transistor (HEMT) concept [7.23] and quantum-well lasers with extremely low threshold current [7.24]. In addition, the search for new possibilities in optoelectronics and high-speed large-scale integration continues. From both a theoretical and a practical viewpoint, a most intriguing feature is the variability of boundary conditions on a length scale comparable to or even smaller than the electron mean free path (e. g., for the electron-phonon interaction). In fact, many of the effects discussed above are a consequence of unusual boundary conditions, be they periodic or nonperiodic. It is clear that complicated boundaries can introduce new and interesting features into otherwise simple and well-defined potential problems. For the engineer, in particular, the opportunity arises to *engineer* new forms of semiconductors and the corresponding Fourier transforms, thus influencing both the real-space and wavevector-space properties.

This chapter is devoted to the problem area of *hot electrons* in heterostructures and superlattices. We do not intend here to give a complete review but will concentrate on effects which we found theoretically interesting and promising for new types of applications. Section 7.1 describes the band structure and scattering mechanisms in multilayer structures and Sect. 7.2, heterolayer-electronic transport at low and intermediate electric fields. Section 7.3 deals with real-space transfer effects, Sect. 7.4 with hot-electron emission over large potential barriers, and Sect. 7.5 with lateral two- and three-dimensional superstructures. Section 7.6 gives a summary of the chapter.

7.1 Band Structure and Scattering Mechanisms in Layered Semiconductor Structures

7.1.1 The Band Structure in Semiconductor Heterolayers

Among all possible combinations of semiconductors, the lattice-matched (or at least closely matched) combinations have attracted most interest. The reason is, of course, the possibility of ideal heterojunction interfaces if the lattice constant of the constituents is the same. It is interesting to note that perfect lattice match does not guarantee an ideal interface, as can be seen from the results for the GaAs-Ge system [7.25]. In addition to lattice match, we also have to require that both systems be binary (or ternary, quaternary) in order to avoid, for instance, the site allocation problem [7.25]. An overview of some important alloys suitable for heterojunction formation can be found in [7.26].

A picture of the effects of the heterojunctions on the band structure can be obtained by employing the effective-mass theorem. The simplest approach is to view the semiconductor as homogeneous with an additional potential super-imposed on the crystal potential as shown in Fig. 7.1. Then the effective-mass theorem [7.27] tells us that for electrons at band minima (maxima) we can replace the Hamiltonian

$$\hat{P}^2/(2\,m_0) + V(\mathbf{r}) + \phi(z) \tag{7.1}$$

by

$$\hat{P}^2/(2\,m) + \phi(z), \tag{7.2}$$

and therefore transform away the periodic crystal potential $V(\mathbf{r})$ [provided that $\phi(z)$ and its Fourier components fulfill certain conditions]. In other words,

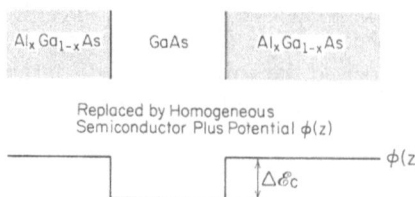

Replaced by Homogeneous
Semiconductor Plus Potential $\phi(z)$

Fig. 7.1. Typical example of a semiconductor heterostructure and model potential replacement

we have reduced the heterojunction problem to a well-defined potential problem for a quasi-free electron, and we can use all the well-known solutions for single (square-well model) and multiple (Kronig-Penney model) quantum wells. The $\mathscr{E}(\mathbf{k})$ relation for a periodic square-well potential is easily derived and given (in k_z direction) by

$$\mathscr{E}_z = \tfrac{1}{2} \mathscr{E}_0 (1 - \cos k_z L_z). \tag{7.3}$$

Here z is the direction perpendicular to the heterolayers, $\mathscr{E}_0$ is a constant energy term, and L_z is the layer thickness.

 Esaki and Tsu [7.4] suggested that, because of the adjustable layer thickness L_z, superlattices may be useful for high-frequency generation. The idea is based on substituting the solution of the equation of motion

$$\hbar \frac{dk_z}{dt} = eE \tag{7.4}$$

into (7.3), which results in an ac current with frequency eEL_z/h, where E is the electric field. It was this idea which provided the original impetus for research on superlattices.

 To complete the effective-mass theory we need to know the potential $\phi(z)$. Although it is difficult to determine $\phi(z)$ relative to the vacuum level, a calculation of $\phi(z)$ relative to the band edge of one component of the heterostructure seems to be easier. Harrison et al. [7.13, 28] proposed that $\phi(z)$ can be determined by the linear combination of atomic orbitals (LCAO) method which gives the valence band edge $\mathscr{E}$, as

$$\mathscr{E}_v = \frac{\mathscr{E}_a + \mathscr{E}_c}{2} - \left[\left(\frac{\mathscr{E}_a + \mathscr{E}_c}{2} \right)^2 + V_{xx}^2 \right]^{1/2}, \tag{7.5}$$

where $\mathscr{E}_a$ ($\mathscr{E}_c$) are the p-state energies of the anion (cation) atoms, and V_{xx} is the interatomic matrix element of the crystal Hamiltonian between p states of adjacent atoms. The valence-band discontinuity $\Delta\mathscr{E}_v$ (relevant for holes) of an AlAs-GaAs heterojunction is therefore

$$\Delta\mathscr{E}_v = \mathscr{E}_v(\text{AlAs}) - \mathscr{E}_v(\text{GaAs}). \tag{7.6}$$

As shown in Fig. 7.1, the conduction-band discontinuity $\Delta\mathscr{E}_c$ in relation to $\phi(z)$ is given by

$$\Delta\mathscr{E}_c = \mathscr{E}_g(\text{AlAs}) - \mathscr{E}_g(\text{GaAs}) + \Delta\mathscr{E}_v. \tag{7.7}$$

Here $\mathscr{E}_g$ is the energy gap of AlAs or GaAs. A comparison of these values with experimental results has been given by Margaritondo et al. [7.15], who found an overall agreement between theory and experiments.

The fact that different potentials $\phi(z)$ are necessary for conduction and valence electrons (holes) hints at the limitations of the above model, which are discussed below. The effective-mass approximation in the form described above makes it necessary to deal with the envelope wave functions which are supposed to match smoothly across the interface. However, the wave functions ψ and their derivatives are continuous, and not necessarily the envelope functions. Typically we have

$$\psi = \exp{(i\mathbf{k}_{\|}\mathbf{r}_{\|})}\,\xi(z)\,u_k(\mathbf{r}_{\|}), \tag{7.8}$$

where $\mathbf{k}_{\|}$ and $\mathbf{r}_{\|}$ are two-dimensional vectors in the plane of the layers and u_k is the Bloch-periodic part of the wave function. Then we have

$$\psi_B^i = \psi_A^i, \quad \text{and} \tag{7.9a}$$

$$\partial\psi_B^i/\partial z = \partial\psi_A^i/\partial z, \tag{7.9b}$$

but the same is not necessarily true for the envelope functions $\xi(z)$. The superscript i in (7.9) means that the values are taken at the interface; the subscripts B and A indicate the different materials, e.g., GaAs $=$ B and AlAs $=$ A. Nevertheless, if the wave function on one side far from the interface is given, then that of the other side is uniquely determined for any junction. This fact can be described according to *Ando* and *Mori* [7.29] by the following relation between the envelopes and their derivatives at the interface

$$\begin{pmatrix} \xi_B^i \\ \dfrac{\partial\xi_B^i}{\partial z} \end{pmatrix} = \tilde{T}_{BA} \begin{pmatrix} \xi_A^i \\ \dfrac{\partial\xi_A^i}{\partial z} \end{pmatrix}, \tag{7.10}$$

where $\tilde{T}_{BA}$ is a complex 2×2 matrix called transfer matrix. If $\tilde{T}_{BA}$ is identical to the unit matrix, the simple effective-mass approach is valid. *Ando* and *Mori* [7.29] found for the GaAs-Al$_{0.1}$Ga$_{0.9}$As system

$$\tilde{T}_{BA} \approx \begin{pmatrix} 1.021 & 0 \\ 0 & 1.101 \end{pmatrix}, \tag{7.11a}$$

which is very close to the unit matrix. Therefore, the simple effective-mass picture should be excellent for the Al$_x$Ga$_{1-x}$As system with small x. For the GaSb-InAs system, on the other hand, where a GaAs monolayer interface naturally exists, *Ando* and *Mori* [7.29] found

$$\tilde{T}_{BA} \approx \begin{pmatrix} 0 & 1.88 \\ 0.24 & 0.06 \end{pmatrix}, \tag{7.11b}$$

which means that the simple effective-mass picture breaks down and the transfer matrix corrections are essential.

In some instances, where details of the band structure away from the minima play a role, or layer widths are very small (of the order of the lattice constant itself), then more powerful methods must be used to calculate the electronic band structure. In both cases, several tight-binding approaches have been used and have shown excellent agreement with experimental data [7.30–32].

During his tight-binding calculations, *Mon* [7.32, 33] noticed an interesting phenomenon for superlattices composed of two materials with indirect energy gap such as Si-GaP. For certain layer widths, of the order of a few lattice constants, the superlattice becomes direct. This means that the X minima along the superlattice directions are folded into the zone center, i.e., the superlattice forms a semiconductor with a direct energy gap. Recent experimental investigations in this direction are discussed in [7.34].

Let us emphasize at the end of this section that superlattices are three-dimensional and not one-dimensional objects. The various subbands (mini-bands) are connected by a two-dimensional continuum (parallel to the layers) and there are no real energy gaps between the minibands. This is an important factor for many investigations and especially important for the following discussion of electronic transport.

7.1.2 Scattering Mechanisms

Many aspects of scattering at interfaces have been reviewed before [7.35, 36]. Therefore, we will only give here a brief description of the most important features which distinguish interface from bulk transport.

Let us first investigate scattering by ionized impurities in heterostructures. There are three major new features compared to impurity scattering in bulk materials:

i)   The electron gas can be two-dimensional at the interface because of size quantization. Various possibilities are shown in Fig. 7.6a–c.
ii) If the dielectric constant of the two materials is different, image-force effects play a major role. In simple cases the dielectric constant can be replaced by the arithmetic mean of the two constants.
iii) There exists the possibility that the electrons and the impurities are separated over considerable distances (e.g., several effective Bohr radii). In the case of modulation doping, this distance is intentionally introduced in order to raise the mobility by the reduction of impurity scattering [7.7]. This is a very important effect. Mobilities higher than $10^6 \text{ cm}^2/\text{Vs}$ have been observed in modulation-doped GaAs-Al$_x$Ga$_{1-x}$As layers under illumination by Bell Laboratory scientists.

All the above phenomena can to a certain extent be described by a "model matrix element" which is obtained for a two-dimensional sheet of electrons

$(x-y$ direction) and an impurity distribution with density $N_I(z)$ where z is the coordinate perpendicular to the electron sheet which is located at $z=0$. The absolute square of the matrix element for the transition of an electron from $k_\parallel$ to $k_\parallel'$ is then given by [7.20]

$$|M(k_\parallel, k_\parallel')|^2 = \frac{e^4}{4\,A^2 L_z(\bar\varepsilon\varepsilon)^2(q+S_{low})^2} \times \int_a^b \exp\,(-2\,q|z|)\,N_I(z)\,dz. \qquad (7.12a)$$

Here A is the area of the electron sheet; $\bar\varepsilon$ is the (arithmetic) average of the relative dielectric constant of the two neighboring media, $q=|k_\parallel - k_\parallel'|$; and S_{low} is the two-dimensional screening constant which for high electron density is equal to $2/a_B^*$ where a_B^* is the effective Bohr radius. The two coordinates a and b define the boundaries of the sample in the z direction. The choice of a and b has given rise to confusion in the past: in symmetric structures a can be set equal to 0 and the result for the matrix element later multiplied by 2. Since the samples are usually very large compared to $1/q$, $b=\infty$ is a good choice [7.20]. However, because of the infinite range of the Coulomb force, the scattering diverges for $S_{low}=0$ (no screening) and infinite sample width.

Many of the investigated modulation-doped structures consist of one $Al_xGa_{1-x}As$ $(x \lesssim 0.33)$ layer on top of GaAs with the homogeneous intentional doping in the $Al_xGa_{1-x}As$ layer starting at a distance z_0 from the electron gas (Fig. 7.2). The scattering rate is then found to be [7.20, 38, 39]

$$\frac{1}{\tau} = \beta \int_0^\pi \exp\,(-4kz_0 \sin\,\theta)\, \sin\,\theta(2k \sin\,\theta+S_{low})^{-2}d\theta, \qquad (7.12b)$$

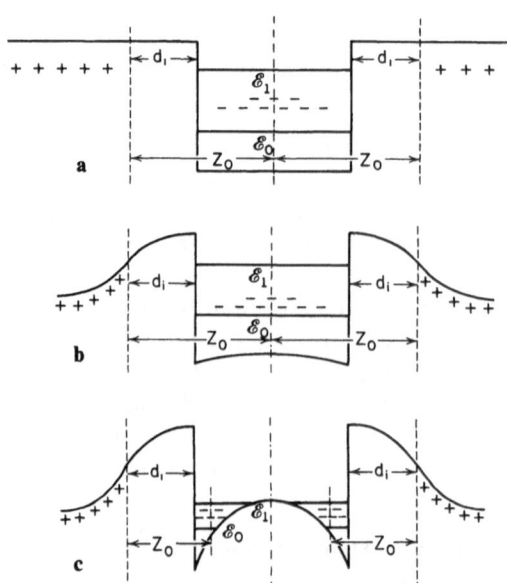

Fig. 7.2a–c. Schematic diagram of the conduction band edge in modulation-doped GaAs-$Al_xGa_{1-x}As$ layers including (a) no band bending, (b) moderate band bending, (c) strong band bending. Here $\mathscr{E}_0$ and $\mathscr{E}_1$ are the lowest and first excited quantum states which are above the GaAs conduction-band edge. The distance d_i is the spacing in the $Al_xGa_{1-x}As$ layer which is free of intentional doping. Notice the narrowing in electron-impurity distance Z_0 when going from (b) to (c)

$$\beta = e^4 m N_R (8 \pi \hbar^3 (\bar{\varepsilon}\varepsilon)^2 k)^{-1}. \tag{7.13}$$

Here m is the effective mass and N_R the remote impurity density. If the screening is strong, scattering occurs only for small values of $\theta (kz_0 > 1)$, and $\sin \theta$ can be replaced by θ in the exponent of (7.12). Then

$$\frac{1}{\tau} \simeq (2\beta/S_{low}^2)\,(16\,k^2 z_0^2 + 1)^{-1}. \tag{7.14}$$

This equation can also be used for calculating the scattering by the background impurities by setting $z_0 = 0$ and $N_R = N_B$, where N_B is the background density. If the impurities are on both sides, the right-hand side of (7.14) has to be multiplied by 2. Eq. (7.14) clearly illustrates the advantages of modulation doping in increasing the electron mobility:

i)   the distance z_0 reduces the scattering rate;
ii) the background is screened effectively since the local electron density (S_{low} is proportional to the electron concentration at low densities) is larger than the density of background impurities.

There are many details which need to be understood in the treatment of remote impurity scattering. Some are described in [7.37–39].

The electron-phonon interaction in heterolayers is more difficult to understand since, for example, the confinement of the electrons to a layer does not mean that the phonons are also confined and vice versa. Limiting cases have been discussed before [7.21, 38]. There do not seem to be very large deviations of the scattering rate (more than a factor of ~ 2) from the bulk values, and there are no major new physical concepts involved except for the possibility of remote phonon scattering [7.40] which, however, is not expected to influence the mobility drastically either. For details, we refer the reader to [7.38, 40]. The effect of intersubband scattering is described in [7.21, 41]. A comparison of some experimental and theoretical results has been given in [7.39]. The good agreement found shows that, overall, the effects are understood. In its ultimate limitations, however, there are still many open questions. It is not known which scattering mechanism will be the limiting one at low temperature. Candidates include not only acoustic-phonon scattering and interface-charge scattering, but also effects such as boundary-alloy scattering. The boundary of the GaAs is usually the ternary alloy $Al_x Ga_{1-x}As$, and since the electron wave function penetrates into the $Al_x Ga_{1-x}As$, the electron should experience some alloy scattering.

The above considerations apply mainly to single layers. The scattering mechanisms of superlattices have not been discussed in detail yet. It has always been assumed that superlattices behave more or less like bulk semiconductors with respect to scattering rates. It is clear, however, that differences can be expected, especially in effects which involve symmetry. (For example, optical-deformation-potential scattering is negligible in GaAs at Γ point because of symmetry, and the consequences of the new superlattice symmetry have not yet been investigated.)

Finally, let us mention that the quasi–two-dimensionality of the electron gas in heterolayers can give rise to dramatic effects in high magnetic fields, as demonstrated in numerous papers on the quantum Hall effect [7.42, 43].

7.2 Electronic Transport at Low and Intermediate Fields

The extremely high mobilities which have been observed in modulation-doped structures at low temperatures persist only at very low electric fields ($E \lesssim 1$ V/cm). At intermediate fields, pronounced warm- and hot-electron effects can be observed.

In contrast, the mobility in bulk GaAs with a high donor concentration (high electron densities) is very low and therefore warm-electron effects (below the Gunn threshold) are not usually observed.

Selectively doped (Al, Ga)As/GaAs heterostructures combine the unique features of high mobilities and large electron concentrations. Warm-electron effects in these structures should therefore be treated with a model which includes strong electron-electron interactions. The simplest way of including electron-electron scattering is the assumption of a Fermi distribution at elevated temperature T_e for the spherical symmetrical part of the distribution function.

The calculation is then straightforward, using the two-dimensional model for the mobility as described above. To calculate the carrier temperature T_e, we can use the power balance equation as done in [7.44], where at low temperatures we

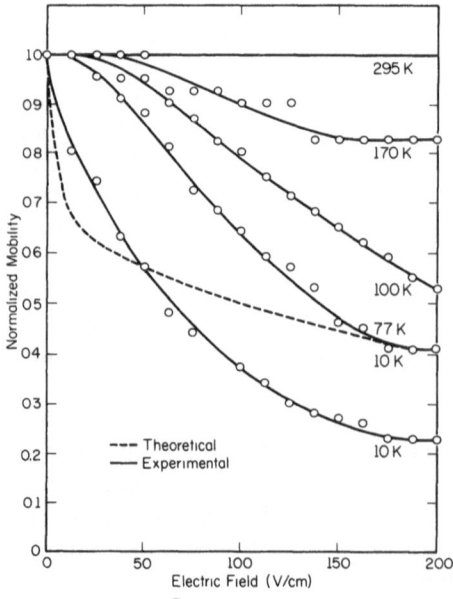

Fig. 7.3. Experimental (O O O) and theoretical (– – –) results for the normalized mobility of modulation-doped structures at moderate electric fields. (———) are only guides for the eye to the experimental results [7.46]

have to include the power loss to acoustic modes [7.45]. Experimental curves for a very high mobility ($\mu \simeq 2 \times 10^5$ cm^2/Vs at 10 K) modulation-doped structure are illustrated in Fig. 7.3. A theoretical curve is shown for comparison for the case of the lowest temperature (10 K).

In comparing experimental and theoretical results, two aspects are apparent. At very low fields the theory predicts a deviation from Ohm's law, which is steeper than the experimentally observed deviation. This discrepancy can be accounted for by assigning a larger energy-loss rate due to the phonon interaction. At higher electric fields, the slope of the electron mobility versus field curve decreases at a slower rate, which may be an indication of a lower energy-loss rate than that used in our model.

Recently, more direct measurements of the electron temperature in modulation-doped structures have been reported [7.47]. These measurements can be used to obtain directly the rate of energy loss P_e per electron to the phonons (which, unlike the mobility, obviously does not depend on elastic scattering processes). The experimentally determined P_e values at a given carrier temperature T_e are approximately seven times smaller than the theoretical values [7.45]. This is consistent with the findings from Fig. 7.3 and is typical for large deviations of the distribution function from the Fermi shape at energies above the optical-phonon energy [7.48]. At first glance, one would think that the electron-electron interaction should be strong enough to establish a Fermi-like distribution function. However, detailed investigations show that even at an electron concentration of about 10^{18} cm^{-3} the deviations of the distribution function from the Fermi-like shape above the optical-phonon energy can be substantial [7.49]. Since the range of intermediate electric fields is very important for the operation of HEMT devices [7.50] (dominated by low-source resistance), these effects deserve further investigation. Especially since screening of the phonons may also play a major role.

7.3 Hot-Electron Thermionic Emission over Small Potential Barriers: Real-Space Transfer

Emission of thermal electrons over heterojunction barriers has been investigated in great detail, e.g., in connection with Schottky barriers [7.51]. For heterostructures the current density $j_{A \to B}$ from the semiconductor A to B is obtained from the density of electrons with energies sufficient to overcome the potential barrier (assumed to be in the z direction):

$$j_{A \to B} = e \sum_{k_x k_y} \sum_{k_z > k_{z_l}} v_{k_z}. \tag{7.15}$$

Here k_{z_l} is the smallest possible component of the wavevector in the z direction which allows the electron to escape. Equation (7.15) results in the relation

$$j_{A \to B} = A^* T_0^2 \exp\left[-(\mathscr{E}_F - \Delta\mathscr{E}_c)/K_B T_0\right]. \tag{7.16}$$

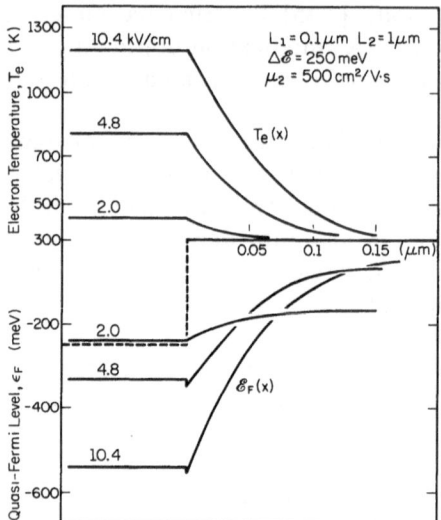

Fig. 7.4. Electron temperature and quasi-Fermi level as functions of distance. The GaAs potential well is shown by (---). The interface between the GaAs and the $Al_xGa_{1-x}As$ is set at $Z=0$ [7.52]

Here A^* is the effective Richardson constant $[A^* = 120\,(A/cm^2/K^2)m/m_0]$; m is the effective mass (for anisotropic mass see [7.51]); T_0 is the lattice temperature; and $\mathscr{E}_F$ is the quasi-Fermi level. It is clear that T_0 has to be substituted by the actual carrier temperature T_e (as deduced from the average carrier energy) if the electron distribution function is not at equilibrium.

Inspection of (7.16) shows that the current can be greatly increased if $T_e > T_0$, i.e., for hot-electron conditions. The carrier temperature can be raised by various means, the simplest being an electric field applied parallel to the interface. This is illustrated in Fig. 7.4, which shows the electron temperature and the quasi-Fermi level as a function of the distance perpendicular to a GaAs-$Al_xGa_{1-x}As$ double heterojunction. This case was first discussed by *Hess* in 1979 [7.53]. Notice that the quasi-Fermi level changes in the z direction although the external electric field is applied in a perpendicular direction. In all conventional semiconductor devices the differences in the quasi-Fermi level arise in the direction of the applied field. The reason for this unconventional result follows immediately from (7.16), if the metal is replaced by $Al_xGa_{1-x}As$ and the semiconductor is GaAs, as shown in Fig. 7.4. The current $j_{GaAs \rightarrow Al_xGa_{1-x}As}$ is then enhanced because T_0 has to be replaced by a high electron temperature T_e. This high T_e arises from the huge energy gained by electrons (at a rate $e\mu E^2$), which is proportional to the mobility. In the $Al_xGa_{1-x}As$ the mobility is usually low, especially under conditions of modulation doping, and therefore the current $j_{Al_xGa_{1-x}As \rightarrow GaAs}$ is not enhanced but stays practically at its equilibrium value. This means that electrons transfer from the GaAs to the $Al_xGa_{1-x}As$ until the number of electrons in the $Al_xGa_{1-x}As$ is sufficiently increased so that $j_{Al_xGa_{1-x}As \rightarrow GaAs}$ increases and balances the outflowing current $j_{GaAs \rightarrow Al_xGa_{1-x}As}$. Of course, the actual balance also depends on the doping and, in particular, on the distribution of the donors,

Details are discussed in [7.52–54]. In the limiting case of small electron densities, with large mobility in the GaAs and small mobility in the $Al_xGa_{1-x}As$, all electrons will transfer out of the GaAs as the applied electric field parallel to the layers increases.

This real-space-transfer effect can be generalized in the following way. As electrons or holes are accelerated (e. g., by dc or ac electric fields) in the three- or two-dimensional continuum of a semiconductor (or insulator), their average energy increases. It is therefore possible that the electrons (holes) reach states at such high energies that propagation to a neighboring semiconductor is possible even if a substantial band-edge discontinuity has to be overcome. In the case of a superlattice, the electrons can be energized from the lowest confined states to higher minibands in which they may then propagate in any direction. In a single quantum well, electrons can be accelerated in the two-dimensional continuum and then at high energies propagate out of the well region after a scattering event which supplies the necessary momentum. Of course, the effect can also occur in quasi-one-dimensional systems where the electrons can be accelerated in a one-dimensional continuum and then leave the one-dimensional "wire" if they reach high enough energies. The consequence of this transfer can give rise to negative differential resistance, switching, and storage. ,

Figure 7.5 shows some of the inherent possibilities of these switching effects for complicated geometrical structures. Figure 7.5a refers to the simplest case: for a small value of the voltage V_1 (so that $eV_1 < \varDelta\mathscr{E}_c$), current flows only in the GaAs to Contact 1. For a larger value of V_1 (so that $eV_1 > \varDelta\mathscr{E}_c$), there is also a current channel through the AlGaAs and Contact 2. Figure 7.5b shows the same arrangement but with a floating GaAs region which, for example, can store electrons. All kinds of geometrical arrangements can be thought of, and the figure is merely for illustrative purposes. A large variability is also given by the use of various band-edge discontinuities; superlattices; single wells, coupled by tunneling or separated; etc. In simple systems, the possibility of real-space-transfer oscillators and electron storage devices has been demonstrated and is described below.

Negative differential resistance arises if the mobility in the $Al_xGa_{1-x}As$ is much lower (impurity scattering) than in the GaAs. Various experimental results on this effect have been described before [7.55] and will not be repeated here since they are similar in nature to the following results [7.56] of storage and switching.

A switching and storage test structure shown in Fig. 7.6 was fabricated using standard photolithographic and chemical etching techniques. Four ohmic contacts were formed by vacuum evaporation of Au-Sn and subsequent alloying at 450 °C for 30 s in flowing H_2. The resulting structure had a dumbbell-shaped center region containing nine heterostructure periods (each period consisting of a $GaAs/Al_xGa_{1-x}As$ layer pair) contacted at each end by an ohmic contact (labeled "side contact" in Fig. 7.6) that was diffused down through most of the layers. The areas containing the main contacts (Fig. 7.6) were etched down so that the top three $GaAs-Al_xGa_{1-x}As$ layer pairs were removed, leaving only the lower six periods contacted by the main contacts. If voltages are applied only to

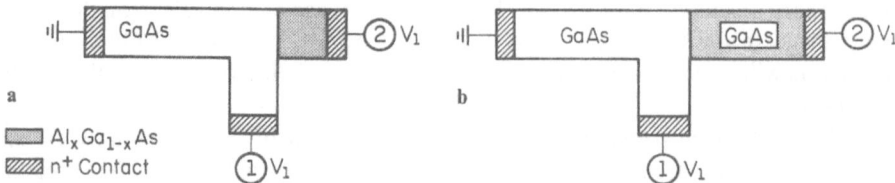

Fig. 7.5a, b. Possible geometries to demonstrate switching effects due to real-space transfer in GaAs-$Al_xGa_{1-x}As$ heterostructures

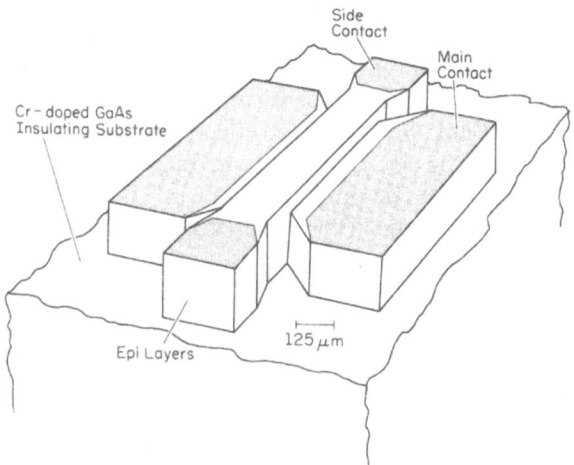

Fig. 7.6. Device test structure used for measuring switching and storage effects. When a high electric field is applied, hot electrons can propagate from the bottom high-field layers to the upper electrically isolated layers. [7.56]

the main contacts (no applied voltage on side contacts), then only the GaAs layers in the bottom six periods will have high electric fields. Some of the electrons that escape from the GaAs layers in these bottom six periods can propagate into the upper electrically isolated layers nearer to the surface in the dumbbell bridge. If the sample is cooled to cryogenic temperatures, the electrons that lose energy in these top isolated layers will become trapped in the GaAs regions and will not be able to gain enough energy from the crystal lattice or fringing electric fields to escape by thermionic emission. Thus, as soon as the electric field between the main contacts becomes high enough to cause electrons to be emitted from the lower GaAs layers, a fraction of these electrons can propagate to the upper isolated GaAs layers. There they are trapped, provided that the side contacts do *not* connect through to the lower layers interconnected by the main contacts. Consequently, the resulting current between the main contacts will be reduced for all following measurements of the current at lower fields. Now, suppose a voltage is then applied only to the side contacts so that the resulting electric field in the dumbbell bridge region is high enough to allow thermionic emission of the electrons in all the upper GaAs layers as well as of the electrons between the main

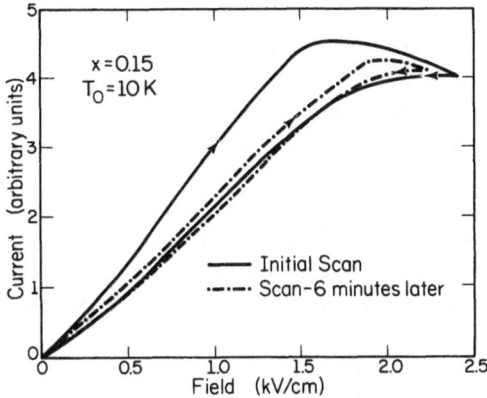

Fig. 7.7. Current-field characteristics measured between the main contacts of the test structure in Fig. 7.6 at $T_0 = 10$ K. The prolonged reduction in the current flow following the initial voltage scan is attributed to the switching and storage of electrons. The initial (virgin) scan characteristics are restored by applying a high electric field between the side contacts. [7.56]

contacts. The initially higher concentration of electrons stored in the top three periods will then be redistributed to the other periods, thus restoring the original carrier distribution of the sample before any voltages were applied. The current-field characteristics between the main contacts at low fields will then also return to their original values.

The results of the current-field measurements between the main contacts on a sample cooled to 10 K are shown in Fig. 7.7. Measurements were performed using 700 ns voltage pulses at low repetition rates to reduce sample heating. Measurements of the current versus field shown in the graphs were taken at 600 ns after the beginning of each pulse, although no time dependence was observed in the pulses between 1 and 700 ns. The sample was mounted on a temperature-controlled cold finger in an evacuated sample chamber, and standard 50Ω sampling oscilloscope and $x - y$ recorder techniques were used. All measurements were performed with the sample in the dark. The upper solid line in Fig. 7.7 shows the virgin current-field characteristics between the main contacts when the first voltage was initially applied to the main contacts (no voltage between side contacts). After the electric field exceeded a threshold value necessary to cause a real-space transfer of electrons, the voltage was returned smoothly to zero, resulting in the lower solid curve.

The significantly lower current values measured as the voltage was returned to zero indicated that a fraction of the electrons had been transferred and stored and could no longer contribute to the current. Repeating the same voltage scan several seconds later revealed only an insignificant change in the reduced current. Another scan of the voltage several minutes later (dot-dashed curve in Fig. 7.7) exhibited only a small further increase in the reduced current, indicating the number of stored electrons had changed very little. Immediately thereafter, a high field was applied between the side contacts (no applied voltage on main contacts) to redistribute the sorted electrons. The following voltage scan between the main contacts resulted in a retracing of the initial virgin current-field curve (solid line in Fig. 7.7). The above pattern of voltage applications was repeated many times with the same results. The experiments are still inconclusive as to whether the electrons are stored in the upper GaAs layers, in deep traps in the $Al_xGa_{1-x}As$, or in surface states, since a complete electrical separation of the layers (side contacts) is difficult to achieve. Also, the density of deep traps is not well known in these samples and more research needs to be done. However, the effect not only proves the concepts of real-space transfer as discussed before, but also shows once again the potential for device applications.

More recent developments in the general area of real-space transfer concentrate on the use of strained superlattices such as $GaAs-In_xGa_{1-x}As$ and $GaAs-GaAs_{1-x}P_x$. Current-voltage characteristics very similar to the $Al_xGa_{1-x}As$-GaAs system are measured for these materials. This shows the universality of the effect [7.57].

Another set of recent experiments [7.58] makes use of transient capacitance measurements. It has been shown that quantum wells behave like deep traps and therefore can be investigated by deep-level transient capacitance spectroscopy. An important result of these experiments is, e.g., the value of the band-edge discontinuity. Finally, we would like to mention that in spite of the exciting new possibilities, there remain many problems from the point of view of device applications. These are mainly connected with the GaAs surface and with deep levels in the heterolayer or at the heterolayer interfaces. This is true for the $Al_xGa_{1-x}As$ system and even more for the strained superlattices.

7.4 Hot-Electron Emission and Capture in Heterojunction Lasers and Field Effect Transistors

The real-space transfer of *hot* electrons over heterobarriers also occurs in more conventional semiconductor devices. Therefore, we will discuss in the following some aspects of this effect in heterojunction lasers, HEMT, and finally in metal-oxide-semiconductor (MOS) devices. We start with heterojunction lasers.

If excess carriers are photogenerated or injected in the confining layers of a double heterostructure within a diffusion length of the narrow-gap active region, the carriers will diffuse to the (quantum) well, be collected, and recombine in the well (GaAs) if it is not too narrow compared with the mean free path for electron-phonon interaction ($L_z \geq 80$ Å). For simple well size, $L_z \leq 80$ Å, a cutoff of the confined-particle recombination is observed. The dynamics of carrier collection in (quantum) wells depends on details of the band structure and the electron-phonon interaction. Monte Carlo simulations of the electrons percolating down in energy have been performed [7.59] which give a detailed account of the history of the electrons in the Γ, X, and L valley of GaAs. They show that cutoff features for confined-particle recombination depend sensitively on the valley type in which the electron resides when approaching and traversing the quantum well. The transient energy distribution also reflects a characteristic structure with respect to the phonon energy, which can be related to a similar structure in the laser spectrum of the quantum-well heterojunction. Typical results for the

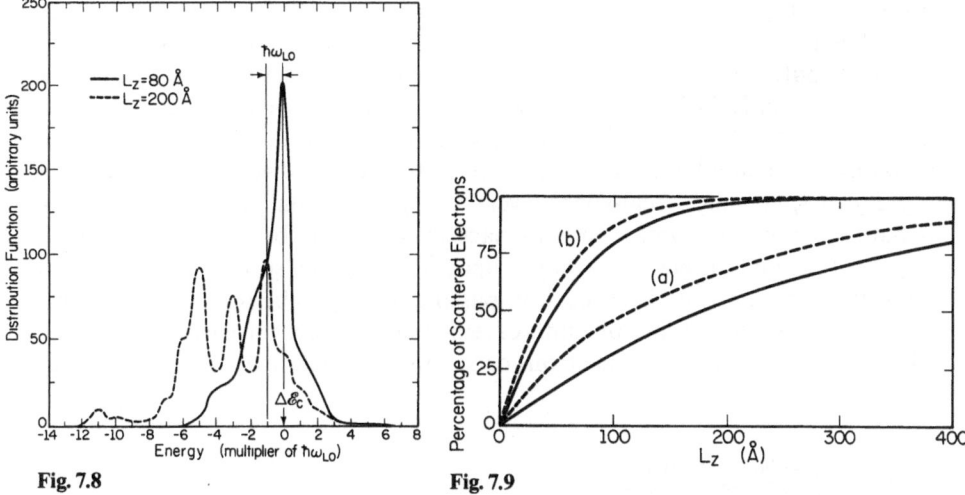

Fig. 7.8. Energy distribution of quantum-well electrons versus energy (in multiples of $\hbar\omega_{LO}$) for well sizes $L_z = 80$ and 200 Å. $\Delta\mathscr{E}_c \sim 0.53$ eV is the injection energy. ($T_0 = 300$ K)

Fig. 7.9. Percentage of electrons scattered into a single GaAs quantum well from two $Al_xGa_{1-x}As$ confining layers by phonon emission. (——) are calculated for one transit, and (– – –) include multiple reflections by the well boundaries. Curves (a) are calculated for electrons at Γ and curves (b) for electrons at X in both the $Al_xGa_{1-x}As$ confining layers and the GaAs quantum well

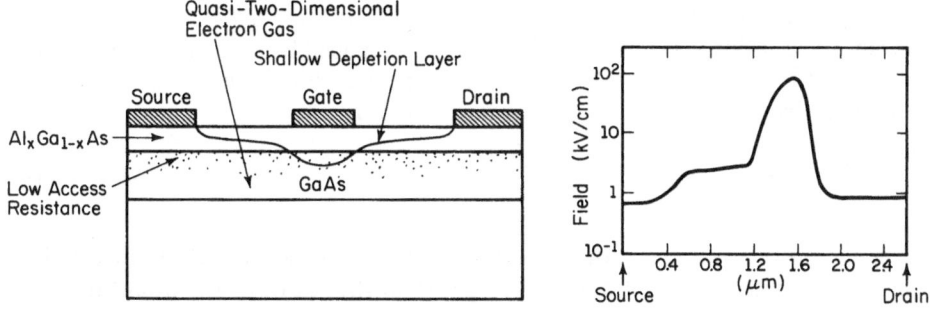

Fig. 7.10. Typical geometry and electric field distribution for a high electron mobility transistor (HEMT)

energy distribution of injected electrons and their collision-free path are given in Figs. 7.8, 9. Notice that with increasing well width, multiple phonon scatterings occur which shift the peak to lower energies (for details see [7.59]). Note also that the electron-electron interaction has been neglected in these computations.

In the above example, the electrons diffuse perpendicular to the layers. In HEMT devices the electrons are accelerated by electric fields parallel to the layers. For high drain voltages, however, their average energy becomes so high that real-space transfer occurs. A typical HEMT structure and a typical electric field distribution for 3 V drain voltage are shown in Fig. 7.10. At the source side the field is relatively low and probably not sufficient for real-space transfer. Below the gate, the electric field reaches values close to 10^5 V/cm. This field is certainly high enough for real-space transfer to occur. However, the field varies rapidly in space and "real-space overshoot phenomena" may be important (Chap. 8). In other words, the real-space transfer may be incomplete since it requires time and the electrons may propagate through a substantial part of the high-field region within this time, thus passing the region before transfer occurs.

The characteristic time constants of real-space transfer can easily be deduced. Thermionic emission of hot electrons occurs in a characteristic "switching time" t_s given by

$$t_s \approx \frac{eN_c L_z m_0}{A^* T_e^2 m} \exp \frac{\Delta \mathscr{E}_c}{K_B T_e}. \tag{7.17}$$

Here N_c is the effective density of states in the GaAs.

Assuming the applied electric field is such that $K_B T_e = \Delta \mathscr{E}_c = 0.20$ eV in a GaAs layer of width $L_z = 500$ Å, one obtains $t_s \approx 2.5 \times 10^{-14}$ s. However, the time required for heating of the electrons is about $1–5 \times 10^{-12}$ s, which means that for high electric fields, the energy relaxation time limits the process. For lower electric fields, the switching time may be limited by (7.17).

The carriers can propagate back into the well due to their thermal motion and energy. If the potential barrier in the $Al_x Ga_{1-x}As$ due to the field of the donors is large, an equation similar to (7.17) (with $\Delta \mathscr{E}_c$ replaced by the barrier height) will describe the backtransfer. Otherwise, the time constant for backtransfer can be obtained from the equation appropriate for diffusion

$$t_s^{back} \approx 4 L_b^2/(\pi^2 D), \tag{7.18}$$

where L_b is the barrier width and D the diffusion coefficient. When the electron mean free path due to phonons is larger than the width of the quantum well, correction factors have to be introduced.

In cases of complete transfer, the average electron-drift velocity will decrease since the saturation velocity in $Al_x Ga_{1-x}As$ is smaller than in GaAs. In this operation range, the speed advantages of HEMT devices are therefore restricted to the source region and are due to the low access resistance. In any case, the real-space-transfer effect has negative consequences for HEMT devices. The above considerations bring us to hot-electron transfer in MOS transistors, where it causes threshold shifts and device instabilities. The conduction-band-edge discontinuity between silicon and silicon dioxide is about 3.1 eV. Electrons, therefore, barely overcome this barrier and larger numbers of electrons are emitted in time spans of months and years only. The high energies involved in the emission process, however, also introduce band-structure effects and effects of collision broadening, which usually are not important.

A detailed investigation of electron emission from silicon into silicon dioxide was performed by *Tang* and *Hess* [7.60]. This investigation was based on a Monte Carlo simulation of electrons propagating in high electric fields from silicon to the SiO$_2$ interface, subject to the conditions of the *Ning* et al. experiments [7.61, 62]. A two-conduction-band model with absolute minima near X points and upper minima at L points, as calculated from the pseudopotential method, are accounted so that the Monte Carlo method includes all the known phonon scattering mechanisms in silicon plus X-L scattering. The inclusion of the second conduction band is of utmost importance since there are almost no states in the first band at energies of 3.1 eV. It should also be noticed that even impact ionization plays an important role in the simulations.

In Fig. 7.11, we show the trajectory of an electron moving from the silicon substrate toward the interface in a MOS transistor with large reverse bias (the smooth curve is the conduction-band edge). The electron sometimes luckily escapes the scattering events and gains significant kinetic energy. It loses energy

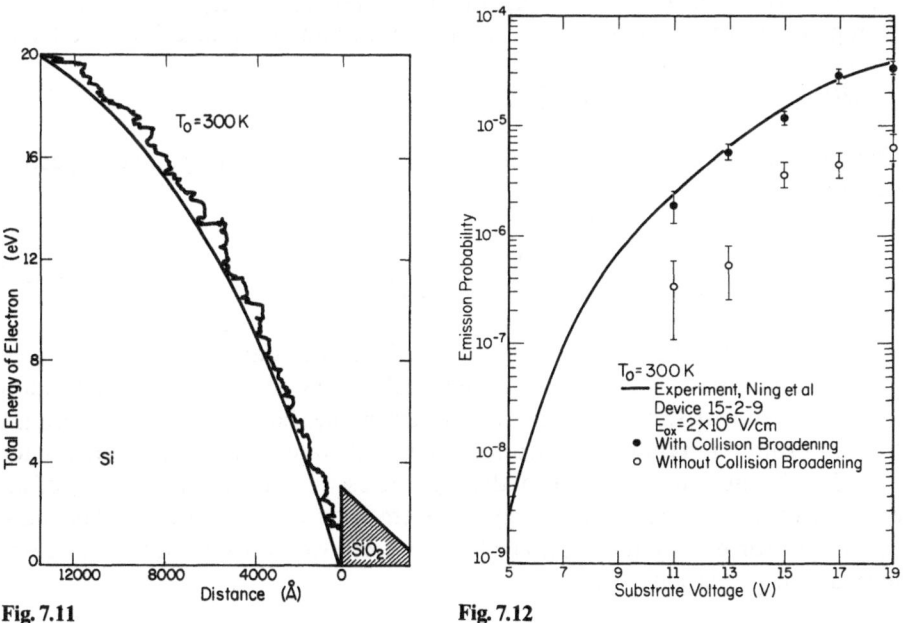

Fig. 7.11   **Fig. 7.12**

Fig. 7.11. Energy histogram of an electron propagating toward the SiO$_2$ in a MOS transistor with large back bias

Fig. 7.12. The calculated emission probability as a function of substrate voltage. The system is modeled for the experimental conditions of *Ning* [7.61]

by emissions of phonons, by traveling against the field, and by impact ionization. Clearly, this electron does not enter into the silicon dioxide layer.

The hot-electron emission probability calculated in this way (for device 15-2-9 in *Ning*'s experiments [7.61, 62]) is plotted in Fig. 7.12 as a function of the substrate voltage. The solid line gives the experimental results of *Ning* et al. [7.61]. The open circles, nearly an order of magnitude below the experimental results, represent the emission probability as obtained from the Monte Carlo simulation. Collision broadening was not included in obtaining these results. If the collision broadening effect due to the electron-phonon interaction is taken into account, the full circles shown in Fig. 7.12 are obtained. These results agree extremely well with the experiments for substrate voltage above 11 V. The computation is too costly for substrate voltages below 11 V because emission events are very rare.

The effect of collision broadening in increasing the emission probability can easily be understood. For a noninteracting system, the wave function of a particle in state k evolves as

$$\psi(t) \sim \exp[-i\mathscr{E}(k)t/\hbar], \tag{7.19}$$

where $\mathscr{E}(k)$ is the energy of the free-particle state. The Fourier transform of (7.19) is the spectral density function $A(\omega, \mathscr{E}) \sim \delta(\omega - \mathscr{E}(k))/\hbar$. For an interacting system (in our case, electron phonon and impact ionization), we have

$$\psi(t) \sim \exp\{-i[\mathscr{E}(k) + \Sigma(k)]t/\hbar\}. \tag{7.20}$$

The self-energy $\Sigma(k)$ is a complex number $[\Sigma(k) = \Delta(k) + i\Gamma(k)]$ and the spectral density has a Lorentzian shape

$$A(\omega, \mathscr{E}) \sim \frac{1}{\pi} \frac{\Gamma(k)}{\{\hbar\omega - [\mathscr{E}(k) + \Delta(k)]\}^2 + [\Gamma(k)]^2}, \tag{7.21}$$

which determines the broadening (line width) of the state.

The quantities Δ and Γ have been calculated for the electron-phonon interaction and are shown in Fig. 7.13. The exact line shape also needs to account for pair production (impact ionization) and final-state broadening, which is not included. Because of the line broadening, the escape probability $P_{esc}(\mathscr{E})$ of an electron over a barrier with energy $\mathscr{E}_0$ is smeared out. If we have classically $P_{esc}^c(\mathscr{E}) = 1$ for $\mathscr{E} > \mathscr{E}_0$ and $P_{esc}^c(\mathscr{E}) = 0$ for $\mathscr{E} < \mathscr{E}_0$, we now obtain the quantum mechanical probability P_{esc}^Q:

$$P_{esc}^Q(\mathscr{E}) = \hbar \int_{\mathscr{E}_0}^{+\infty} A(\omega, \mathscr{E}) d\omega. \tag{7.22}$$

This equation has an important implication: since $A(\omega, \mathscr{E})$ is Lorentzian, P_{esc}^Q does not decrease exponentially when the energies (voltages) are scaled down,

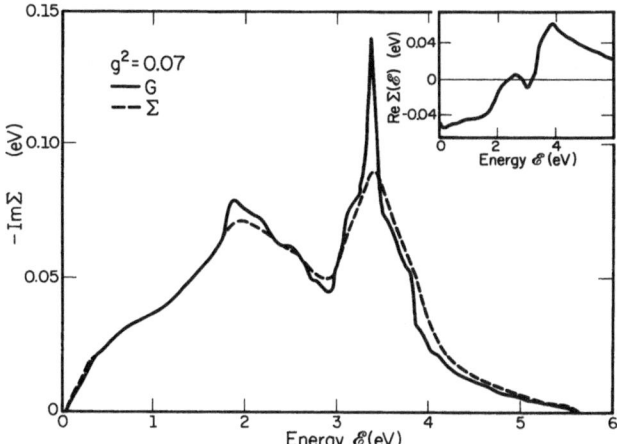

Fig. 7.13. The imaginary part of the self-energy (proportional to the scattering rate) as a function of energy. (——) is the first-order self-energy (the Golden Rule in the Born approximation) and (– – –) is the full-order self-energy. The inset shows the real part of the self-energy

i.e., to avoid the hot-electron emission the voltages need to be scaled down *much below* the barrier height (~ 3.1 V).

Emission of *hot* electrons over potential barriers is also important in many other devices, e.g., buried-channel structures [7.63], field effect transistors (hot-electron substrate current), and novel devices such as planar doped barriers [7.64–66]. However, the basic effects are covered by the above treatment.

7.5 Lateral Superlattices for Millimeter-Wave and Microelectronic Applications

Surfacing repeatedly, a key question concerns the role of electrical contacts in submicrometer device technology, not only from a fabrication and processing point of view, but also from the standpoint of charge injection and extraction to and from submicrometer devices. A consensus indicates that the "contact" problems such as arise from ill-defined boundaries, high resistance scattering centers, and the like are difficult, if not impossible, to overcome in the scaling down of *conventional* devices and, as such, might well be a major barrier to scaling down submicron devices to below one-tenth micrometer feature size. Therefore, it is necessary that new directions be established for device technology to penetrate the ultrasubmicrometer (less than one-tenth micrometer) regime. A prime methodology for establishing practical first-generation ultrasubmicrometer devices is available through MBE and MOCVD techniques: these methods utilize modulation doping to produce well-controlled charge reservoirs adjacent to the electrically active regions, thereby circumventing the problems of

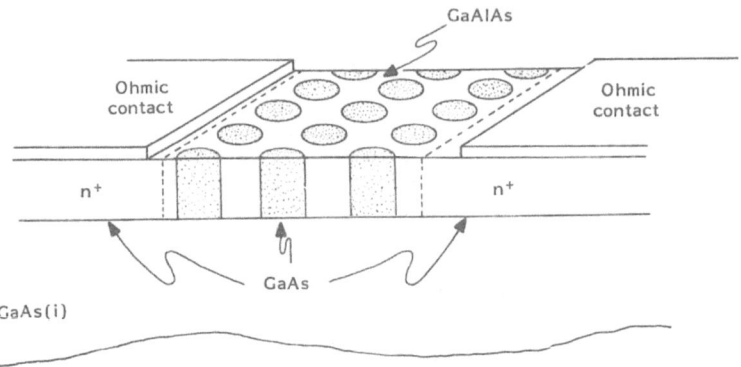

Fig. 7.14. Supermatrix quantum-well FET radiator

conventional contacting. To date, the primary emphasis has been on elucidating the transport and spectroscopic properties of multilayered epitaxial films (heterostructure superlattices). In this section, we look beyond the present state of the art and initiate innovative concepts concerning the role of multidimensional superlattices and electron confinement in future generation electronic devices. These concepts suggest the possibility of multidimensional superlattices and heterostructures with extrapolation to three-dimensional microelectronic circuits.

We have conceived and studied the properties of a variety of two- and three-dimensional quantum-well superlattice structures with novel microelectronic and optical properties. Most of the configurations studied are lateral super-lattices in which the superstructure lies in the surface of a heterostructure layer [7.67]. A typical prototype configuration consists of a thin GaAlAs epitaxial layer with ultrasmall (less than 250 Å) cylindrically shaped periodic regions of GaAs embedded (Fig. 7.14). Such a configuration can be fabricated by high-resolution lithographic patterning, selective anisotropic etching, and selective area epitaxial growth. Other configurations consist of a lateral silicon-FET superlattice (Fig. 7.15), a Ge-GaAs three-dimensional amorphous superlattice (Fig. 7.16), and a solid-state radiator (Fig. 7.17). We predict that these structures will exhibit a host of relevant transport properties [7.68], electron confinement effects, and dielectric charge instabilities such as synergetic switching under the condition of population inversion. We also predict that these structures will offer the promise of exciting new millimeter-wave sources, as well as ultrahigh-density memories based on charge storage and optical addressing.

From the fabrication standpoint, MBE, MOCVD, and grapho-epitaxy provide a method for selective epitaxial growth of ultrasubmicron two- and three-dimensional structures. These methods generally make use of compound semiconductors such as GaAs and GaAlAs because of their natural epitaxial growth characteristics, but Si/SiO_2 metal composites (Fig. 7.15) are material possibilities for some applications as well. In addition, scanning transmission

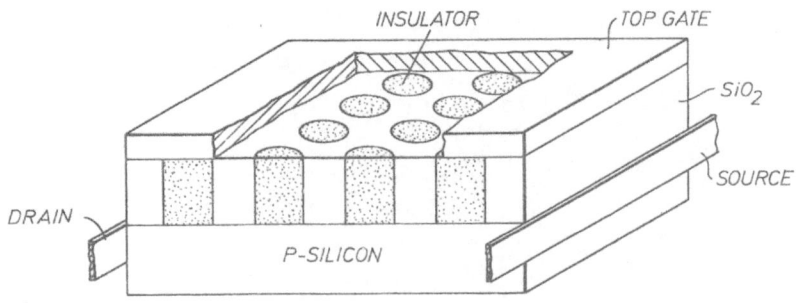

Fig. 7.15. Lateral silicon-FET superlattice

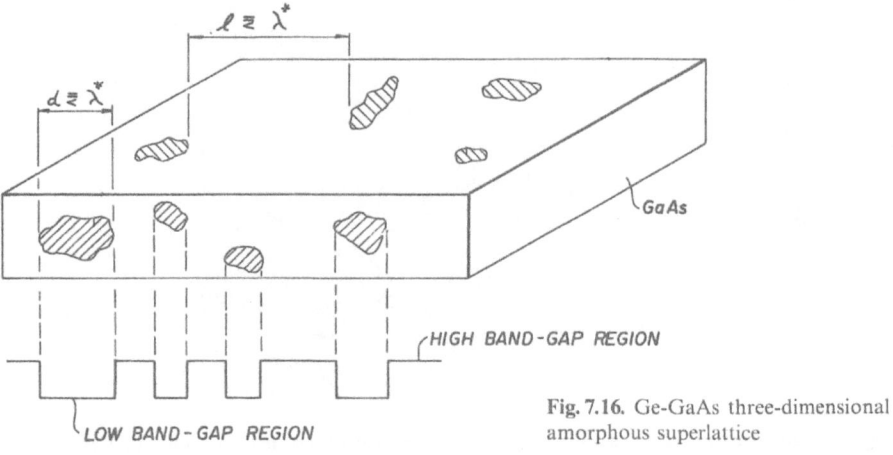

Fig. 7.16. Ge-GaAs three-dimensional amorphous superlattice

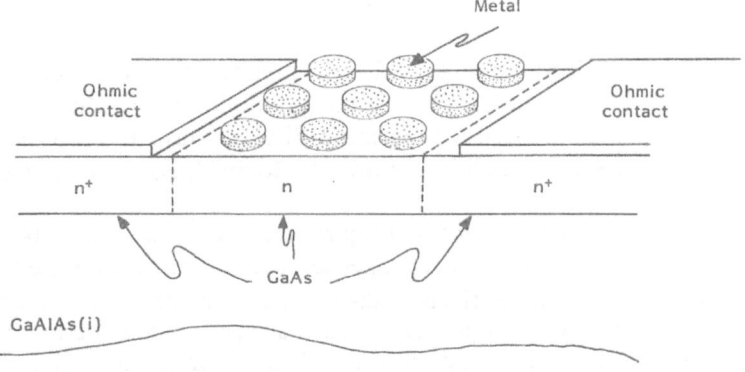

Fig. 7.17. Solid-state radiator

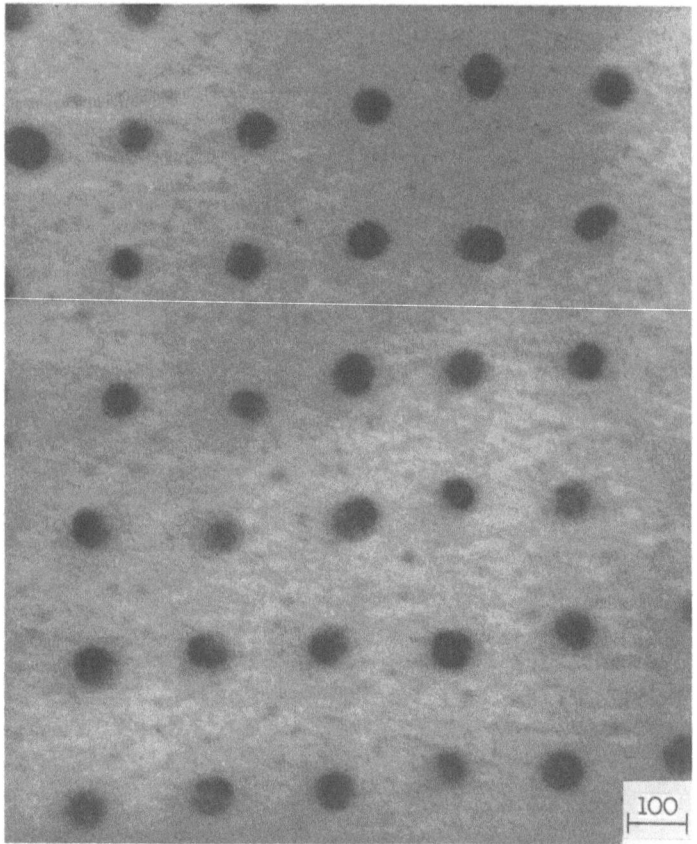

Fig. 7.18. Dot array of Ag disks on a carbon film. The array covers an area 1 μm² and has a "lattice constant" of ~250 Å. The bar in the photograph represents 100 Å

electron microscopy (STEM) optics and ion milling/anisotropic etching provide methods for writing and developing ultrasmall dot array. Dot arrays having 100 Å feature size and 200 Å separation distance have recently been fabricated (Fig. 7.18) at AT & T Bell Laboratories (Holmdel, NJ) by *Craighead* and *Mankiewich* [7.69].

The two-dimensional superlattice configuration shown in Fig. 7.14 is adapted to manifest significant transport properties; in addition, it lends itself to a possible solid-state source of relatively high-power millimeter-wave coherent emission. With regard to transport properties, such a superlattice structure has two-dimensional superlattice miniband splittings which are predicted to give rise to strong "Bloch oscillation" dissipation [7.70] above a certain threshold electric field, thereby leading to a strong negative differential resistance in such structures. This opens up the possibility of high-speed (10^7 cm/s) negative differential resistance devices and millimeter-wave oscillators from planar devices. In addition, such structures can be tailored to switch from a nonconducting state to a conducting state due to the resonant transfer of charge from one well to an adjacent well.

With respect to optical properties of a device such as shown in Fig. 7.14, depending upon the configuration and feature size of the array, energy-level spacings can be tailored to be in the spectroscopic range covering infrared (100 meV) to millimeter-wave (approximately 1 meV) frequencies. Although such frequencies are also realizable in conventional one-dimensional superlattice structures, transitions between the energy levels of one-dimensional superlattices are weak because the superlattice energy spacing is smeared due to the continuum of states available in the directions perpendicular to the one-dimensionality. In the structural configuration disclosed in Fig. 7.14, however, electrons are truly confined in all dimensions, thus giving rise to strong emission between superlattice bands. Radiation can be emitted from such a structure by either exposing the array to radiation with frequency above the GaAs band gap, or by injecting charge into the upper state of the quantum well through the contacts placed on the sides of the active region of the structure (lateral quantum-well laser). In addition, population inversion and resonant charge transfer are accomplished by applying a uniform electric field across the device and adjusting its value, so that the difference in potential between two closely spaced adjacent wells corresponds to the occupied lower level of one well and the upper unoccupied level of the adjacent well. In this way, a single injected electron, in traversing the array from one side to another, can emit radiation many times by tunneling, at constant energy, between adjacent wells. In each tunneling step, electrons fall to a new energy level, thereby emitting radiation.

In Fig. 7.17 we show a conceptually workable "solid-state radiator" that comprises a high-speed GaAs electron confinement channel in close proximity to a conductivity grating. In such a configuration, electrons, when interacting with the grating, will radiate "Smith-Purcell" [7.71] radiation in a frequency range determined by the speed of the electrons and the periodicity of grating. Although the structure shown in Fig. 7.17 can be tailored to radiate in the millimeter range, the power output may be quite low due to the weak coupling of the channel current to the grating. Nonetheless, other configurations, which make use of quantum-well heterojunction and real-space transfer effects, have been conceived that offer strong current-grating coupling.

In summary, a variety of practical two- and three-dimensional supermatrix structures with novel millimeter-wave and microelectronic applications has been described. The material technology required to fabricate these novel structures is within the state of the art yet meaningful implementation will require innovative materials and fabrication effort.

7.6 Conclusion

We have described in this chapter some of the transport effects which have been observed in recent years in semiconductor heterolayers. Special emphasis has been placed on real-space-transfer effects, i.e., the separation (and reunion) of electrons and donors at heterojunctions with and without the presence of external fields.

We have shown that these effects present interesting opportunities for new device applications. The basis of the new phenomena is the control of boundary conditions on the scale of the de Broglie wavelength of the electrons. This finely tuned control is possible because of the great advances of MOCVD and MBE crystal growth capabilities.

Of course, the effects described in this chapter represent only a beginning. The controllability of the boundaries should enable us, for the first time in history, to design a certain function from basic physical considerations and to engineer truly "functional" devices.

Although a variety of effects is already offered by superlattice structures, the control of the other two dimensions, as described in Sect. 7.5, will open yet another door for new possibilities and may ultimately contribute to a renewal of some areas of electronics as it proceeds from physics to function.

Acknowledgements. The work was supported by the Office of Naval Research and the Army Research Office.

References

7.1   R.Dingle: "Confined Carrier Quantum States in Ultrathin Semiconductor Heterostructures", in *Festkörperprobleme 15* (Vieweg, Braunschweig 1975) pp. 21–47
      K.Hess, N.Holonyak: Comments Solid State Phys. **10**, 67–84 (1981)
7.2   J.R.Arthur: J. Appl. Phys. **39**, 4032 (1968);
      A.Y.Cho: J. Vac. Sci. Technol. **8**, 531 (1971)
      A.Y.Cho: Appl. Phys. Lett. **19**, 467 (1971)
7.3   H.M.Manasevit: J. Electrochem. Soc. **118**, 647 (1971);
      R.D.Dupuis, P.D.Dapkus: Proc. Symp. on GaAs and Related Compounds, St. Louis, (1978), ed. by C.M.Wolfe: Conf. Ser. 45 (Institute of Physics, London 1979), pp. 1–9
7.4   L.Esaki, R.Tsu: IBM J. of Res. **14**, 61 (1970)
7.5   L.Esaki, L.L.Chang: Phys. Rev. Lett. **33**, 495 (1974)
7.6   More recently, interest has been focused on Type II systems such as InAs-GaSb; *see*, e.g., G.A.Sai-Halasz, L.L.Chang, J.M.Weter, C.-A.Chang, L.Esaki: Solid State Commun. **27**, 935 (1978)
7.7   R.Dingle, H.L.Störmer, A.C.Gossard, W.Wiegmann: Appl. Phys. Lett. **33**, 665 (1978)
7.8   N.Holonyak, Jr., R.M.Kolbas, R.D.Dupuis, P.D.Dapkus: Appl. Phys. Lett. **33**, 73 (1978)
7.9   W.T.Tsang: Appl. Phys. Lett. **39**, 786 (1981)
7.10 R.J.Malik, J.R.AuCoin, R.L.Ross, K.Board, C.E.Wood, L.F.Eastman: Electron. Lett. **16**, 836 (1980)
7.11 K.Hess: J. Physique **C7**, C7-3 (1981)

7.12 Work on the GaAsP-GaAs and InGaAs-GaAs system is just starting on a larger scale
7.13 W.A.Harrison: J. Vac. Sci. Tech. **14**, 1016 (1977)
7.14 W.R.Frensley, H.Kroemer: Phys. Rev. **B16**, 2642 (1977); J. Vac. Scie. Technol. **13**, 810 (1976)
7.15 G.Margaritondo, A.D.Katani, N.G.Stoffel, R.R.Daniels, Te-Xiu Zhao: Solid State Commun. **43**, 163 (1982)
7.16 N.Holonyak, Jr., W.D.Laidig, B.A.Vojak, K.Hess, J.J.Coleman, P.D.Dapkus, J.Bardeen: Phys. Rev. Lett. **45**, 1703 (1980)
7.17 P.M.Petroff, A.Y.Cho, F.K.Reinhart, A.C.Gossard, W.Wiegmann: Phys. Rev. Lett. **48**, 170 (1981)
7.18 W.D.Laidig, N.Holonyak, Jr., M.D.Camras, K.Hess, J.J.Coleman, P.D.Dapkus, J.Bardeen: Appl. Phys. Lett. **38**, 776 (1981)
7.19 P.Voisin, G.Bastard, C.E.T.Concalves da Silva, M.Voos, L.-L.Chang, L.Esaki: Solid State Commun. **39**, 79 (1981)
7.20 K.Hess: Appl. Phys. Lett. **35**, 484 (1979)
7.21 P.J.Price: Ann. of Phys. New York **133**, 217 (1981)
7.22 S.Mori, T.Ando: J. Phys. Soc. Jpn. **48**, 865 (1980)
7.23 T.Mimura, S.Hiyamizu, T.Fujii, K.Nanbu: Jpn. J. Appl. Phys. **19**, L225 (1980)
7.24 D.R.Scifres, R.D.Burnham, W.Streifer: Appl. Phys. Lett. **41**, 1030 (1982)
7.25 H.Kroemer, K.J.Polasko, S.C.Wight: Appl. Phys. Lett. **36**, 763 (1980)
7.26 H.C.Casey, Jr., M.B.Panish: *Heterostructure Lasers*, Part B: Materials and Operating Characteristics (Academic, New York 1978)
7.27 P.T.Landsberg: *Solid State Theory* (Oxford University Press 1969)
7.28 W.A.Harrison: *Electronic Structure and the Properties of Solids* (Freeman, San Francisco 1980) p. 68
7.29 T.Ando, S.Mori: Surf. Sci. **113**, 124 (1982)
7.30 J.N.Schulman, T.C.McGill: Phys. Rev. **B19**, 6341 (1979)
7.31 J.N.Schulman, Y.C.Chang: Phys. Rev. **B24**, 4445 (1981)
7.32 K.K.Mon: Solid State Commun. **41**, 699 (1982)
7.33 K.K.Mon: private communication (October 1981)
7.34 N.Holonyak, Jr., K.Hess: *Quantum Well Heterostructure Lasers* (Academic, New York 1983) and references therein
7.35 K.Hess: Aspects of High-Field Transport in Semiconductor Heterolayers and Semiconductor Devices, in *Advances in Electronics and Electron Physics* **59**, (Academic, New York 1982) pp. 239–291
7.36 D.K.Ferry, K.Hess, P.Vogl: Physics and Modeling of Submicron Insulated-Gate Field-Effect Transistors. II., in *VLSI Electronics* ed. by N. Einspruch (Academic, New York 1981) p. 68
7.37 J.Lee, H.N.Spector, V.K.Arora: Appl. Phys. Lett. **42**, 363 (1983)
7.38 K.Hess: See [Ref. 7.35; p. 258]
7.39 T.J.Drummond, H.Morkoc, K.Hess, A.Y.Cho: J. Appl. Phys. **52**, 5231 (1981)
7.40 K.Hess, P.Vogl: Solid State Commun. **30**, 807 (1979)
7.41 H.L.Störmer, A.C.Gossard, W.Wiegmann: Solid State Commun. **41**, 707 (1982)
7.42 K. von Klitzing, G.Dorda, M.Pepper: Phys. Rev. Lett. **45**, 494 (1980)
7.43 D.C.Tsui, H.L.Störmer, A.C.Gossard: Phys. Rev. **B25**, 1405 (1982)
7.44 K.Hess, N.Holonyak, Jr., W.D.Laidig, B.A.Vojak, J.J.Coleman, P.D.Dapkus: Solid State Commun. **34**, 749 (1980)
7.45 K.Hess, Solid State Commun. **25**, 191 (1978)
7.46 T.J.Drummond, M.Keever, W.Kopp, H.Morkoc, K.Hess, B.G.Streetman, A.Y.Cho: Electron. Lett. **17**, 545 (1981); M.Keever, W.Kopp, T.J.Drummond, H.Morkoc, K.Hess: Jpn. J. Appl. Phys. **12**, 1489 (1982)
7.47 J.Shah, A.Pinczuk, H.L.Störmer, A.C.Gossard, W.Wiegmann: Appl. Phys. Lett. **42**, 55 (1983)
7.48 K.Hess, C.T.Sah: Phys. Rev. **B10**, 3375 (1974)
7.49 J.P.Leburton, K.Hess: Phys. Lett. **99**A, 335 (1983)
7.50 T.Mimura, S.Hiyamizu, T.Fujii, K.Nanbu: Jpn. J. Appl. Phys. **19**, L225 (1980)
7.51 S.M.Sze: *Physics of Semiconductor Devices* (Wiley, New York 1981)
7.52 H.Shichijo, K.Hess, B.G.Streetman: Solid State Electron. **23**, 817 (1980)

7.53 K.Hess: Physics of Nonlinear Electron Transport, Proc. NATO Adv. Study Inst., Sogesta Conference Centre, Italy 1980, ed. by D.K.Ferry, J.R.Barker, C.Jacoboni (Plenum, New York 1980) p. 32
7.54 T.H.Glisson, J.R.Hauser, M.A.Littlejohn, K.Hess, B.G.Streetman, H.Shichijo: J. Appl. Phys. **51**, 5445 (1980)
7.55 K.Hess: J. Physique **C7**, Suppl. 10, C7-3 (1981)
7.56 M.Keever, K.Hess, M.Ludowise: IEEE **EDL3**, 297 (1982)
7.57 M.Ludowise, W.T.Dietze, C.R.Lewis, P.Gavrilovic, T.C.Hsieh, K.Hess: J. Appl. Phys. **54**, 6771 (1983)
7.58 P.A.Martin, K.Meehan, P.Gavrilovic, K.Hess, N.Holonyak, Jr., J.J.Coleman: J. Appl. Phys. **54**, 4689 (1983)
7.59 J.Y.Tang, K.Hess, N.Holonyak, Jr., J.J.Coleman, P.D.Dapkus: J. Appl. Phys. **53**, 6043 (1982)
7.60 J.Y.Tang, K.Hess: IEEE Trans. Elec. Dev. (to be published)
7.61 T.H.Ning, C.M.Osburn, H.N.Yu: J. Appl. Phys. **48**, 286 (1976)
7.62 T.H.Ning: Solid State Electron. **21**, 273 (1978)
7.63 K.Hess, H.Shichijo: IEEE Trans. **ED-27**, 503 (1980)
7.64 R.J.Malik, T.R.AuCoin, R.L.Ross, K.Board, C.E.C.Wood, L.F.Eastman: Electron. Lett. **16**, 836 (1980)
7.65 J.M.Shannon: Appl. Phys. Lett. **35**, 63 (1979)
7.66 T.Wang, J.P.Leburton, K.Hess: Proc. 3rd Workshop on the Physics of Submicron Structures, Urbana, Illinois (1982) (to be published)
7.67 G.J.Iafrate, D.K.Ferry, R.K.Reich: Surf. Sci. **113**, 485 (1982)
7.68 R.K.Reich, R.O.Grondin, D.K.Ferry, G.J.Iafrate: Phys. Lett. **91A**, 28 (1982)
7.69 H.G.Craighead, P.M.Mankiewich: J. Appl. Phys. **53**, 7186 (1982)
7.70 R.K.Reich, R.O.Grondin, D.K.Ferry, G.J.Iafrate: Electron. Dev. Lett. **3**, 381 (1982)
7.71 S.J.Smith, E.N.Purcell: Phys. Rev. **92**, 1069 (1953)

8. Non-Steady-State Carrier Transport in Semiconductors in Perspective with Submicrometer Devices

Eugene Constant

With 22 Figures

Progress in the microelectronics industry is strongly coupled with the ability to make an ever-increasing number of smaller devices on a single chip. The advent of high-resolution electron and x-ray lithographic techniques is leading toward an era in which individual features' sizes might well be fabricated on the scale of 0.01–0.1 µm. It will then become feasible to develop very small device structures characterized by higher and higher operating frequencies or shorter and shorter switching delays. In this type of device, carriers are often submitted to electric fields characterized by fast time variations and/or by strong spatial nonuniformities. Due to these conditions, a "non-steady-state transport" will occur and the characteristics of such a transport can be very different from those obtained in the usual steady transport. Consequently, it becomes obvious that we must now ask whether classical device modeling may be extrapolated down to the very small space and time scales usually met in submicrometer devices. New phenomena may occur and it is the purpose of this chapter to study the specific features of electron transport in these conditions.

First, the basic physics of carrier transport in semiconductors is briefly recalled. At the smallest size (feature sizes down to 0.01 µm), one does not know the extent to which the conventional effective-mass concept and band theory can be scaled down. In addition, the collisions acting on the carriers cannot be assumed to occur instantaneously either in space or in time [8.1]. At the most usual size (in the range 0.1–0.5 µm), which will be the only size studied in detail, the transport physics may still be based on the Boltzmann transport equation, but it will be shown that, even for this case, the features characterizing carrier transport can be very different from those related to steady-state and bulk transport. These new features occur either when the semiconductor samples are submitted to very fast time variation of the electric field E or when the electric field is characterized by small spatial scale.

Time-dependent phenomena will be studied in Sects. 8.2–4. First, the case of high-frequency sinusoïdal electric field superimposed to an applied steady field will be investigated for materials such as Si and III-V compounds. Transient phenomena when a time-pulse or a time-step configuration of E are applied to the whole semiconductor will then be analyzed. Drift and diffusion mechanisms will be discussed with particular attention to conditions in which the specific features of carrier transport could be used strongly to increase the average velocity of the carriers over the distance of the sample.

Space-dependent phenomena will then be described in Sect. 8.5. In this case all the characteristics of carrier transport cannot be deduced from the results obtained by applying a time configuration of the electric field to a semiconductor, since additional phenomena caused by spatial nonuniformity occur simultaneously. To compare clearly the electron dynamics features of a uniform bulk semiconductor submitted to a transient electric field with those of a semiconductor characterized by small space nonuniformity, we will first neglect the effect of space charge due to electrons (via Poisson's equation) on the value of the applied electric field. Then, by taking into account the effect of space charge, a few examples which occur in submicron devices will be illustrated.

Finally, the problems concerning the experimental investigation of these transient hot-electron effects will be briefly discussed.

8.1 Carrier Transport and Physical Scale

8.1.1 Theoretical Background

To study the new phenomena which may occur in transient conditions, it is necessary to recall briefly the basic physics of carrier transport in a semiconductor.

Transport properties are usually described by the Boltzmann transport equation (BTE),

$$\frac{\partial f}{\partial t} + v\,\frac{\partial f}{\partial r} + eE\,\frac{\partial f}{\partial k} = \int_{k'} dk'\,[f(k')\,W(k',k) - f(k)\,W(k,k')], \tag{8.1}$$

which with the same meaning of symbols has been discussed in Chap. 2. It allows us to obtain the carrier distribution $f(r, k, t)$ in k space and geometrical r space. In the above equation the carrier velocity v can be obtained from the band structure $\mathscr{E}(k)$ characterizing the semiconductor material, through the classical equation

$$v = \hbar^{-1}\,\frac{\partial \mathscr{E}}{\partial k}. \tag{8.2}$$

Consequently, for a given electric field $E(r, t)$ the parameters which enter the BTE will refer to the band structure and scattering mechanisms [represented by the functions $W(k, k')$]. If all these quantities are known, the BTE can be solved to obtain the distribution function from which all ensemble-average quantities, such as the average velocity $\langle v \rangle$ and the mean energy $\langle \mathscr{E} \rangle$, can be deduced. This is achieved practically using either quite direct methods as the iterative techniques [8.2, 3] or more indirect methods such as the Monte Carlo simulation [8.4–7]. We shall mention that a wider interest for this last method has been recognized in recent years for two main reasons: it has proven to be a powerful tool for the study of high-field transport, and it has provided a clear physical description of the electron dynamics in a semiconductor.

In this method, thoroughly described in Chap. 2, the stochastic motion of the carrier representative points in k space is studied numerically in order to obtain at each given time step Δt complete information on the state of the carrier (e.g., its energy and its velocity). The kinetic coefficients are thus calculated taking into account at each time step the determinist effect of the electric field, which in absence of collisions will modify k according to the equation of motion:

$$\Delta k = \left(\frac{E}{\hbar}\right)\Delta t, \tag{8.3}$$

and the stochastic effect of the scattering mechanisms, which will change the position of the representative points with a probability

$$\lambda(k)\Delta t = \Delta t \int_{k'} W(k, k')\, dk', \tag{8.4}$$

$\lambda(k)$ being the scattering rate. In practice, suitable random trials will be used to find out whether an interaction has occurred during Δt, and, if so, to determine which specific scattering mechanism has occurred.

Since, for each Δt, the energy and velocity of the carrier are known, the average quantities $\langle \mathscr{E} \rangle$ and $\langle v \rangle$ can be obtained in general from a simulation of an ensemble of carriers.

Consequently, the Monte Carlo method enables us to solve the BTE whatever the scattering mechanisms and band structure may be, and to obtain useful results on carrier transport in semiconductors in almost all cases. However, significant results will be obtained only in cases where the BTE can be applied, and so it is necessary to point out the three main assumptions which are needed for the BTE to be valid:

1) effective mass and band model hold,
2) collisions are instantaneous in both space and time,
3) scatterings are independent of the electric field.

8.1.2 Electron Transport and Physical Scale

Based on this theoretical background, the influences of the size and of the operating conditions of the device on the characteristics of carrier transport will now be studied in Sects. 8.1.3–5.

8.1.3 Large Devices, Low-Frequency Operation

This is the classical case met in most devices which are characterized by active lengths generally much larger than a few tenths of a micron and operate at frequencies lower than 30 GHz. For this type of device, no major theoretical problems occur: band-structure theory and the BTE are valid. In addition, the time and space variation of the electric field during the carrier mean free time and along the carrier mean free path can be neglected. A steady-state carrier transport is then achieved (even in nonsteady regime). Thus when the BTE is used, two main simplifications are usually introduced.

i) One generally assumes that in the device the space and time variations of the electric field and the distribution function are so weak that at each point of the semiconductor a steady state is attained. This results because of a balance between the effects produced by the perturbation due to the electric field and the scattering mechanisms (practically, this assumption means that the term $\partial f/\partial t$ in the BTE can be neglected).

ii) The term $v(\partial f/\partial r)$ in the BTE is taken into account only by introducing a diffusion current described by a diffusion coefficient D.

Using these two simplifications, it can be easily shown that the main characteristics of carrier transport in a large device (i.e., drift velocity v_d, average energy $\langle \mathscr{E} \rangle$, and diffusivity D) depend only on the instantaneous local electric field, regardless of the values of the electric field in the past and all around the point being studied in the semiconductor material. In this case, the current equation can be used assuming that v_d and D only depend on the local electric field.

8.1.4 Large Devices, Very-High-Frequency Operation

This is the case when very-high-frequency or fast transient voltage (i.e., frequency higher than 30 GHz or switching time shorter than a few picoseconds) are applied to bulk semiconductors or to relatively large devices. The variation of the electric field during a mean free time cannot be neglected and consequently in the Boltzmann equation the term $\partial f/\partial t$ has to be taken into account. Therefore, the characteristics of this non-steady-state electron transport can be very different from those obtained in the low-frequency operation.

8.1.5 Submicron Devices

Very small devices are at present generally characterized by active lengths smaller than 0.5 µm. In most cases, the properties of carrier transport in such devices may still be based on the BTE and the band-structure model [with a few special reservations concerning quantum effects on the band structure of inverted layers or of heterostructure used in the high electron mobility transistor (HEMT), (see Chap. 7)]. However, no other simplification is justifiable; the BTE has to be solved exactly taking into account the small spatial scale and often, in operating condition, the short-time scale of both field E and distribution f. As a consequence, the usual concept of mobility (even depending on the electric field) and diffusion should be generalized. Since the variation of the electric field along a mean free path cannot be neglected, new features, even in steady-state regime, will characterize this non-steady-state electron transport. These will be studied in the following.

To obtain a satisfactory understanding of all these phenomena which occur either in large devices or in submicron devices, we shall successively study: (i) the case of a bulk semiconductor submitted to short-time variations of the electric field, and (ii) the case of a submicron device characterized by strong spatial variations of the electric field.

8.2 Time-Dependent Phenomena in Uniform Bulk Semiconductors

8.2.1 How to Study Them

The problem is to study the motion of an ensemble of electrons in a uniform bulk semiconductor subject to various time configurations of a uniform electric field. Thus an ensemble average will enable us to obtain the time dependence of all the dynamic parameters such as the mean carrier energy and the average carrier velocity. It is to be noted that, in the case of a uniform semiconductor and electric field, the carrier concentration remains uniform and diffusion phenomena do not occur; as a consequence, the average velocity $\langle v \rangle$ reduces to the drift velocity v_d:

$$\langle v \rangle = v_d . \tag{8.5}$$

However, it must be pointed out that (8.5) is only valid for a uniform semiconductor.

The Monte Carlo method, described in Chap. 2, is well suited to carry out this study and consequently most of the results described here will be obtained by this method. (The microscopic models to be used for Si, GaAs, and CdTe are given in Tables 8.1, 2, 4.) Nevertheless, simplified methods can also be useful not only to save computation time but also, owing to their flexibility, to help the physical interpretation of the phenomena. An interesting example is the balance-equation method which makes use of the momentum and energy relaxation time approximation to obtain the drift velocity v_d and the average energy $\langle \mathscr{E} \rangle$. Under this approximation the following equations are obtained:

$$\frac{d}{dt}[m^*(\langle \mathscr{E} \rangle)v_d] = eE - \frac{m^*(\langle \mathscr{E} \rangle)v_d}{\tau_m(\langle \mathscr{E} \rangle)} . \tag{8.6}$$

$$\frac{d\langle \mathscr{E} \rangle}{dt} = eEv_d - \frac{\langle \mathscr{E} \rangle - \mathscr{E}_0}{\tau_\mathscr{E}(\langle \mathscr{E} \rangle)} , \tag{8.7}$$

where e is the unit charge; m^* is a mean effective mass at the given average energy; $\mathscr{E}_0$ is the thermal energy $3/2\, K_B T_0$ of the lattice; τ_m and $\tau_\mathscr{E}$ are the momentum and energy relaxation times, respectively.

As discussed by *Nougier* et al. [8.8], the above equations can be deduced from a suitable integration of the BTE when the scattering mechanisms are taken into account through the momentum and energy relaxation times τ_m and $\tau_\mathscr{E}$, assumed to depend on the average energy of the carrier. As a result, it should be noted that the transient motion, which will be determined by using (8.6, 7), will depend only on the instantaneous value of the electric field E and of the average energy $\langle\mathscr{E}\rangle$ of the carriers. This means that in (8.6, 7) it is implicitly assumed that the instantaneous distribution function depends only on the average energy of the carriers and, in a few cases, this assumption can be very rough. Now the main difficulty met in using (8.6, 7) is to know "a priori" the dependence of τ_m, $\tau_\mathscr{E}$ and m^* against $\langle\mathscr{E}\rangle$. This difficulty is practically overcome by using the values of the static characteristics $v_s(E_s)$, $\mathscr{E}_s(E_s)$, and $m^*(E_s)$ relating stationary drift velocity, average energy, and effective mass to the electric field (as obtained, for example, by Monte Carlo calculations in steady-state conditions).

In this case, by equating $d(m^* v_d)/dt$ and $d\langle\mathscr{E}\rangle/dt$ to zero in (8.6, 7), very simple relations for τ_m and $\tau_\mathscr{E}$ are obtained as

$$\tau_m(\mathscr{E}_s) = \frac{m v_s}{e E_s}, \tag{8.8}$$

$$\tau_\mathscr{E}(\mathscr{E}_s) = \frac{\mathscr{E}_s - \mathscr{E}_0}{e E_s v_s}. \tag{8.9}$$

Table 8.1. Model used in Monte Carlo calculations of electrons in Si [8.26, 27]

Density	$\varrho_0 = 2.329\ \text{gr cm}^{-3}$
Longitudinal sound velocity	$u_{lo} = 9.037 \times 10^5\ \text{cm s}^{-1}$
Relative static dielectric constant	$\varepsilon_0 = 11.7$
Longitudinal effective mass	$m_{lo} = 0.9163\, m_0$
Transverse effective mass	$m_{tr} = 0.1905\, m_0$
Nonparabolicity parameter	$\alpha = 0.5\ \text{eV}^{-1}$
Acoustic-deformation-potential parameter	$E_1 = 9\ \text{eV}$

Intervalley Scattering

Type	Equivalent temperature [K]	Coupling constant [eV cm^{-1}]
f	210	1.5×10^7
	500	2.5×10^8
	630	4×10^8
g	140	5×10^7
	210	8×10^7
	700	3×10^8

Table 8.2. Model used in Monte Carlo calculations of electrons in GaAs [8.7]

Density		$\varrho_0 = 5.36$ gr cm^{-3}	
Longitudinal sound velocity		$u_{lo} = 5.24 \times 10^5$ cm s^{-1}	
Relative static dielectric constant		$\varepsilon_0 = 12.90$	
Relative high-frequency dielectric constant		$\varepsilon_\infty = 10.92$	
Longitudinal optical-phonon energy at $q=0$		$\hbar\omega_{op} = 35.36$ meV	
Piezoelectric constant		$p = 0.16$ C/m^2	

		Γ	L	X
Number of equivalent valleys		1	4	3
Effective mass	m/m_0	0.063	0.222	0.58
Nonparabolicity parameter	[eV^{-1}]	0.610	0.461	0.204
Intervalley energy gap (measured from the top of the valence band)	[eV]	1.439	1.769	1.961
Acoustic deformation potential	[eV]	7.0	9.2	9.27
Optical deformation potential	[eV/cm]	0	3.0×10^8	0
Optical-phonon energy	[meV]	0	34.3	0
Intervalley deformation potential [eV/cm] $\{\,\Gamma$		0	10^9	10^9
L		10^9	10^9	5×10^8
X		10^9	5×10^8	7×10^8
Intervalley phonon energy [meV] $\{\,\Gamma$		0	27.8	29.9
L		27.8	29.0	29.3
X		29.9	29.3	29.9

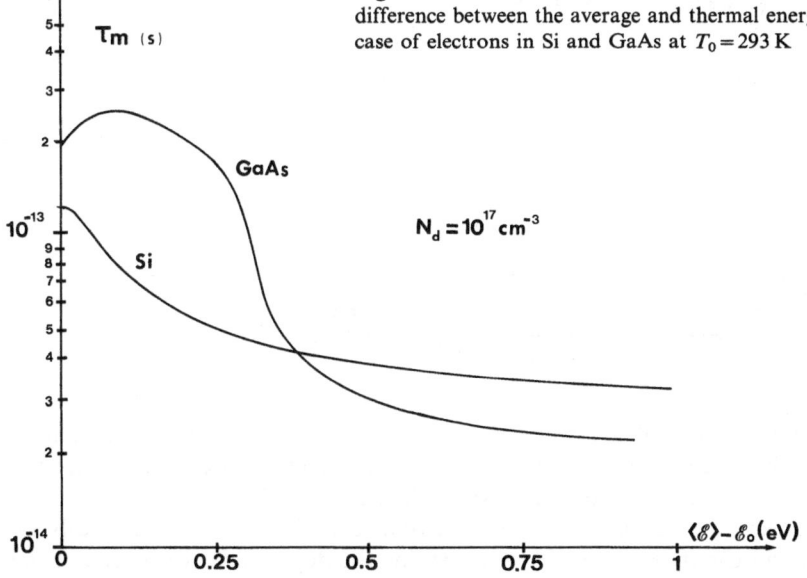

Fig. 8.1. The momentum relaxation time as a function of the difference between the average and thermal energies for the case of electrons in Si and GaAs at $T_0 = 293$ K

τ_m (s)

GaAs

Si

$N_d = 10^{17}$ cm^{-3}

$\langle \mathcal{E} \rangle - \mathcal{E}_0$ (eV)

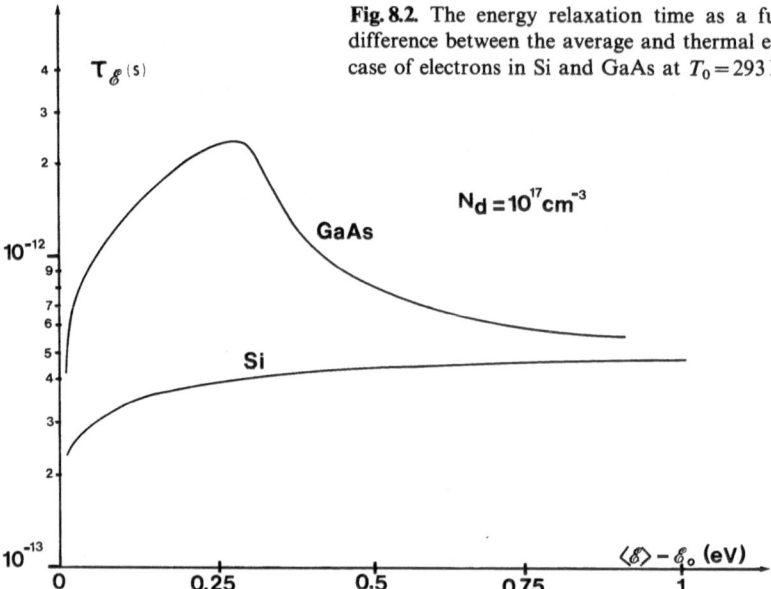

Fig. 8.2. The energy relaxation time as a function of the difference between the average and thermal energies for the case of electrons in Si and GaAs at $T_0 = 293\,\mathrm{K}$

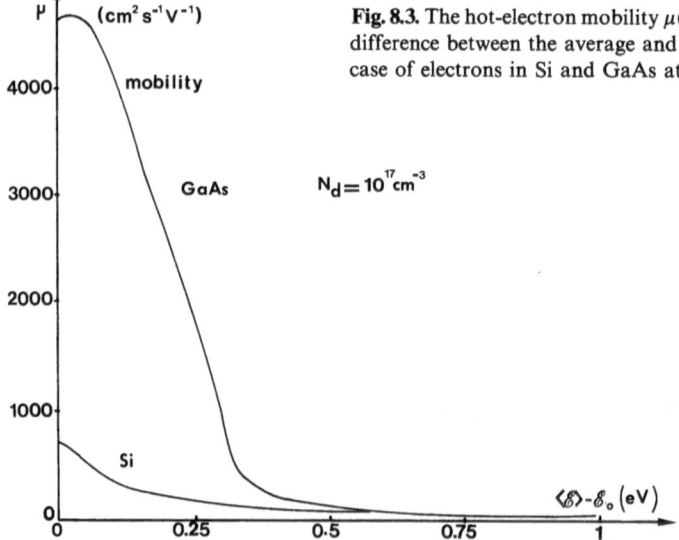

Fig. 8.3. The hot-electron mobility $\mu(\langle \mathscr{E} \rangle)$ as a function of the difference between the average and thermal energies for the case of electrons in Si and GaAs at $T_0 = 293\,\mathrm{K}$

Typical variations of τ_m, $\tau_{\mathscr{E}}$ and mobility $\mu(\mathscr{E}) = e\tau_m m^*$ have been calculated (8.8, 9) for the case of Si and GaAs using the microscopic models summarized in Tables 8.1, 2, and the results are reported in Figs. 8.1–3.

The conservation equations (8.6, 7) have already been used successfully by *Shur* [8.9] and *Cappy* et al. [8.10] to determine the dependence with time of the

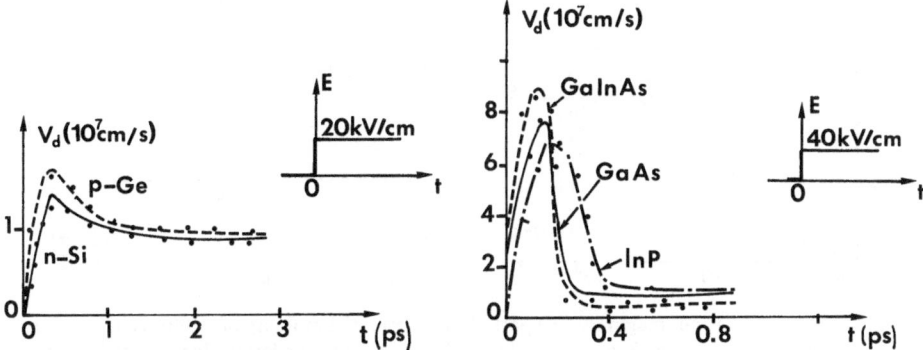

Fig. 8.4. Transient drift velocity against time when a time step of the electric field is applied to an ensemble of electrons initially at thermal equilibrium. (——, – – –, – · –) correspond to the relaxation time description of (8.6, 7) and (● ● ●) to Monte Carlo calculations (electrons in Si, GaInAs, InP, GaAs) or results obtained from iterative methods (holes in Ge)

drift velocity and the average energy when a given time configuration of the electric field is applied to the semiconductor, and we shall point out that good agreement with more exact calculations has been generally observed. This is clearly shown in Fig. 8.4 where a comparison between the results obtained with both the relaxation time description and the Monte Carlo or the iterative method has been carried out. It can be seen that, for the five semiconductor materials studied, results obtained on the drift velocity by the relaxation time method agree very well with more exact calculations. Consequently, owing to its simplicity, the balance-equation method with the relaxation time approximation appears to be very useful for the study of time-dependent phenomena.

8.2.2 Periodic Field

First, let us consider the case when superimposed to a steady-state field E_0 a high-frequency sinusoïdal electric field $E_1 \sin(\omega t)$ is applied to the whole semiconductor material,

$$E(t) = E_0 + E_1 \sin(\omega t), \tag{8.10}$$

and the transient drift velocity, which results from the applied electric field $E_1 \sin(\omega t)$, has to be determined. The results concerning the time-dependent drift velocity $v_d(t)$, caused by the sinusoïdal electric field, can be analyzed by introducing a complex differential mobility μ^* through a linear-response relation of the usual type:

$$v_\omega = \mu^* E_1 = (\mu' + i\mu'') E_1. \tag{8.11}$$

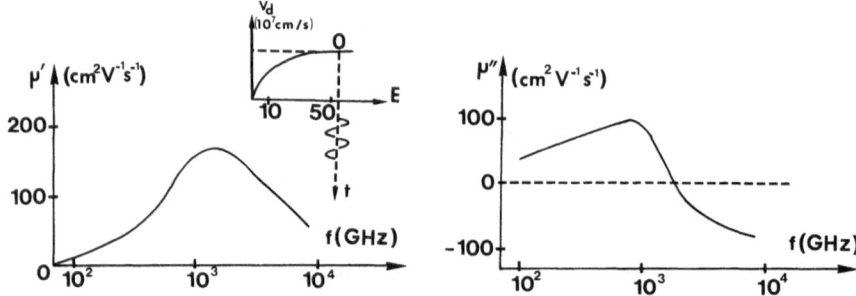

Fig. 8.5. The real (μ') and imaginary (μ'') part of the differential mobility of electrons in silicon as a function of frequency at 293 K as obtained from Monte Carlo calculations for $E_0 = 50\,\text{kV/cm}$ and $E_1 = 5\,\text{kV/cm}$ ($E \| \langle 111 \rangle$)

where v_ω is the Fourier transform of $v_d(t)$, and μ' and μ'' are the real part and the imaginary part of the complex differential mobility.

At low frequency the differential mobility can easily be obtained as the derivative $(dv_d/dE)_{E=E_0}$ of the static velocity field characteristics because, as a result of the balance between the effect of the electric field and of the scattering mechanisms, it can be assumed that a steady state always exists. At very high frequency this is no longer the case because the carriers can be strongly accelerated or decelerated between two successive collisions. Therefore, differential mobility values very different from low-frequency values can be obtained.

The case of electrons in Si and GaAs, as typical semiconductor materials, will be successively considered.

An example of the results obtained in Si [8.6], using a Monte Carlo procedure, is reported in Fig. 8.5. Here the real and imaginary parts of the differential mobility are plotted against the frequency of the applied sinusoïdal electric field. For the value of the steady-state electric field studied ($E_0 = 50\,\text{kVcm}^{-1}$), the static differential mobility dv_d/dE vanishes. As a consequence, a very low value of the real part of the differential mobility is obtained in the centimeter and millimeter wavelength region below about 100 GHz. At increasing frequency a velocity modulation due to E_1 comes into play. Accordingly, the real part of the differential mobility increases, and, in turn, even the imaginary part increases, as can be seen from Fig. 8.5. This phenomenon will occur and has to be taken into account in devices operating in the millimeter wavelength region such as IMPATT and TUNNET diodes [8.11]. In fact, at these high frequencies (between 100 and 400 GHz), owing to a $\mu'(\omega) \neq 0$ the device semiconductor material will become absorbing even if it is operating in the saturated drift velocity range [i.e., $\mu'(0) = 0$]. In turn, a decrease of the output microwave power as well as of the efficiency of the oscillating structures will be observed.

An effect of the same type is the noninstantaneousness of the transferred electron mechanism between the minimum of the conduction band and upper valleys. It will also determine a frequency limit of the real part of the negative differential mobility of most of the III-V compound semiconductors. An example of the results obtained in [8.12] is given in Fig. 8.6. Here the value of E_0 is chosen in order to obtain a negative value of the static differential mobility, and the frequency dependence of the real part of the differential mobility is calculated for two values of E_1 and for two semiconductor materials (GaAs and InP). It can be noted that the value of the cut-off frequency can be increased significantly in the case of large-signal operation.

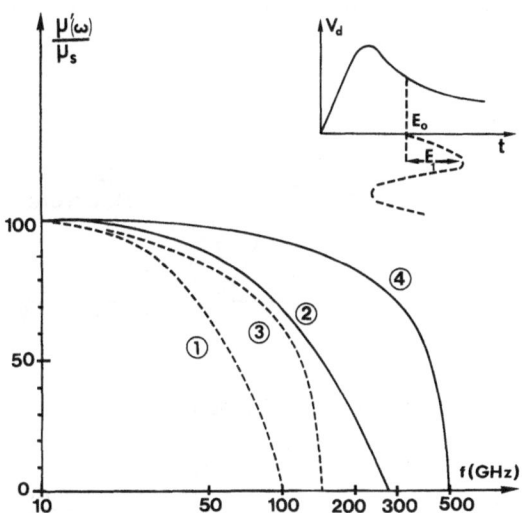

Fig. 8.6. The real part of the negative differential mobility normalized to the static value versus frequency at $T_0 = 293$ K. Curves (1) and (2) refer to the case of GaAs; (---): $E_0 = 7$ kV/cm, $E_1 = 3$ kV/cm; (——): $E_0 = 20$ kV/cm, $E_1 = 15$ kV/cm. Curves (3) and (4) refer to the case of InP; (---): $E_0 = 15$ kV/cm, $E_1 = 6$ kV/cm; (——): $E_0 = 55$ kV/cm, $E_1 = 45$ kV/cm

8.2.3 Transient Phenomena

All the phenomena studied in Sect. 8.2.2 occur only in microwave devices operating under periodic-field conditions. The case of a time-pulse or a time-step configuration of the electric field appears to be even more interesting. In fact, this is the simplest way to simulate roughly the electric field applied to a carrier when it travels through the active layer of a submicron semiconductor device in the dc operating conditions. Consequently drift and diffusion phenomena, which occur in this case, will be successively studied in Sects. 8.3, 4.

8.3 Drift Phenomena

8.3.1 Ballistic, Overshoot Motion

To illustrate the main phenomena, let us first consider the case of a "single-valley" semiconductor (appropriately modeled to the case of Si with $E \| \langle 111 \rangle$) subjected to a simple time-step configuration of the electric field. The corresponding transient drift velocity and average energy can be determined by using an ensemble Monte Carlo method, and an example of the results is shown in Fig. 8.7a. It can be noted that, just after the application of the field step, the drift velocity increases linearly with time: the motion of the carriers is freely accelerated. This resembles the classical concept of ballistic transport [8.13]. Then the drift velocity exhibits a maximum value which is much higher than the steady-state value: this is the so-called overshoot phenomenon [8.14]. As time

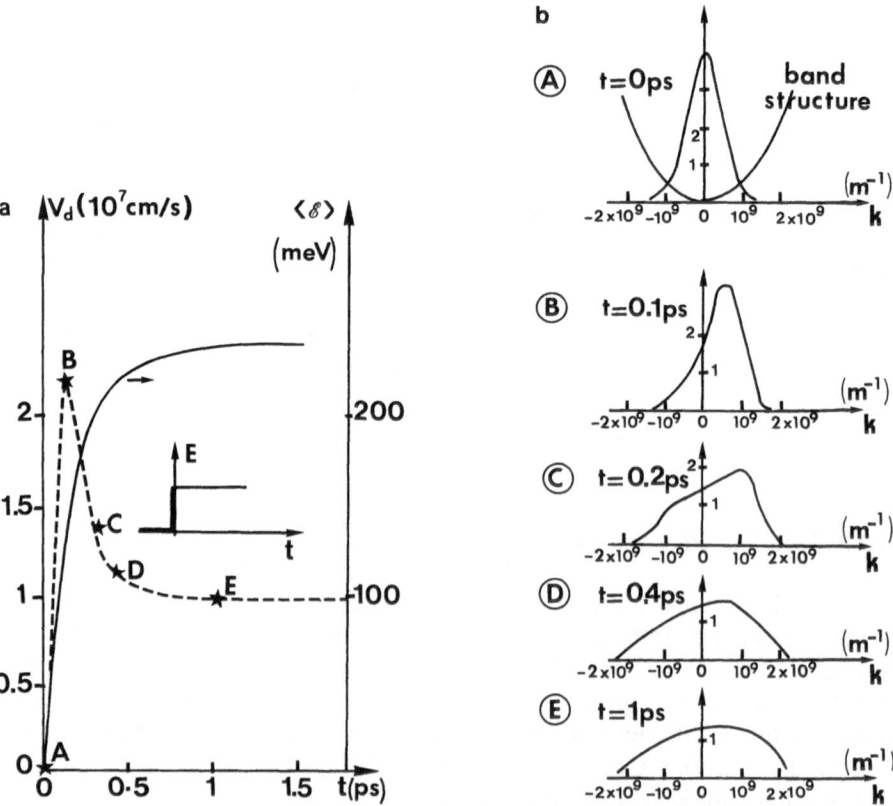

Fig. 8.7. (a) Transient drift velocity $v_d(t)$ and average energy $\langle \mathscr{E} \rangle$ versus time resulting from a time-step configuration of the electric field for the case of electrons in Si with $E \| \langle 111 \rangle$ at $T_0 = 293$ K. **(b)** Evolution of the corresponding momentum distribution function

increases further, the drift velocity decreases until reaching its stationary value, and the average energy of the carriers is found to increase progressively with time until a steady-state value is obtained.

For better understanding of these phenomena, Fig. 8.7b reports the corresponding momentum distribution functions, as obtained by Monte Carlo calculations at different values of time after the application of the time step [8.6]. At the initial time (Case A) the symmetric distribution function of the equilibrium condition is shown together with the band structure. Obviously, in this symmetric case the drift velocity is zero, while the value of the average energy is equal to $3(K_B T_0)/2$. The distribution function calculated 0.1 ps after the electric field has been applied is reported as Case B: the scattering mechanisms have not yet occurred, and consequently only the perturbation due to the electric field is observed. Each carrier, or more exactly, each representative point in the band structure, is shifted the same amount in the direction of the field. Then, only a rigid shift of the initial distribution function, proportional to the elapsed time and to the electric field value (8.3), will occur. The velocity of all carriers along the field direction, and in turn the drift velocity, increases linearly with time so that a high value of the drift velocity can be achieved. We say that a ballistic transport condition is obtained, which occurs during the first tenth of a picosecond in the case here analyzed of Si at room temperature. As time increases, however, the scattering mechanisms progressively occur. Owing to the progressive velocity randomization introduced by the scattering efficiency, the regions of k states opposite to the field direction become more populated. The peak of the distribution function, initially rigidly shifted in the field direction, relaxes toward its steady-state position (Cases C, D, E) and the drift velocity tends to attain its stationary value (Case E). During the transient (Cases B, C, D), drift velocity values higher than the steady-state value are observed (overshoot phenomenon), until the steady state is reached. There (Case E) a balance between the energy gained from the electric field and the energy lost through scattering mechanisms is finally achieved.

The relaxation time equations can be of valid help in obtaining a simple but often useful understanding of these phenomena, thus in the following we shall illustrate their application.

For time $t < \tau_m$, the scattering term $m^* v_d / \tau_m$ can be neglected and (8.6) can be written as

$$\frac{d}{dt} (m^* v_d) = eE. \tag{8.12}$$

This equation characterizes a ballistic motion where the velocity only depends on inertial effects. For a constant value of the electric field and for a parabolic valley, so that the effective mass does not depend on energy, the drift velocity should increase proportionally with time; this is the result which has been already observed in Fig. 8.7a just after the application of the electric field step.

For time $t \gg \tau_m$, on the other hand, we can neglect the "inertial" term $d(m^* v_d)/dt$ and (8.6) reduces to

$$v_d = \frac{e\tau_m}{m^*} E = \mu(\langle \mathscr{E} \rangle) E. \tag{8.13}$$

It can be noted that the drift velocity not only depends on the value of E but also on the instantaneous average energy of the carriers $\langle \mathscr{E} \rangle$. This energy can be obtained from the energy conservation equation (8.7). In the case of a time-step configuration of E, $\langle \mathscr{E} \rangle$ increases progressively from the thermal energy $\mathscr{E}_0$ to the steady-state value $\mathscr{E}_s$ achieved when $d\langle \mathscr{E} \rangle/dt = 0$.

In most semiconductors $\mu(\langle \mathscr{E} \rangle)$ decreases at increasing $\langle \mathscr{E} \rangle$ (Fig. 8.3). As a result, the instantaneous drift velocity will reach higher values at the beginning of the motion when $\langle \mathscr{E} \rangle$ will be close to $\mathscr{E}_0$ and thus $\mu(\langle \mathscr{E} \rangle)$ will keep a value higher than that of the steady state. The overshoot phenomena is then obtained. In fact, by neglecting the interial term, the maximum value of the instantaneous drift velocity is $\mu(\mathscr{E}_0) E$ (8.13), and this value can be much higher than the steady-state value, $\mu(\mathscr{E}_s) E$, particularly when $\mu(\langle \mathscr{E} \rangle)$ is a strongly decreasing function of $\langle \mathscr{E} \rangle$. As can be seen in Fig. 8.3, this is particularly the case with GaAs for which, due to intervalley transfers, at energy higher than the intervalley energy gap $\mathscr{E}_{\Gamma L}$, the mobility $\mu(\langle \mathscr{E} \rangle)$ is strongly reduced. Consequently, transient effects for this type of semiconductor appear most interesting to study.

8.3.2 Undershoot Motion, Rees Effect

Here we shall consider the case of GaAs and report the results concerning overshoot phenomena as well as additional related effects usually referred to as undershoot phenomena and Rees effect. In this example, a GaAs semiconductor sample is submitted to a time-pulse configuration of the electric field and simultaneously to a steady field. The results obtained for the transient drift velocity and average energy are illustrated in Fig. 8.8. They can be roughly understood by using (8.13), where it should be noted that the drift velocity depends not only on the electric field but also, via the mobility $\mu(\langle \mathscr{E} \rangle)$, on the instantaneous value of the average energy. State A, just before the application of the pulse, will be described first. The steady electric field value is only 2 kV/cm and all carriers remain in the Γ valley; the average energy value is close to the thermal equilibrium value and the mobility value $\mu(\langle \mathscr{E} \rangle)$ (plotted in Fig. 8.3) is still close to its high ohmic value. As a result, relatively high values of the drift velocity can be achieved. State B refers to a few tenths of a picosecond after the application of the high electric field pulse (20 kV/cm): the instantaneous energy is not yet increased and the mobility remains very high, in spite of the high value of the electric field, and a very high overshoot velocity is observed. These phenomena will occur whenever the instantaneous energy is lower than its steady value. State C refers to 1 ps after the application of the pulse;

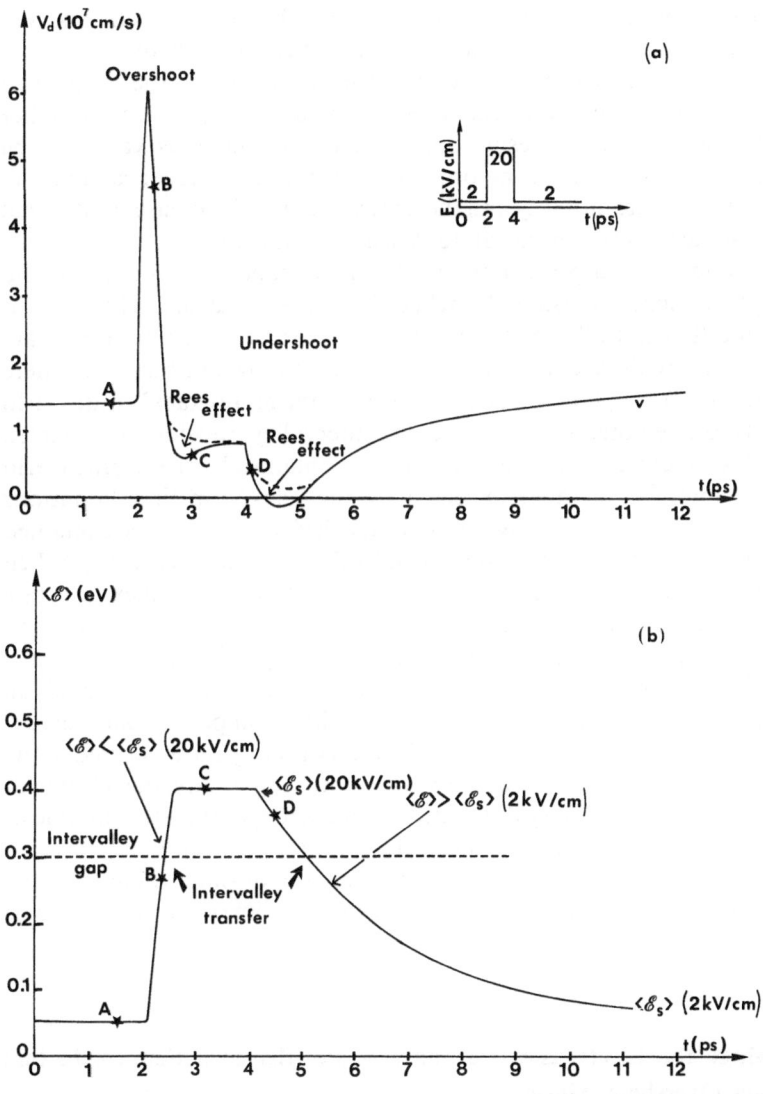

Fig. 8.8a, b. Transient drift velocity (**a**) and average energy (**b**) versus time when a time-pulse configuration of the electric field and simultaneously a steady-state field are applied to a multivalley semiconductor (GaAs, 300 K). Results are obtained with a Monte Carlo procedure; (– – –) refers to the relaxation time equations

the energy has now reached its steady-state value and a great number of carriers are in the X valleys. Owing to the increased scattering efficiency, mainly due to intervalley transitions and to an increased effective mass, the mobility is very low and the drift velocity is reduced, although the electric field remains very high. State D refers to a few tenths of a picosecond after the application of the pulse:

the instantaneous energy has not yet been reduced; the mobility remains very low; and, since the electric field value is now also very low, there is a drastic decrease in the drift velocity. Such phenomenon can be called the *undershoot velocity effect*; it will be achieved whenever the instantaneous energy is higher than its steady-state value corresponding to the instantaneous value of the electric field. However, at increasing time the energy will decrease; carriers will return to the Γ valley; the mobility and the drift velocity will increase; and a final steady state identical to the initial State A will be attained.

From the above, it appears possible to understand the overshoot and undershoot phenomena by using the relaxation time equations. However, it should be noted that not all the features of the transient carrier transport are taken into account by these equations. This is clearly shown in Fig. 8.8a where the results obtained from (8.6, 7) are compared with the more exact Monte Carlo calculations. It can be noted that just after the intervalley transfer has occurred (Fig. 8.8a, b) the exact values of the drift velocity, obtained by the Monte Carlo procedure, are lower than those expected by the relaxation time equations (dashed lines in Fig. 8.8a). These low values of the drift velocity can be explained by the following mechanism, as first suggested by *Fawcett* and *Rees* [8.15]. When intervalley transfer occurs, an electron scattered into the central valley in a state with a large k wavevector component antiparallel to the electric field E (i. e., with a negative velocity fluctuation) will remain in the Γ valley where it will perform a long flight before suffering another intervalley scattering. On the other hand, an electron scattered into the central valley in a state with a component parallel to E (i. e., with a positive velocity fluctuation) will be very rapidly retransferred to the original valley. Consequently, when intervalley scattering occurs, the lifetime of an electron with negative velocity fluctuation is much longer than the lifetime of an electron starting with a positive velocity fluctuation. Therefore, the transient drift velocity can be reduced with respect to its final steady-state value, with the possibility to obtain even transient negative values at times shorter than 5 ps (Fig. 8.8a).

8.3.3 How Make the Electrons Go as Fast as Possible in a Semiconductor: Ballistic Versus Overshoot Motion

It has been shown above that, due to overshoot or ballistic motion, drift velocity values much higher than those corresponding to the final steady-state conditions could be achieved. These phenomena, occurring under non-steady-state conditions, might be achieved in submicron devices. Thus they can be used to increase strongly the velocity of the carriers with the objective to reduce further their transit time along the active region. This appears to be an interesting goal since it is well known [8.16] that to increase cut-off frequency of microwave devices and to reduce the propagation delay of logic devices, this transit time must be as short as possible.

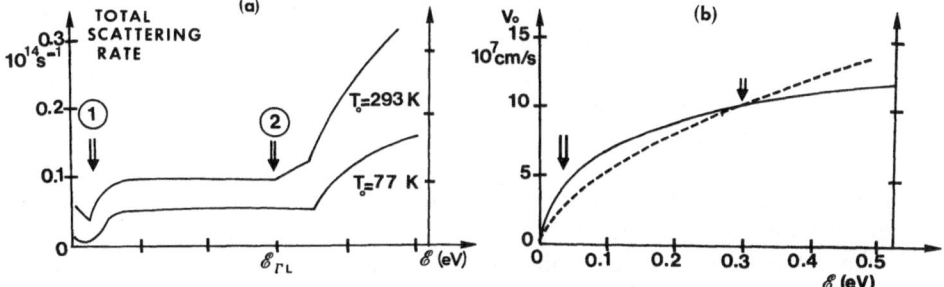

Fig. 8.9. (a) Total scattering rate in the Γ valleys of GaAs as a function of electron energy. Impurity scattering is neglected. (b) Maximum velocity which can be achieved in the Γ valley as a function of electron energy; values are obtained from the classical equation $v_0 = \hbar^{-1}(\partial \mathscr{E}/\partial k)_\mathscr{E}$. (---) corresponds to a parabolic valley characterized by a maximum velocity v_0, (8.2), for $\mathscr{E} = \mathscr{E}_{\Gamma L}$. *Case 1* corresponds to an energy slightly smaller than the polar optical-phonon energy. *Case 2* corresponds to an energy slightly smaller than the intervalley energy gap

Consequently, an interesting problem to investigate is the research of the optimum electric field configuration which makes electrons go as fast as possible in the device, i.e., over a given distance d. As has been already noted in this chapter, the simplest way to try to solve this problem (not always the most reliable) is to study the motion of an ensemble of electrons in a uniform bulk semiconductor subject to various time configurations of the electric field. From the knowledge of the dependence with time of the drift velocity $v_d(t)$, the average distance d traveled by the carriers over a time t, and in turn the average velocity v_a over a distance d, can be obtained from the following equations:

$$d = \int_0^t v_d(t')dt', \tag{8.14}$$

$$v_a(d) = d/t. \tag{8.15}$$

Then the optimum electric field time configuration, which for a given distance d enables us to obtain the maximum value of v_a, can be found and discussed.

Most of the phenomena involved can be understood by considering Fig. 8.9b where, for the case of GaAs, we report the maximum velocity[1] achievable along one direction in space as a function of the energy of the carrier (velocity distribution and velocities along the perpendicular directions are assumed to be zero). In Fig. 8.9a, the dependence of the total scattering rate on energy is also shown. From these data, used in the Monte Carlo calculations described in this

---

1 This maximum velocity is determined by using (8.2) with a nonparabolic energy wavevector relation which is modeled on the Γ minimum as used in classical Monte Carlo calculations. It should be noted that using a realistic band structure (e.g., as obtained from pseudopotential calculations) the maximum velocity should depend slightly on the crystallographic axis.

paper, it can be noted that, with an energy slightly smaller than the polar optical-phonon energy (Case 1) or the intervalley energy gap (Case 2), it is possible to obtain very high velocities (3.5×10^7 and 10^8 cm/s, respectively) with reasonably low values of the scattering rate. Thus these high velocities can survive for a relatively long time.

We shall note that for attaining the conditions of Case 1, low-temperature operations (e. g., 77 K) must be used so that initial carrier energies lower than the optical-phonon energy are easily obtained. This is not required for Case 2 which is characterized by a much higher velocity, and therefore appears to be much more interesting than Case 1.

The main point is now to find the best way to attain this condition, typified by Case 2, where the electron energy is just a little bit lower than the intervalley energy gap and where the velocity and energy distributions have to be as sharp as possible.

Two opposite or limiting cases are considered in Fig. 8.10, where a few results obtained by a Monte Carlo simulation in GaAs are reported [8.17].

In the first case (Fig. 8.10a), which will be now referred to as the overshoot case, a constant electric field is suddenly applied to an ensemble of electrons initially at thermal equilibrium until the average energy reaches a value close to $\mathscr{E}_{\Gamma L}$. Since the time duration is short (about 2 ps), a steady state is not attained and high drift velocities can be achieved. However, it should be noted that, due to scattering which broadens the velocity distribution, the value of the drift velocity (6×10^7 cm/s) which can be achieved for an energy close to $\mathscr{E}_{\Gamma L}$ is lower than the value expected without scattering (10^8 cm/s). In addition, one can note that, owing to inertial effect, the velocity overshoot cannot be reached instantaneously. Even if immediately after the application of the electric field scattering

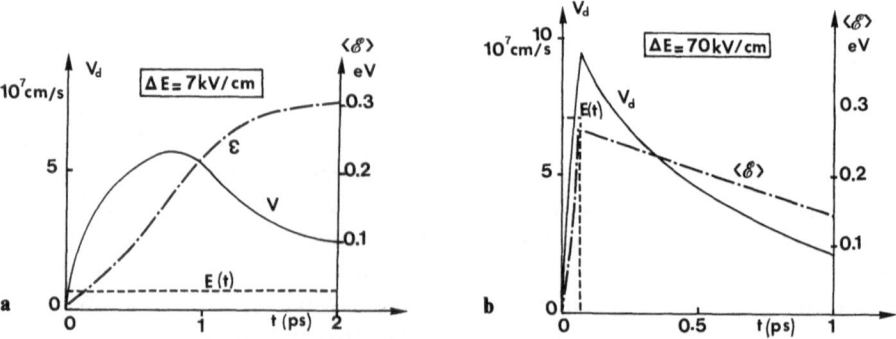

Fig. 8.10. (a) *Overshoot Case.* Average velocity and average energy of electrons subject to a time-step configuration of the electric field as functions of time (GaAs: $N_d = 0$, $T_0 = 77$ K). The applied electric field step ($E = 7$ kV/cm) is also plotted in the same figure. (b) *Ballistic Case.* Average velocity and average energy of electrons subject to a time-pulse configuration of the electric field as functions of time (GaAs: $N_d = 0$, $T_0 = 77$ K). The applied electric field pulse characterized by an amplitude $\Delta E = 70$ kV/cm and a duration $\Delta t = 0.06$ ps is also plotted in the same figure

mechanisms have not yet occurred and the motion is of ballistic type, one can remark that the velocity attainable during this type of motion remains quite small. Consequently, this type of ballistic behavior is not usefully exploited with this field configuration.

To take full advantage of the ballistic behavior of the carriers a second procedure can be suggested. This corresponds to the other limiting case (Fig. 8.10b) and will be referred to as the ballistic case. Here a very short time pulse of the electric field configuration, characterized by a very high amplitude, is used[2]. The pulse duration is chosen in order to obtain a final energy slightly lower than the intervalley energy gap. For this very short pulse (0.06 ps in Fig. 8.10b), scattering events have no time to occur and consequently very high drift velocity can be achieved together with a sharp velocity distribution. The electric field being switched off, a really ballistic motion of all the carriers is then achieved, a little like a ballistic rocket where all the motion is achieved without driving field, i.e., with the engine stopped. Consequently, the very high values attained by the velocity decrease very slowly after the field is switched off. As a result, quite long distances can be achieved within very short times.

The problem is now to determine which, between these two limiting cases, is the most advantageous for obtaining the maximum value of the average velocity over a given distance.

To answer this question, very simple analytical calculations can be carried out using an approximate, but in most cases a quite realistic, model [8.15]. To this end, the classical relaxation time equations (8.6, 7) are used, under the assumption that the momentum relaxation time τ_m and the effective mass m^* remain constant for energies lower than the intervalley energy gap. In addition, the energy relaxation time τ_ε will be considered much longer than the momentum relaxation time τ_m. Using these assumptions, the distance d traveled by the carrier over a time T, just before its energy reaches at time T the value $\mathscr{E}_{\Gamma L}$, can easily be obtained.

In the *overshoot case*, the distance d traveled by the carrier over the time T is given by

$$d_{\text{overshoot}} = \frac{d_0}{\sqrt{2}}\left[\frac{T}{\tau_m} - (1 - e^{-T/\tau_m})\right]^{1/2}. \tag{8.16}$$

In the ballistic case, d is given by

$$d_{\text{ballistic}} = d_0(1 - e^{-T/\tau_m}). \tag{8.17}$$

---

2 Obviously, due to the displacement current, it appears very difficult to apply such very short time pulse. Nevertheless, by using heterojunction, it appears possible to apply in a semiconductor a very small space pulse [8.16] which could give rise to the same type of effect.

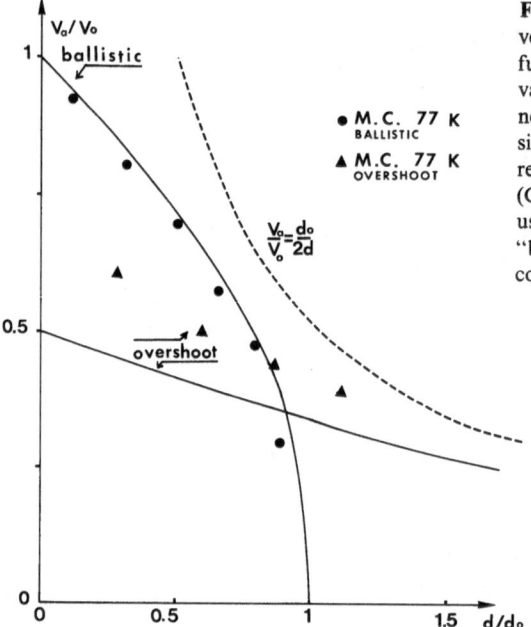

Fig. 8.11. Maximum transient average velocity of electrons over a distance d as a function of the distance d (normalized values are reported for the sake of generality). (——, ----) correspond to the simplified theory of (8.16–22). (▲, ●) correspond to Monte Carlo simulations (GaAs: $N_d = 0$, $T_0 = 77\,\mathrm{K}$) carried out using in the "overshoot case" and in the "ballistic case" the type of electric field configurations reported in Fig. 8.10

In these two equations v_0 is the maximum value of the carrier velocity as obtained, for $\mathscr{E} = \mathscr{E}_{\Gamma L}$, from

$$v_0 = \hbar^{-1} \left(\frac{\partial \mathscr{E}}{\partial k} \right)_{\mathscr{E} = \mathscr{E}_{\Gamma L}},\tag{8.18}$$

and d_0 is a characteristic distance given by

$$d_0 = v_0 \tau_m.\tag{8.19}$$

From (8.10, 11), it can easily be shown that for $T \ll \tau_m$

$$d_{\text{ballistic}} = 2\, d_{\text{overshoot}} = v_0 T.\tag{8.20}$$

Consequently, we argue that over small distances the transient average velocity in the ballistic case can be twice as high as in the overshoot case. Fig. 8.11 reports a set of universal curves which can be used for any semiconductor to determine the maximum transient average velocity $v_a = d/T$ achievable by a carrier traveling over a given distance d. This quantity, calculated by using either (8.16) or (8.17), has been plotted versus d using the dimensionless variables v_a/v_0 and d/d_0. From Fig. 8.11, it can be noted that the average velocity obtained in the ballistic case is higher (maximum advantage is twice) than that obtained in the overshoot case only for distances smaller than

d_0. Thus, for distances larger than d_0, purely ballistic motion can no longer be of use and, to achieve high average velocities, velocity overshoot phenomena should be exploited. In this case for $T > \tau_m$ inertial effect can be neglected [i.e., the term $m^* dv/dt$ in (8.6)] and the average velocity is given by the following equation:

$$\frac{v_a}{v_0} = \frac{d_0}{2d} \quad \text{or} \tag{8.21}$$

$$v_a = \frac{\mathscr{E}_{\Gamma L} \mu}{ed}. \tag{8.22}$$

The curve corresponding to (8.21) has been plotted using a dashed line (Fig. 8.11). The velocity values obtained are higher than those determined from (8.16) taking inertial effect and ballistic behavior at the beginning of the motion into account. As a consequence, it should be noted that ballistic or inertial effects reduce the velocity overshoot phenomenon (at least in the case where a constant electric field is applied throughout the motion).

From these results it clearly appears that, to maximize the value of the average velocity, the most suitable semiconductor materials should be characterized by the highest values of the following quality factors: v_0, d_0, and $\mu \mathscr{E}_{\Gamma L}$ (8.22). Table 8.3 summarizes the values of these three quantities for the most common III-V semiconductors.

It should, however, be noted that rough assumptions are needed to obtain (8.16, 17) where the three previous quality factors take a part. Consequently, a comparison with more exact results, as those obtained from a Monte Carlo simulation, looks necessary. To this end two cases which have been studied with such a simulation [8.17] will be presented; the "overshoot case", where an ensemble of thermal equilibrium electrons are subjected to a time-step configuration of the electric field; and the "ballistic case", where an ensemble of thermal equilibrium electrons are subjected to a very short time pulse (in the

Table 8.3. Typical values of maximum drift velocity v_0, maximum ballistic distance d_0, and product of mobility and intervalley energy gap $\mu \cdot \mathscr{E}_{\Gamma L}$ for intrinsic material. Calculation of d_0 has been carried out assuming that the momentum relaxation time can be obtained from the value of the steady-state mobility $\mu = v_d/E = e\tau_m m^*$ when the average energy of carriers is $\mathscr{E} = \mathscr{E}_{\Gamma L}/2$

		v_0 [10^8 cm/s]	d_0 [μm]	$\mu \mathscr{E}_{\Gamma L}$ [10^3 eV cm^2 s^{-1} V^{-1}]
GaAs	(370 K)	1.0	0.2	0.8
GaAs	(293 K)	1.0	0.3	2.4
GaAs	(77 K)	1.0	0.6	12.0
InP	(293 K)	1.1	0.2	2.4
GaInAs[a]	(293 K)	1.2	0.25	10.0

[a]   Alloy composition lattice matched to InP.

Monte Carlo calculations the pulse duration is 0.06 ps). In both cases the electric field values are chosen in order to obtain the maximum value of the average velocity over a given distance d. Taking the values of v_0 and d_0 given in Table 8.1, the results obtained for v_a/v_0 as a function of d/d_0 are also plotted in Fig. 8.11. The agreement obtained is found to be quite satisfactory in the ballistic case while, in the overshoot case, Monte Carlo calculations give velocity values systematically higher than those obtained using the simplified model. This discrepancy results because in the Monte Carlo simulation, the nonparabolicity of the Γ minimum is taken into account. Therefore, at the beginning of the motion it is possible to obtain values of the velocity higher than those expected assuming, in the analytical calculation, a parabolic valley characterized by a maximum velocity v_0 for an energy equal to the intervalley gap (Fig. 8.9b).

Additional results of the overshoot case obtained by Monte Carlo calculations are shown in Fig. 8.12. Here it can be noted that the average velocity can be

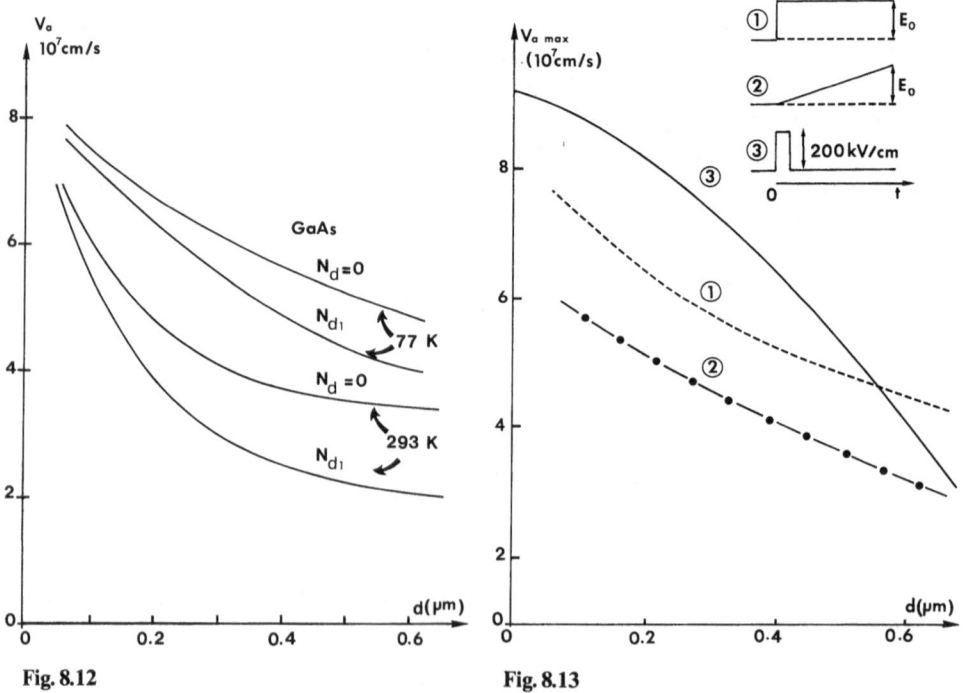

Fig. 8.12   Fig. 8.13

Fig. 8.12. Maximum transient average velocity of electrons over a distance d calculated with a Monte Carlo simulation for two operating temperatures (77 K, 293 K) and two impurity concentrations (GaAs : $N_d = 0$, $N_d = 3 \times 10^{17}$ cm^{-3}). A time-step electric field configuration is applied and, in all cases, the electric field amplitude E_0 is chosen to obtain a maximum value of v_d for a given distance d

Fig. 8.13. Monte Carlo calculations of the maximum transient average velocity of electrons (GaAs : $N_d = 0$, $T_0 = 293$ K) over a distance d as a function of the distance d for various time configurations of the electric field (see inset). In 1 and 2, which refer to overshoot case, E_0 is chosen to obtain a maximum value of v_d for a given distance d

increased by using very small distance (in the submicron range), low impurity concentrations, and low temperature operations. However, at very small distances ballistic phenomena progressively occur regardless of impurity concentration or temperature. This can be easily understood, since at these distances scattering events will not occur and the average velocity which a carrier will achieve only depends on the energy wavevector relationship of the band structure. In addition, it should be pointed out that the optimum value of the electric field amplitude $\dot{E}_0$ needed to achieve a maximum average velocity is roughly inversely proportional to the distance d, the product (edE_0) being very close to the intervalley energy gap $\mathscr{E}_{\Gamma L}$ as is expected from the assumptions made in the simple theory based on (8.16, 17).

Numerical results obtained in the ballistic case are reported in Fig. 8.13 under Label 3. For comparison, in the same figure the results obtained in the overshoot case (Label 1), and those obtained by using a field configuration where E increases proportionally to time (Label 2), are also reported. It must be pointed out that in this latter case, which may roughly simulate the electric field applied to the carriers when traveling through the active region of a common submicron device [field effect transistor (FET), n^+-n-n^+ structures], the average velocity remains lower than that obtained by using a simple time step in the whole region of d values considered.

8.4 Diffusion Phenomena

The study of diffusion phenomena under transient regime condition is of peculiar interest, mainly because of its implication in the analysis of transport properties in submicron devices [8.18]. Under non-steady-state conditions the value of diffusivity can be very different from its steady-state value and, in order to illustrate such a phenomenon, transient diffusion will now be studied in a uniform semiconductor submitted to a time-step configuration of the electric field.

To this end we shall use the generalized definition of the longitudinal diffusion coefficient at short time, as reported in Chap. 2:

$$D_{lo}(t) = \frac{1}{2} \frac{d}{dt} \langle (x - \langle x \rangle)^2 \rangle, \tag{8.23}$$

where x is the distance covered by a carrier at time t along the x direction parallel to the field.

Results will be presented for the cases of CdTe and GaAs as typical examples [8.19]. The microscopic models are reported in Tables 8.2, 4. Fig. 8.14 reports the results obtained in CdTe at $T_0 = 300$ K and $E = 40.5$ kV/cm for an ensemble of 2×10^3 electrons which start at the same point $x = 0$ and have an initial equilibrium-Maxwellian velocity distribution. The second central moment of this ensemble is calculated at intervals of the order of 10^{-14} s, with an accuracy of

Table 8.4. Model used in Monte Carlo calculations of electrons in CdTe [8.20]

			Γ	L
Average longitudinal elastic constant			$\varrho_0 u_{lo}^2 = 6.97 \times 10^{11}$ dyn cm^{-2}	
Relative static dielectric constant			$\varepsilon_0 = 10.6$	
Relative high-frequency dielectric constant			$\varepsilon_\infty = 7.13$	
Longitudinal optical-phonon energy at $q=0$			$\hbar\omega_{op} = 21.37$ meV	
Number of equivalent valleys			1	4
Effective mass	m/m_0		0.0963	0.50
Intervalley energy gap (measured from the top of the valence band)	[eV]		1.5	3.0
Acoustic deformation potential	[eV]		0	9.5
Intervalley deformation potential	[eV/cm]	Γ	0	1×10^9
Intervalley phonon energy	[meV]	Γ	0	18.96

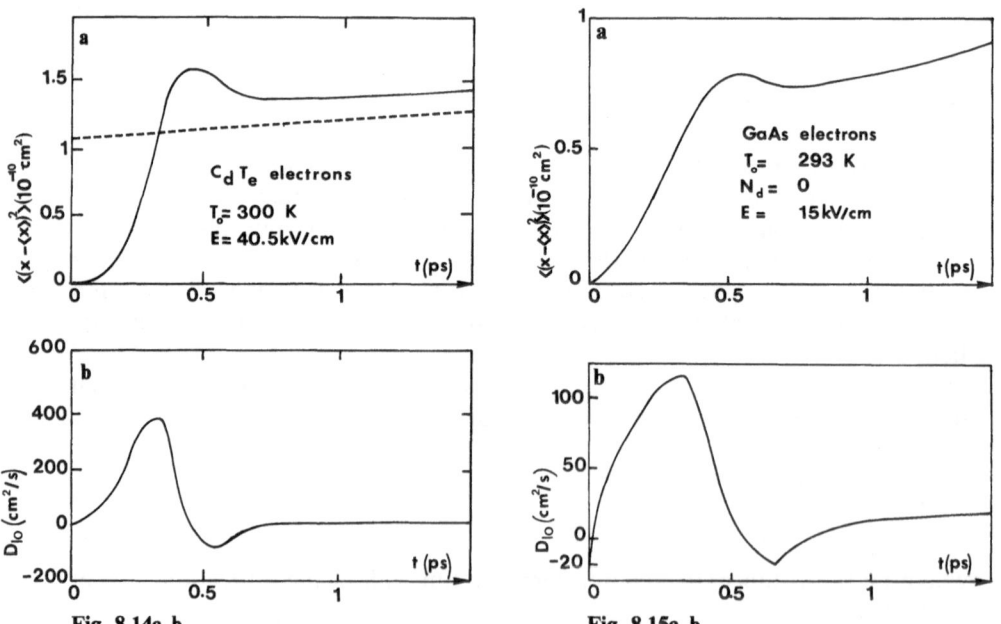

Fig. 8.14a, b   Fig. 8.15a, b

Fig. 8.14. (a) Second central moment and (b) longitudinal diffusion coefficient as functions of time for the case of electrons in CdTe at $T_0 = 300$ K and $E = 40.5$ kV/cm. (–––) in (a) is an extrapolation at short times of the long-time limit of the second central moment which gives $D_{lo} = 6.6$ cm^2/s

Fig. 8.15. (a) Second central moment and (b) longitudinal diffusion coefficient as functions of time for the case of electrons in GaAs at $T_0 = 293$ K and $E = 15$ kV/cm

about 3%. As shown in Fig. 8.14a, this quantity from the very beginning increases quadratically with time, in agreement with the limiting behavior at short times predicted by the Langevin equation which describes the Brownian-like motion of the particles; it achieves a maximum value at 0.45 ps, then it decreases, attains a minimum value at about 0.8 ps, and finally increases monotonically. The asymptotic behavior, as obtained from the simulation in the range of times from 2.4 to 15 ps, is given by the dashed line in the Fig. 8.14a and corresponds to the steady-state value of $D_{lo} = 6.5$ cm^2/s. In the range of times between 0.45 and 0.8 ps the negative slope exhibited by the second central moment versus time can be interpreted as a negative diffusivity region, and from a numerical evaluation of (8.16) it is seen that a minimum value for D_{lo} of about -100 cm^2/s is found at $t = 0.55$ ps (Fig. 8.14b).

Figure 8.15 reports the results obtained in GaAs at $T_0 = 293$ K and $E = 15$ kV/cm making use of the microscopic model reported in Table 8.2 for an ensemble of 2×10^3 electrons starting at the same point with an initial equilibrium-Maxwellian velocity distribution. The main features observed are very close to those obtained for CdTe. In the range of times between 0.5 and 0.7 ps the negative slope exhibited by the second central moment can again be interpreted as a negative diffusivity region. In addition, it should be noted that times longer than 1 ps are needed for the steady-state value of the diffusion coefficient to be obtained.

It appears interesting to try to understand why a negative diffusion coefficient can be obtained in a transient condition, i.e., why a narrowing of the electron bunch can be achieved just a short while after a time-step configuration of E has been applied. This can be explained in the following way: when the electric field is applied to an ensemble of electrons starting at the same point, due to the velocity distribution, a widening of the electron bunch and a positive value of the diffusion coefficient can first be observed. However, carriers which are in front of the electron bunch are characterized by positive and high value of the velocity. Consequently, they can be heated very rapidly by the electric field and they can transfer more rapidly in the L valleys, where they will strongly slow down. On the contrary, carriers which are behind the electron bunch are characterized at the initial time by negative or low values of the velocity. Therefore, they can remain for a longer time in the Γ valley where their velocity will progressively increase. This latter type of carrier will be traveling much faster than the former one a short while after the application of the electric field. Consequently, a narrowing of the electron bunch and a negative value of D will be observed.

In addition, a microscopic interpretation of the value of the time τ_{min} at which the minimum of the simulated second central moment occurs has been given in [8.19]. When intervalley scattering occurs, an electron scattered into the central valley to a state with large k wavevector component antiparallel to E has the possibility to perform a long flight before suffering another intervalley scattering (Rees effect described in Sect. 8.3.2). As illustrated in Fig. 8.16, under ideal conditions (i.e., absence of intravalley scattering mechanisms and intervalley

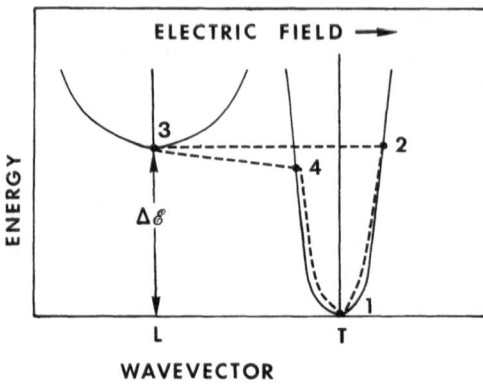

ELECTRIC FIELD →

ENERGY

WAVEVECTOR

Fig. 8.16. Schematic representation of the two-valley band model of CdTe illustrating an ideal long flight in k space; 1–2: free acceleration under the field, 2–3: intervalley scattering from the central to a satellite valley through an optical-phonon emission, 3–4: intervalley scattering from a satellite to the central valley, with a final k wavevector antiparallel to the field direction, through an optical-phonon emission, 4–1: free deceleration

scattering occurring immediately at threshold) these long flights would occur periodically in both real and reciprocal space, and bring the electron gas to its initial conditions within a cycle of period τ_0 given by

$$\tau_0 = \frac{2\sqrt{2m^*\mathscr{E}_{\Gamma L}}}{eE}. \tag{8.24}$$

It is worthwhile noting that the values of τ_0 obtained from (8.24) have been found [8.19] to be very close to the values of $\tau_{\min}$ obtained from the Monte Carlo simulation.

All the phenomena described in Sects. 8.2–4 show the rather complex features which can characterize the transport of electrons in transient conditions and which must be taken into account in most submicron devices.

8.5 Space-Dependent Phenomena in Submicron Devices

Now we shall illustrate two methods which are suitable for studying space-dependent phenomena at small distances: the Monte Carlo procedure and the balance equation.

8.5.1 Monte Carlo Procedure

Monte Carlo simulation appears to be the most useful method to study space-dependent phenomena. In such a simulation, the motion of an ensemble of particles, representative of the carriers in the device, is studied simultaneously in k and r space. Accordingly, a description of the scattering process in a three dimensional k space and a description of the electric field acting on the carriers in the device in a one-, two-, or three-dimensional r space can be used.

At the beginning, the Monte Carlo simulation is used to obtain the velocity of each particle from which the instantaneous position of the carriers in the device is deduced. From the space description of the charge carriers, it then becomes possible to determine the electric field acting on each carrier and to carry out a further simulation with the proper self-field accounted for.

The space dependence of the electric field is generally calculated from Poisson's equation through the knowledge of the carrier density. Two main methods can be used to calculate the carrier density; we shall call them the single-carrier and the multicarrier methods.

In the former method, the motion of only one particle is simulated and the carrier density is determined progressively by recording the total time ΔT spent by the particle in an elementary volume Δr of the device through the relation

$$n(r) = \frac{M \Delta T}{T \Delta r},$$

(8.25)

where T is the observation time during which the motion has been studied, and M is a normalization constant determined from the total number of carriers in the device. This method does not allow us to obtain the time dependence of n because, by using (8.25), we implicitly assume ergodicity of the system, that is, steady-state regimes. This method can therefore be used only to obtain stationary values of the various parameters of the device.

In the latter method, the carrier density is determined from the number of simulated carriers observed at each instant t in an elementary volume Δr. In this case the time dependence of the carrier density can be obtained, and the high-frequency behavior of the device eventually studied.

8.5.2 Balance Equations

An alternative method for studying space-dependent phenomena makes use of the balance equations as obtained by integration of the Boltzmann equation over k space [8.21]. In the single-valley case, by using a one-dimensional treatment this integration leads to three fundamental macroscopic equations which express respectively the conservation of the number, momentum, and energy of the particle ensemble:

$$\frac{\partial n}{\partial t} + \frac{\partial n \langle v \rangle}{\partial x} = 0,$$

(8.26)

$$\frac{\partial}{\partial t} (nm^* \langle v \rangle) = enE - \frac{\partial}{\partial x} (nK_B T_e) - \frac{\partial}{\partial x} nm^* \langle v \rangle^2 - \frac{nm^* \langle v \rangle}{\tau_m(\langle \mathscr{E} \rangle)}.$$

(8.27)

$$\frac{\partial}{\partial t} (n \langle \mathscr{E} \rangle) = en \langle v \rangle E - \frac{\partial}{\partial x} [n \langle v \rangle (\langle \mathscr{E} \rangle + K_B T_e)] - n \frac{(\langle \mathscr{E} \rangle - \mathscr{E}_0)}{\tau_\mathscr{E}(\langle \mathscr{E} \rangle)},$$

(8.28)

where T_e is the electronic temperature (the other symbols have already been defined).

In the case of a multivalley semiconductor, these equations must be written for each valley together with the coupling continuity equation, since an additional term $(\partial n/\partial t)_c$ has to be introduced in the second member of (8.26) to take into account the intervalley transfers. A rough but useful approximation [8.20, 22] consists of averaging all the relevant quantities over all valleys of the considered band structure. Doing so, (8.26–28) can still be applied to a multivalley semiconductor. This procedure means that again we assume that the distribution function is fully determined by the knowledge of the instantaneous energy $\langle \mathscr{E} \rangle$, and obviously this assumption can, in a few cases, be very rough. Nevertheless, following *Shur* [8.9], the energy and momentum relaxation times, the effective mass, and the electronic temperature can be all determined from Monte Carlo simulations under static steady-state conditions. Thus, at least this method has the advantage of providing steady-state results which are exact within the band model used since all the operating parameters are analytically fitted with Monte Carlo calculations.

As a result, a good agreement between Monte Carlo findings and results based on (8.26–28) is generally obtained. This is shown in Fig. 8.17 which illustrates the spatial dependence of the drift velocity obtained by the two aforementioned methods for a GaAs n^+–n–n^+ structure under static conditions at 77 K.

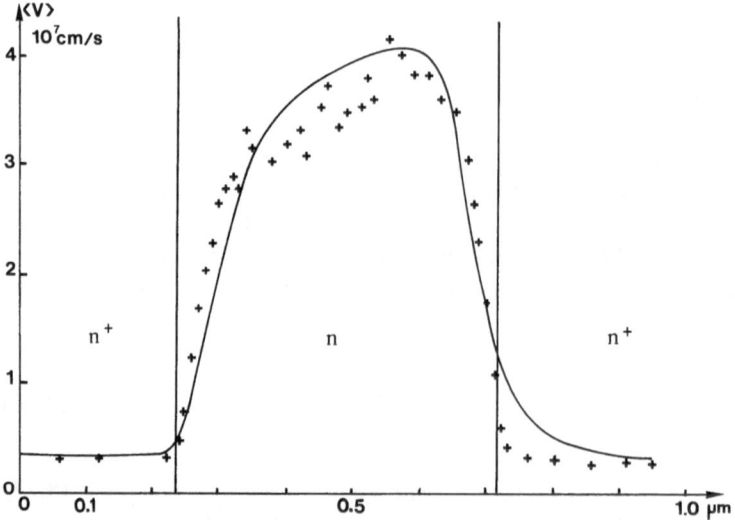

Fig. 8.17. Spatial dependence of the electron drift velocity for a n^+–n–n^+ GaAs structure at 77 K obtained from a Monte Carlo simulation and an analytical formulation under static conditions. (——) analytical formulation; (+ + +) Monte Carlo simulation

8.5.3 Transient Dynamics in Semiconductors Subject to Space Configurations of the Electric Field Characterized by Small Spatial Scales

In this case, some of the characteristics of the carrier transport can be deduced from the results obtained in Sects. 8.2, 3, where a time-pulse configuration of E is applied to the whole semiconductor, by using the usual relation:

$$dr = v_{\mathrm{d}}dt. \tag{8.29}$$

However, from the beginning it is very important to note that the spatial nonuniformity of the free-carrier concentration, energy, and distribution function is not taken into account by using (8.29) and this nonuniformity generates additional phenomena. The most important are

1) diffusion current due to the nonuniformity of n,
2) heat conduction due to the nonuniformity of $\langle \mathscr{E} \rangle$,
3) thermoelectronic current due to the nonuniformity of $\langle v \rangle^2$.

These phenomena can be taken into account in the balance equations (8.27, 28) by the additional terms compared to (8.6, 7).

To illustrate some of the above phenomena, a Monte Carlo simulation (8.17, 18) is carried out when a very small space-pulse configuration of the electric field is applied to an ensemble of electrons in a semiconductor. For the sake of simplicity, the effect of the space charge due to electrons on the value of electric field (via Poisson's equation) has not been accounted for; this is clearly to illuminate the additional phenomena due to a space configuration of the electric field. The results obtained when a stationary state is attained are reported in Fig. 8.18 where in part (b) they are compared with those obtained when a very short time-pulse configuration of the electric field is applied. One can note (and this appears to be the most striking feature) that the average velocity of the carriers is now, in general, much lower than in the previous case, which refers to a time analysis. This difference can be partly explained by the diffusion current which arises since the carrier concentration is no longer uniform.

The highest values of the velocity, which one should have expected at the end of the space pulse, are in fact not obtained since the steep increase of the carrier concentration is at the origin of diffusion velocity with direction opposite to the drift velocity. On the other hand, in front of the space pulse, the velocity of carriers is slightly higher than in the previous case since the carrier concentration decreases so that the diffusion velocity adds to the drift velocity. Moreover, one can note a pronounced spatial nonuniformity of the average energy; this will give rise to heat conduction phenomena since an increase in front of the space step of the average energy is observed. The low value of the carrier average velocity so obtained in stationary conditions can also be explained by considering the way in which the stationary state is reached in the structure. To this end, in Fig. 8.18 the electron concentration profiles are plotted at various times after the injection of an electron bunch (with a δ-function concentration profile) at the entrance of the

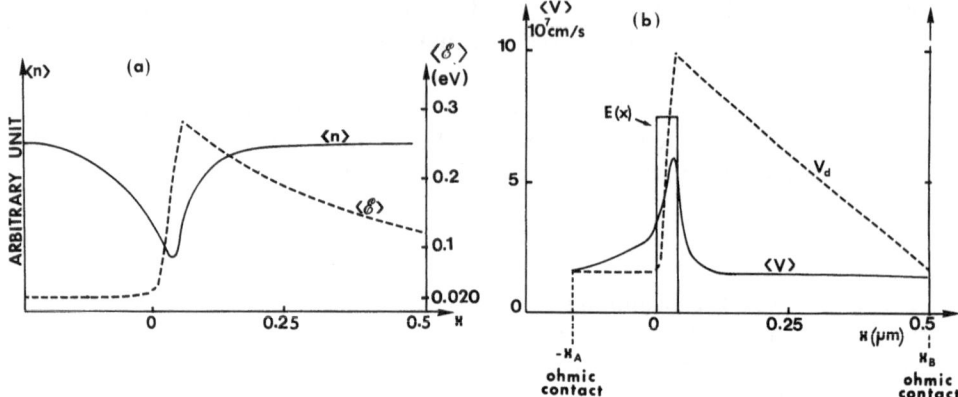

Fig. 8.18. (a) Average concentration and average energy profiles of electrons in GaAs at 293 K when stationary state is reached in the structure characterized by a space-pulse configuration of the electric field. (b) Comparison between the average velocity of electrons submitted to a space pulse (——) or a time pulse (– – –) of the electric field. The value of the electric field pulse is 70 kV/cm over a distance of 0.05 μm. The results obtained by the study versus time are plotted versus the distance x traveled by the electron by using the relation:

$$x = \int_0^t \langle v(t') \rangle \, dt'.$$

In the study versus space, the Monte Carlo simulation is carried out until a steady regime is reached with a nonhomogeneous semiconductor where the space pulse of the electric field (plotted in the figure) is obtained from the study versus time. Ohmic contacts are assumed on each side of the nonhomogeneous semiconductor: this means that any carrier crossing one of the contacts (at $x = -x_A$ or $x = x_B$) is immediately reinjected at the other contact with the same k state and energy (carrier position is simply displaced from $-x_A$ to x_B or from x_B to $-x_A$)

space pulse. It can be noted that as the drift-velocity of the electron bunch (assumed to coincide with the velocity of the peak of the electronic concentration) is very high (close to 10^8 cm/s) during the first picosecond, a widening in the tail of the pulse is very rapidly observed. Such an effect is due to the carriers which, subject to collisions, come back towards the region of the space pulse and are not able to cross over it. Consequently, the number of electrons located near the end of the electric field space pulse increases versus time. These electrons are often characterized by negative values of the velocity, as shown by the dashed curves in Fig. 8.19, and this can explain the low values observed for the average velocity attained in stationary conditions.

In Fig. 8.20 the average velocity of the carriers in the structure is plotted versus time just after the injection of the δ-like electron bunch. The average velocity at the very beginning steeply increases and reaches a very high value (about 9×10^7 cm/s) but, unfortunately, it suddenly decreases finally to reach a quite low stationary value (close to 1×10^7 cm/s). Consequently, from the above analysis we conclude that for the structure considered a ballistic motion seems to be difficult to realize under steady-state conditions.

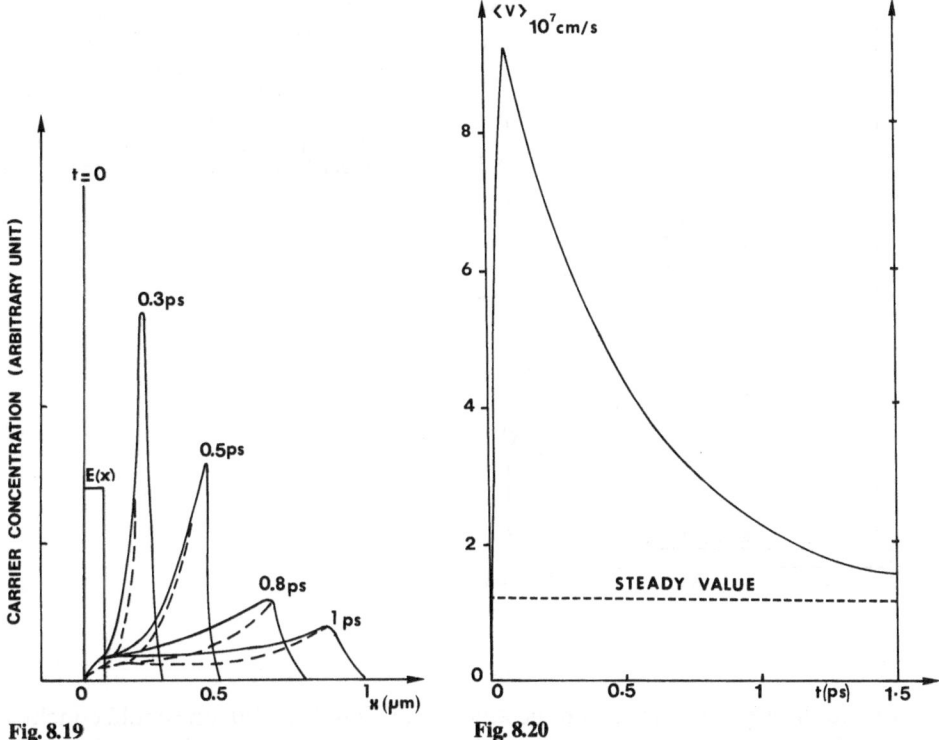

Fig. 8.19   Fig. 8.20

Fig. 8.19. Transient behavior of the structure studied in Fig. 8.17: time evolution of an electron bunch injected at initial time at the entrance of the space-pulse configuration of the electric field. (——): total electron concentration; (– – –): electron concentration with positive velocity: GaAs: $N_d = 0$, $T_0 = 77$ K

Fig. 8.20. Transient behavior of the structure studied in Fig. 8.17: average velocity of the electrons as a function of time when an electron bunch is injected at time 0 at the entrance of the space-pulse electric field; GaAs: $N_d = 0$, $T_0 = 77$ K

Overshoot phenomena have also been studied when a wide space-pulse configuration of the electric field is applied to an ensemble of electrons in the semiconductor. The results so obtained are reported in Fig. 8.21 where they are compared with those obtained when a long time-pulse configuration of the electric field is applied. One can note that the values of the overshoot velocity corresponding to the space-pulse configuration are close (just a little bit lower) to those obtained in the case of a time configuration. In this "overshoot" case, the space variation of the velocity $\langle v \rangle$ and, in turn, of the carrier concentration (in stationary condition the quantity $n\langle v \rangle$ is constant) is smooth and consequently diffusion effects are almost negligible. In addition, one can remark that within about a length of 0.5 μm the velocity increases and carrier concentration decreases; therefore, in this region the diffusion velocity adds to the drift velocity

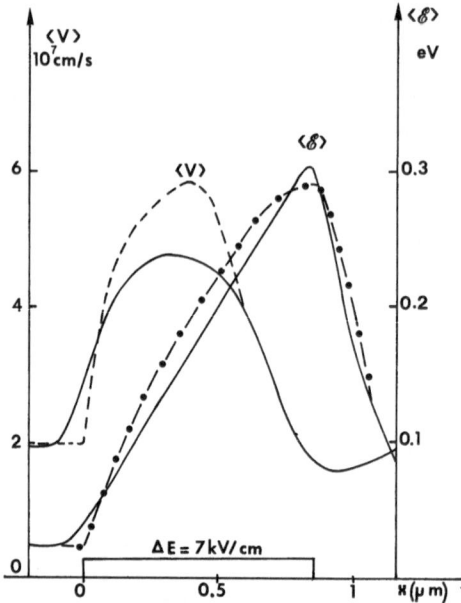

Fig. 8.21. *Overshoot Case.* Average velocity and average energy of electrons as functions of the distance traveled by the carriers (GaAs: $N_d = 0$, $T_0 = 77$ K). The results obtained from a Monte Carlo simulation when a wide space-pulse configuration (——) and a long time-pulse configuration (– – –) are reported for comparison

due to the electric field. From the analysis of Figs. 8.18, 20, one could conclude that the wide space configuration of E should be preferred with respect to the spike pulse to optimize the performances of submicron devices. Nevertheless, it should be noted that in the two examples here studied, the effect of the space charge due to electrons is not taken into account. Thus, additional, more realistic Monte Carlo simulations carried out in submicrometer devices are needed to support fully such a conclusion.

Finally, an example of a microscopic simulation by Monte Carlo methods taking into account all phenomena (diffusion, space-charge, and additional effects due to space nonuniformity) is reported in Fig. 8.22 for the case of an $n^+ - i - n^+$ structure under steady-state conditions. It can be noted that, even in this case, overshoot phenomena exist and rather high values of the velocity can be obtained in the active region of the sample. These results appear to be important since they clearly show that overshoot velocity can be attained in practical devices.

8.6 Conclusion

In this chapter we have analyzed the carrier transport under non-steady-state conditions occurring in a semiconductor submitted either to a very short-time or to a very small-space configuration of the electric field. New features then characterize the carriers' dynamics and some interesting phenomena, such as

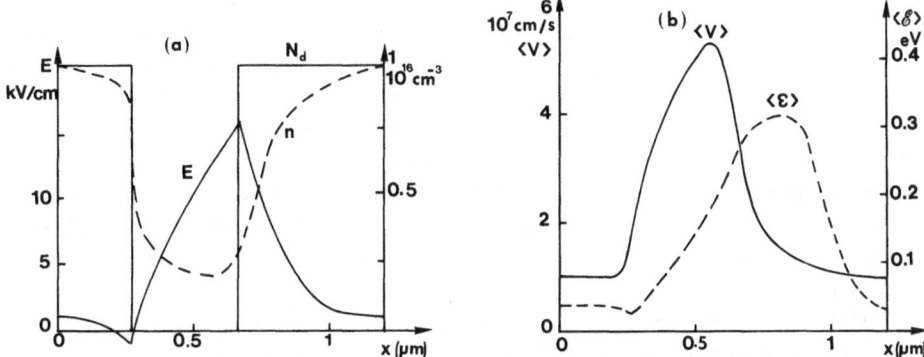

Fig. 8.22a, b. Monte Carlo simulation of a unidimensional GaAs n^+-i-n^+ structure with an applied voltage of 0.5 V at $T_0 = 77$ K: (a) Electric field, impurity, and electron concentration profiles; (b) Drift velocity and average energy profiles. Results refer to steady-regime conditions

ballistic motion, overshoot or undershoot velocity, and negative diffusivity, can be predicted.

These features should be taken into account in the modeling of submicron devices. The classical electrokinetic equations cannot be used directly, and new methods, based, for example, on the Monte Carlo procedure or the relaxation time approximation, have to be employed. With suitable time configurations of the electric field, it seems possible, by using ballistic or overshoot phenomena, to increase strongly the carrier velocity during very short times, which in the spatial scale correspond to submicrometer distances. Nevertheless, owing to space-charge reaction, diffusivity, and heat conduction, the characteristics of the phenomena achieved in submicrometer devices can differ significantly from those obtained from a simple time analysis of a uniform sample. Consequently, a careful study that takes into account all the effects due to the spatial nonuniformity must be carried out to determine the carrier velocities which really can be achieved in practical submicrometer devices.

Only theoretical results have been reported in this chapter because so far the experimental investigation of these phenomena is still at the beginning. Due to the very short time scale involved (1 ps), any experiment appears to be inherently difficult. Despite its importance, the particular problem of determining the drift velocity attainable in submicrometer devices still remains an unsolved problem since classical or even microwave-time-of-flight techniques have not reached such a degree of time resolution.

Methods based on current voltage measurements in typical submicrometer $n^+ - n - n^+$ structures could also be used, and a few experiments have already been described [8.23]. For very small active region length, scattering mechanisms should not occur and ballistic motion could be achieved. In this case current-voltage characteristics similar to those obtained for the macroscopic vacuum diode should be obtained. However, no direct determination of the drift velocity can be achieved since the current depends not only on the carrier velocity, but also on the number of carriers injected in the active region. In addition, the presence of parasitic elements, such as series resistances, can introduce spurious effects which are difficult to take into account.

Consequently, alternative methods have to be devised. In this field we cite recent experiments using picoseconds lasers [8.24] or photo-excitation of submicrometer devices [8.25]. So far the results which have been obtained are not very accurate from a quantitative point of view. Nevertheless, they clearly support the existence of overshoot velocity phenomena in submicrometer structures.

With a careful design – one that takes into account space charge, non-uniformity, and also the possible influence of edge and parasitic effects — it should soon be possible to use the transient phenomena here described to increase significantly the velocity of the carriers so that high-performance logic and microwave submicrometer devices could be realized.

Acknowledgements. I would like to thank Profs. J. P. Nougier, R. Castagne, G. Salmer, L. Reggiani, Drs. B. Boittiaux, A. Cappy, R. Fauquembergue, P. A. Rolland, and J. Zimmermann for many useful discussions.

References

8.1   D.Ferry: *Physics of Nonlinear Transport in Semiconductors*, NATO Adv. Study Inst. Ser. B, Physics **52**, 577 (Plenum, New York 1980)
8.2   H.Budd: Phys. Rev. **158**, 798 (1967)
8.3   J.P.Nougier, M.Rolland: Phys. Rev. **B8**, 5728 (1973)
8.4   H.D.Rees: J. Phys. Chem. Sol. **30**, 643 (1969)
8.5   W.Fawcett, A.D.Boardman, S.Swain: J. Phys. Chem. Sol. **31**, 1963 (1970)
8.6   J.Zimmermann: These d'Etat, Université de Lille (1980)
8.7   M.A.Littlejohn, J.R.Hauser, T.H.Glisson: J. Appl. Phys. **48**, 4587 (1977)
8.8   J.P.Nougier, J.C.Vaissiere, D.Gasquet, J.Zimmermann, E.Constant: J. Appl. Phys. **52**, 825 (1981)
8.9   S.Shur: Electron. Lett. **12**, 615 (1976)
8.10 A.Cappy, B.Carnez, R.Fauquembergue, G.Salmer, E.Constant: IEEE Trans. **ED-27**, 2158 (1980)
8.11 S.M.Sze: *Physics of Semiconductor Devices* (Wiley, New York 1981) Chap. 10
8.12 P.A.Rolland, E.Constant, G.Salmer, R.Fauquembergue: IEEE Trans. **ED-28**, 341 (1981)
8.13 S.Shur: IEEE Electron. Lett. **12**, 615 (1976)
8.14 J.G.Ruch: IEEE Trans. **ED-19**, 652 (1972)
8.15 W.Fawcett, H.D.Rees: Phys. Lett. **A29**, 578 (1969)
8.16 E.Constant: Nato-Asi, *Physics of Submicron Semiconductor Devices*, S. Miniato Italy, July 10–23, 1983 (Plenum, New York, to be published)
8.17 A.Ghis, E.Constant, B.Boittiaux: J. Appl. Phys. **54**, 214 (1983)
8.18 D.K.Ferry, J.R.Barker: J. Appl. Phys. **52**, 818 (1981)
8.19 B.Boittiaux, E.Constant, L.Reggiani, R.Brunetti, C.Jacoboni: Appl. Phys. Lett. **40**, 407 (1982)
8.20 V.Borsari, C.Jacoboni: Phys. Status Solidi **B54**, 649 (1972)
8.21 K.Blotekjaer: IEEE Trans. **ED-17**, 38 (1970)
8.22 P.A.Rolland, M.R.Friscourt, A.Cappy, E.Constant, G.Salmer: IEEE Trans. **ED-30**, 223 (1983)

8.23 M.S.Shur, L.F.Eastman: Electron. Lett. **16**, 522 (1980)
8.24 J.Shah: J. Physique **10**, C7-445 (1981);
      see also C.V.Shank, R.L.Fork, B.I.Greene, F.K.Reinhardt, R.A.Logan: Appl. Phys. Lett. **30**,
      104 (1981)
8.25 S.Laval: Microelectronics J., **13**, 18 (1982)
8.26 C.Canali, C.Jacoboni, F.Nava, G.Ottaviani, A.Alberigi-Quaranta: Phys. Rev. **B12**, 2265
      (1975)
8.27 C.Canali, C.Jacoboni, G.Ottaviani, A.Alberigi-Quaranta: Appl. Phys. Lett. **27**, 278 (1975)

List of Symbols

Page and equation numbers are shown in italics. Equation numbers are given in parentheses and dimensions in square brackets. The Greek symbols are at the end of the list.

a	Length parameter for donor ground state [cm] *(2.212)*
a_B^*	Bohr radius of the impurity ground state [Å] *66*
a_i	Distance between imperfection centers [cm] *72*
a_s	Unit crystallographic slip-distance [cm] *66*
a_0	Lattice parameter [Å] *63*
A	Cross sectional area of a device [cm²] *41*
	Inverse valence-band parameter *(2.180)*
	Characteristic frequency for the spontaneous optical phonon emission [s⁻¹] *(6.1)*
$A(\mathcal{E}, \omega)$	Spectral density function [eV⁻¹] *218*
$[A]$	Matrix for transformation from the main crystallographic axes to the reference frame of the sample *156*
A^*	Effective Richardson constant [A cm⁻² K⁻²] *(7.16)*
$\mathscr{A}$	General physical quantity *(2.66)*
b	Length parameter for donor ground state [cm] *(2.212)*
$\mathbf{B}$	Magnetic field (induction) [T] *153*
B	Inverse-valence band parameter *(2.180)*
$[B^{(\alpha)}]$	Matrix for transformation from the main axes of the α-valley to the main crystallographic axes *156*
$(\Delta B)/2$	Half width of the cyclotron resonance line [T] *(6.14)*
C	Inverse-valence band parameter *(2.180)*
$C(t)$	Autocorrelation function of velocity fluctuations [cm² s⁻²] *(2.61)*
C_A	Fraction of atoms of type A *66*
C_d	Device capacitance [F] *93*
C_i	Input capacitance [F] *93*
C_s	Parasitic capacitance [F] *93*
C'	Parallel capacitance of C_d and C_s [F] *93*
d	Average distance traveled by carriers over a time t [cm] *(8.14)*
d_i	Distance from intentional doping [Å] *206*
d_{ij}	Distance between probes i and j [cm] *153*
d_0	Deformation potential constant for hole optical-phonon interaction [eV] *(2.210a)*
	Maximum ballistic distance [cm] *(8.21)*
D	Diffusion coefficient [cm² s⁻¹] *(1.1)*
D^*	Time-dependent diffusion coefficient independent of initial conditions [cm² s⁻¹] *(4.33)*
$\bar{D}$	Diffusion coefficient averaged over microwave period [cm² s⁻¹] *126*
D_{ij}	Tensorial notation for the diffusion coefficient [cm² s⁻¹] *(2.17)*
$D_{lo, tr}$	Diffusion coefficient longitudinal and transverse with respect to the field direction [cm² s⁻¹] *(2.19)*
D_{jk}	Deformation potential constant for intervalley-phonon scattering with electrons [eV cm⁻¹] *66*

(D_tK)	Deformation potential constant for optical-phonon scattering with electrons [eV cm^{-1}] *66*	
e	Electron charge [C] *(1.1)*	
E	Electric field [V cm^{-1}] *(2.1)*	
$E^{(\alpha)}$	Effective field strength in the α-valley [V cm^{-1}] *(5.2)*	
E_{ac}	ac electric field [V cm^{-1}] *115*	
E_{av}	Average electric field [V cm^{-1}] *(3.3)*	
E_{dc}	dc electric field [V cm^{-1}] *(4.1)*	
E^{ext}	External electric field [V cm^{-1}] *(2.142)*	
E_h	Harmonic mixing field [V cm^{-1}] *117*	
E_{high}	High-field value due to longitudinal domainization [V cm^{-1}] *155*	
E_{low}	Low-field value due to longitudinal domainization [V cm^{-1}] *153*	
E_{max}	Maximum value of an electric field [V cm^{-1}] *92*	
E_{min}	Minimum value of an electric field [V cm^{-1}] *92*	
E_n	Microwave noise electric field [V cm^{-1}] *133*	
$E_{q\omega}$	Fourier component of the electric field [V cm^{-1}] *(2.133)*	
E_s	Static electric field [V cm^{-1}] *(8.8)*	
E_0	Uniform electric field [V cm^{-1}] *(2.71)*	
E^{scf}	Self-consistent electric field [V cm^{-1}] *(2.140)*	
$E_{C,K,S}$	Different type of critical electric field strengths [V cm^{-1}] *160, 163*	
E_ω	Microwave electric field [V cm^{-1}] *191*	
E_ω^{min}	E_{min} for the case of microwave [V cm^{-1}] *195*	
E_1	Small superimposed electric field [V cm^{-1}] *(2.71)*	
E_2	Amplitude of microwave field of the second harmonic [V cm^{-1}] *(4.6)*	
E_i'	ith component of the fluctuational field [V cm^{-1}] *150*	
E_{01}	Electric field in empty waveguide [V cm^{-1}] *117*	
E_1	Acoustic-phonon deformation potential for electrons [eV] *66*	
E_1^d	Deformation potential for scattering with dislocations [eV] *66*	
E_1^0	Acoustic-phonon deformation potential for holes [eV] *80*	
$\mathscr{E}$	Carrier energy [eV] *(2.7)*	
$\mathscr{E}_0$	Carrier thermal energy [eV] *(8.7)*	
	Constant energy term [eV] *(7.3)*	
$\mathscr{E}_{a,c}$	Anion and cation atomic p-states energy [eV] *(7.5)*	
$\mathscr{E}_A$	Affinity energy [eV] *66*	
$\mathscr{E}_b$	Energy associated with the uncertainty broadening [eV] *(2.213)*	
$\mathscr{E}_{cs}$	Chemical shift energy [eV] *(2.212)*	
$\mathscr{E}_e$	Energy of impinging electrons [eV] *(3.14)*	
$\mathscr{E}_F$	Fermi energy [eV] *(7.16)*	
$\mathscr{E}_g$	Energy gap [eV] *55*	
$\mathscr{E}_{i,f}$	Initial and final energy of a carrier free flight [eV] *(2.216)*	
$\mathscr{E}_I$	Binding energy of impurity center [eV] *66*	
$\mathscr{E}_M$	Maximum electron energy [eV] *28*	
$\mathscr{E}_p$	Energy to generate an electron-hole pair [eV] *(3.14)*	
$\mathscr{E}_R$	Energy of the resonant level [eV] *(2.213)*	
$\mathscr{E}_{\Delta_{1c}}$	Energy gap between states of given symmetry [eV] *(2.184)*	
$\mathscr{E}_{\Delta'_{2c}}$	Energy gap between states of given symmetry [eV] *(2.184)*	
$\mathscr{E}_{L_{1c}}$	Energy gap between states of given symmetry [eV] *(2.183)*	
$\mathscr{E}_{L'_{3v}}$	Energy gap between states of given symmetry [eV] *(2.183)*	
$\mathscr{E}_s$	Carrier average energy under steady state conditions [eV] *(8.8)*	
$\mathscr{E}_{so}$	Split-off energy of the lowest valence band [eV] *59*	

$\mathscr{E}_v$	Valence band edge [eV] *(7.5)*
$\mathscr{E}_T$	Trap energy level [eV] *(3.7)*
$\mathscr{E}_\Gamma$	Energy gap at Γ [eV] *(2.182)*
f	Frequency [Hz] *(2.3)*
$f(\boldsymbol{k},\boldsymbol{r},t)$	Single-particle distribution function *(2.5)*
$\bar{f}(\boldsymbol{k},\boldsymbol{r},t)$	Ensemble average of $f(\boldsymbol{k},\boldsymbol{r},t)$ *(2.128)*
$f^{(\alpha)}(\boldsymbol{k})$	Carrier distribution function in the α-valley *156*
$f_0^{(\alpha)}(\mathscr{E})$	Isotropic part of the distribution function in the α-valley *156*
$f_{q\omega}$	Fourier component of the distribution function *(2.134)*
$\boldsymbol{F}$	External force [Nw] *(2.193)*
$g(\boldsymbol{k},\boldsymbol{r},t)$	Normalized single-particle distribution function *(2.20)*
$g_0(\boldsymbol{k},E)$	Normalized single-particle distribution function for space homogeneous and stationary conditions *(2.29)*
$g_1(\boldsymbol{k},\gamma,E)$	Pertubed term to $g_0(\boldsymbol{k},E)$ due to a small concentration gradient *(2.31)*
$[g^{(\alpha)}]$	Matrix for transformation from the main axes of the α-valley to the reference frame of the sample *(5.2)*
$\boldsymbol{g}_{q\omega}$	Langevin random current density [A cm^{-2}] *(2.149)*
G	Green's function of operator $\hat{L}$ [cm^3 s] *(2.32)*
$\mathscr{G}(\boldsymbol{k},\boldsymbol{k}')$	Overlap factor *(2.197)*
h	Planck constant [J s] *(2.6)*
$\hbar$	Planck constant divided by 2π [J s] *(2.6)*
H'	Perturbation Hamiltonian for scattering of electrons [eV] *(2.195)*
i	Imaginary unit $(i=\sqrt{-1})$
I	Current [A] *41*
	Parameter for intervalley scattering at impurities [eV cm^3] *(2.212)*
I_c	Normalized current [A] *(3.14)*
I_0	Average current [A] *41*
$\tilde{I}$	Induced microwave current [A] *(3.10)*
$\tilde{I}_0$	Induced microwave current under bombarded contact [A] *(3.10)*
I_k	Linearized collision operator including electron-electron scattering [s^{-1}] *(2.135)*
I_k^{ee}	Electron-electron collision operator [s^{-1}] *(2.129)*
I_k^{th}	Linear collision operator describing scattering with thermal bath via phonons and imperfections [s^{-1}] *(2.5)*
$\boldsymbol{j}(\boldsymbol{r},t)$	Current density [A cm^{-2}] *(2.12)*
$j_i^{(\alpha)}$	ith component of the density current contribution of th α-valley [A cm^{-2}] *150*
$j_i'^{(\alpha)}$	ith component of the contribution of the α-valley to the fluctuation density current [A cm^{-2}] *150*
j_{dc}	Density of dc current flowing through the sample under simultaneous action of dc and microwave fields [A cm^{-2}] *(4.1)*
j_o	Ohmic dc current density in sample [A cm^{-2}] *(4.3)*
j_h	Harmonic mixing dc current density [A cm^{-2}] *(4.6)*
$\boldsymbol{j}_{q\omega}$	Fourier component of the density current [A cm^{-2}] *(2.148)*
$j_{A\to B}$	Density current flowing from semiconductor A to semiconductor B by thermoionic emission [A cm^{-2}] *(7.15)*
$\boldsymbol{k}$	Wavevector of a carrier [cm^{-1}] *(2.5)*
$\boldsymbol{k}'$	Final wavevector of a carrier [cm^{-1}] *(2.8)*
$\boldsymbol{k}_0$	Given value of $\boldsymbol{k}$ [cm^{-1}] *35*
$\boldsymbol{k}_d$	Average wavevector of a drifted Maxwellian distribution function [cm^{-1}] *(2.131)*
$\boldsymbol{K}$	Reciprocal lattice vector [cm^{-1}] *(2.210a)*
K_B	Boltzmann constant [J K^{-1}] *(1.1)*

K_2	Modified Bessel function of order 2 *(2.188)*
l	Carrier mean free path [cm] *15*
l_{op}	A mean free path for optical-phonon interaction [cm] *(2.226)*
L	$\langle 111 \rangle$ point of the Brillouin zone *55*
L	Length of a device [cm] *(2.114)*
$L_{b,z}$	Width of a quantum well [Å] *(7.3)*
$\hat{L}$	Linear operator to solve perturbatively the Boltzmann equation [s^{-1}] *(2.28)*
m	Carrier effective mass [kg] *(2.131)*
m^*	Mean effective mass at the given carrier average energy [kg] *(8.6)*
m_0	Free-electron mass [kg] *(2.180)*
m_c	Conductivity effective mass [kg] *(2.186)*
m_{de}	Density-of-states effective mass of electrons *(2.188)*
m_{di}	Density-of-states effective mass of holes *(2.189)*
m_h	Density-of-states effective mass of heavy holes *(2.189)*
m_{li}	Density-of-states effective mass of light holes *(2.189)*
$m_{lo,tr}$	Electron longitudinal and transverse effective masses for an ellipsoidal equienergetic surface [kg] *(2.177)*
m_{sat}	Cyclotron mass value [kg] *195*
$[M]$	Dimensionless diagonal tensor for the description of the energy-independent mobility anisotropy *(5.2)*
$M(k_{\parallel}, k_{\parallel}')$	Matrix element for scattering in modulation-doped structures [eV cm$^{-3/2}$] *(7.12a)*
M^m	mth space moment of the carrier distribution [cm^m] *(2.38)*
$n(r, t)$	Carrier density in real space [cm^{-3}] *(2.11)*
$n_{a,b}(k)$	After and before scattering distribution in k-space [cm^3] *(2.69)*
$n^{(\alpha)}$	Electron density in the α-valley [cm^{-3}] *150*
$n_{k,s}$	Number of accumulated (k) and streaming (s) carriers *187*
n_0	Constant average density [cm^{-3}] *(2.46)*
$n_{q\omega}$	Fourier component of the carrier density [cm^{-3}] *(2.141)*
n_1	Amplitude of harmonic carrier density [cm^{-3}] *(2.46)*
	Average electron concentration in valley 1 [cm^{-3}] *(4.21)*
n_2	Average electron concentration in valley 2 [cm^{-3}] *(4.21)*
N	Number of particles *(2.67)*
	Number of electron free flights *(2.70)*
	Numer of fixed-time intervals *(2.73)*
N_A	Number of atoms *66*
N_q	Thermal equilibrium number of phonons with wavevector q *(2.206)*
N_0	Thermal equilibrium number of optical phonons *69*
N_a	Acceptor impurity concentration [cm^{-3}] *92*
N_c	Effective density of states [cm^{-3}] *(7.17)*
N_d	Donor impurity concentration [cm^{-3}] *92*
N_{imp}	Impurity concentration [cm^{-3}] *191*
N_B	Background impurity density [cm^{-3}] *207*
N_I	Net ionized impurity concentration [cm^{-3}] *66*
$N_I(z)$	Net impurity density as a function of z [cm^{-3}] *206*
N_N	Neutral impurity concentration [cm^{-3}] *66*
N_R	Remote impurity density [cm^{-3}] *(7.13)*
N_{sc}	Concentration of centers for space-charge scattering [cm^{-3}] *(2.214)*
N_T	Trap concentration [cm^{-3}] *(3.6)*
p	Appropriate component of piezoelectric tensor [C m^{-2}] *63*
P	Pressure [Nw m^{-2}] *101*

$\hat{p}$	Quantum mechanical momentum operator [kg m s^{-1}]	(7.1)
$P(x;t)$	Probability per unit length that in absence of trapping a particle covers a distance x in a time t [cm^{-1}]	(2.51)
$P[k(t)]$	Probability per unit time that an electron in k at a given time will be scattered between t and $t+dt$ later [s^{-1}]	(2.63)
$P_n(k_0;k,t)$	Probability density that an electron started at $t=0$ with k_0 will pass through k, t during the nth flight [cm^3]	(2.75)
P_{esc}	Escape probability of an electron over an energy barrier	218
$P_{esc}^{c,Q}$	Classical and quantum mechanical escape probability of an electron over phonon	218
P_e	Rate of energy loss of electrons to phonons [W]	209
P_{av}	Available noise power of a two-terminal network for unit bandwidth of frequency [W]	(2.3)
P_k	Power of generated kth harmonic [W]	121
P_m	Microwave power absorbed in a sample per unit volume [W cm^{-3}]	(4.11)
$P_{n\alpha}$	Noise power in α-direction [W]	(4.19)
P_s	Microwave power absorbed by a sample over the microwave period [W]	(4.5)
$\mathscr{P}(t)dt$	Probability that an electron will suffer its next collision during dt around t	(2.63)
q	Wavevector [cm^{-1}]	(2.43)
$q_{a,e}$	Absorption and emission phonon wavevector [cm^{-1}]	(2.209)
Q	Electric charge [C]	(3.1)
	Symbol for a function appropriate to describe MED	(5.9)
Q_{li}	Line charge of a dislocation [C cm^{-1}]	66
$Q_{x,y,z}$	Integrated photocurrent along the respective direction [A]	186
Q_0	Cross-sectional area for space-charge scattering [cm^2]	(2.214)
r	Position in real space [cm]	11
r	Random number evenly distributed in the interval (0,1)	(2.65)
R	Length of the electron-hole pairs generation range [μm]	89
	Symbol for a function appropriate to describe MED	(5.9)
R_0	Input resistance [Ω]	93
R_S	Series resistance [Ω]	93
S	Sample area [cm^2]	(3.14)
$S(x,t)$	Particle generation rate (source term) [cm^{-3} s^{-1}]	(2.43)
S_0	Amplitude of the particle generation rate [cm^{-3} s^{-1}]	(2.43)
S_{low}	Two-dimensional screening constant [cm^{-1}]	(7.12a)
$S_I(\omega)$	Current-noise spectral density [A^2 s]	(2.101)
$S_{j\alpha}$	Spectral density of current density fluctuations in α-direction [A^2 cm^{-4} s]	(4.18)
$S_{j\,int}$	Spectral density of intervalley current density fluctuations [A^2 cm^{-4} s]	(4.21)
$S_{j\parallel,\perp}$	Longitudinal and transverse spectral density of current density fluctuations [A^2 cm^{-4} s]	133
S_v	Spectral density of velocity fluctuations [cm^2 s^{-1}]	(1.2)
$S_{v\parallel,\perp}$	Longitudinal and transverse spectral density of velocity fluctuations [cm^2 s^{-1}]	137
$S_v(\omega)$	Voltage-noise spectral density [V^2 s]	(2.106)
t	Time [s]	(2.5)
t_i	Free-flight duration [s]	(2.66)
t_r	Stochastic free-flight duration [s]	(2.65)
t_s, t_s^{back}	Switching time for forward and backward thermionic emission [s]	216
T	Duration of a single history [s]	(2.66)
	Given time [s]	245
T_R	Carrier transit time [s]	(3.2)
T_R'	Reduced transit time [s]	(3.9)

T_{op}^0	Traveling time required for a carrier to be accelerated up to the optical phonon energy by a dc electric field [s] *(6.2)*
T_{op}	T_{op}^0 in the presence of magnetic field [s] *(6.11)*
T_{op}^ω	T_{op}^0 for the microwave electric field [s] *(6.12)*
T_0	Lattice temperature [K] *(1.1)*
T_e	Electron temperature [K] 8
T_n	White-noise temperature [K] *(1.3)*
$T_{n\alpha}$	Noise temperature in α-direction [K] *(4.18)*
$T_{n\parallel,\perp}$	Longitudinal and transverse noise temperature [K] 133
T_1	Hot contact temperature [K] *(4.15)*
$[T_{ij}]$	Herring and Vogt transformation matrix *(2.190)*
T_{BA}	Transfer matrix between interfaces B and A *(7.10)*
u	Effective sound velocity [cm s^{-1}] 65
$u_{lo,tr}$	Longitudinal and transverse sound velocity [cm s^{-1}] 63
u_0	Dimensionless parameter for intervalley scattering at impurity center *(2.212)*
$u(x-x')$	Unit step function *(2.125)*
$u_k(r)$	Periodic part of Bloch wavefunction [cm$^{-3/2}$] *(2.197)*
U_T	Thermoelectric voltage [V] *(4.14)*
ΔU_{ij}	Potential difference between probes i and j [V] 153
v	Carrier group velocity [cm s^{-1}] *(2.7)*
v_d'	Differential (with respect to the number of carriers) drift velocity [cm s^{-1}] *(2.161)*
v_a	Average velocity over a distance d [cm s^{-1}] *(8.15)*
v_d	Drift velocity [cm s^{-1}] *(2.13)*
v_{do}	Drift velocity in absence of concentration gradient [cm s^{-1}] *(2.15)*
v_{d_1}	Drift velocity perturbation due to a small concentration gradient [cm s^{-1}] *(2.34)*
v_s	Drift velocity under steady state conditions [cm s^{-1}] *(8.8)*
v_d^s	Carrier drift velocity under the streaming motion condition [cm s^{-1}] *(6.4)*
v_{th}	Thermal velocity [cm s^{-1}] *(3.6)*
v_ω	Fourier transform of the time-dependent drift velocity [cm s^{-1}] *(8.11)*
$v_{1,2}$	Sine and cosine Fourier transform of the velocity response to a small ac field [cm s^{-1}] *(2.72)*
v_0	Maximum velocity of a carrier [cm s^{-1}] 243
	Drift velocity associated to a dc field E_0 [cm s^{-1}] *(2.72)*
$V(r)$	Periodic crystal potential energy [eV] *(7.1)*
$\|V(q)\|^2$	Squared Fourier transform of the interaction potential [eV^2] *(2.196)*
V_{xx}	Interatomic energy matrix element [eV] *(7.5)*
V	Voltage [V] *(2.103)*
V_a	Applied voltage [V] *(3.3)*
V_d	Depletion voltage [V] *(3.2)*
V_{dc}	Direct-current voltage [V] 115
V_h	Mixing voltage [V] 118
V_{ko}	Contact potential difference [V] *(4.16)*
V_0	Crystal volume [cm^3] *(2.8)*
V_s	Active volume of sample [cm^3] *(4.5)*
V_{op}	Carrier velocity with which carrier kinetic energy equals the optical phonon energy [cm s^{-1}] *(6.3)*
V_1	Voltage value [V] 211
W	Sample thickness [cm] *(3.1)*
$W(k,k')$	Total transition rate [s^{-1}] *(2.10)*
$W^+(k,k')$	Total transition rate including self-scattering [s^{-1}] *(2.77)*

$W_i(\mathbf{k}, \mathbf{k}')$	Transition rate for the ith process [s^{-1}]	(2.10)
x	Space coordinate [cm] 14	
x_0	Thickness of drifting charge layer [cm]	(3.13)
y	Space coordinate [cm] 14	
$y_{q\omega}$	Langevin random term for the linearized Boltzmann equation [s^{-1}] 49	
z	Space coordinate [cm] 14	
$z(\mathbf{r}, \mathbf{r}', \omega)$	Transfer impedance [Ω cm^{-2}]	(2.109)
z_0	Distance [Å] 206	
$Z(\omega)$	Small signal impedance of a two-terminal device [Ω]	(2.102)
Z_g	Waveguide impedance [Ω] 116	
Z_s	Sample impedance [Ω] 116	
Z_e	Number of charge units of the impurity center 66	
$\langle \, \rangle$	Ensemble average at a given time 13	
$\langle 100 \rangle$	Direction along the indicated crystallographic axes 59	
(110)	Indicated crystallographic plane 152	
α	Nonparabolicity parameter [eV^{-1}]	(2.181)
α^*	Warm electron effective coefficient [V^{-2} cm^2]	(4.11)
α_p	Dimensionless Fröhlich coupling constant	(2.210b)
β	Nonlinear electric field coefficient [V^{-2} cm^2]	(4.6)
γ	Relative gradient of concentration [cm^{-1}]	(2.26)
Γ	Center of the Brillouin zone 55	
$\Gamma(\mathbf{k})$	Total scattering rate including self-scattering [s^{-1}]	(2.77)
δ	Dirac function	(2.18)
Δ	$\langle 100 \rangle$ direction in the Brillouin zone 55	
ε	Free space permittivity [F cm^{-1}]	(2.123)
ε_0	Relative value of the static dielectric constant	(2.123)
ε_∞	Relative value of the high-frequency dielectric constant	(2.210b)
$\bar{\varepsilon}$	Arithmetic average value between relative static dielectric constants	(7.12a)
ζ	Normalized static electric field	(6.9)
ζ^h	ζ for positive holes in silver-halides 188	
$\zeta_{h, li}$	ζ for heavy and light holes in Ge 189	
ζ_ω	Normalized microwave electric field 195	
η	Dimensionless parameter used to adjust the deformation potential constant 164	
θ_{jk}	Equivalent temperature of the intervalley phonon [K] 66	
θ_{op}	Equivalent temperature of the optical phonon [K] 66	
Θ	Noise contribution to diffusion due to electron-electron interaction [cm^2 s^{-1}]	(2.170)
$\varkappa_k$	Distribution semiinvariants [cm^k]	(4.23)
λ	Wavelength [cm] 20	
λ_D	Debye screening length [cm]	(2.211)
$\lambda(\mathbf{k})$	Scattering rate [s^{-1}]	(2.9)
λ_0	Scattering rate for self-scattering process [s^{-1}]	(2.86)
λ_i	Scattering rate relative to the ith mechanism [s^{-1}] 28	
μ	(Chord) mobility [cm^2 V^{-1} s^{-1}]	(1.1)
$\mu^{(\alpha)}$	Mobility tensor for the α-valley [cm^2 V^{-1} s^{-1}]	(5.2)
μ_H	Hall mobility [cm^2 V^{-1} s^{-1}]	(6.5)
$\mu_{lo, tr}$	Components of the mobility parallel and perpendicular to the main valley axes [cm^2 V^{-1} s^{-1}] 156	
μ^*	Complex differential mobility [cm^2 V^{-1} s^{-1}]	(1.2)
μ'	Real part of μ^* (differential mobility) [cm^2 V^{-1} s^{-1}]	(2.112)
μ''	Imaginary part of μ^* [cm^2 V^{-1} s^{-1}]	(8.11)

v	Poisson ratio	72
ξ	Fluctuational electric field [V cm^{-1}]	(2.162)
$\xi(z)$	Envelope wavefunction	(7.8)
ϱ	Resistivity of the material [Ω cm]	89
ϱ_0	Crystal density [gr cm^{-3}]	63
	Zero-field sample resistivity [Ω cm]	117
σ	Conductivity [Ω^{-1} cm^{-1}]	113
σ_0	Zero-field conductivity [Ω^{-1} cm^{-1}]	(4.3)
σ'	Differential conductivity [Ω^{-1} cm^{-1}]	(2.163)
$\sigma'_{\parallel}$	Longitudinal differential conductivity [Ω^{-1} cm^{-1}]	134
σ_c	Capture cross section [cm^2]	(3.6)
σ_d	Real part of the small signal conductivity [Ω^{-1} cm^{-1}]	(4.9)
σ_e	Emission cross section [cm^2]	(3.7)
σ_{li}	Number of dislocation lines per unit area [cm^{-2}]	66
Σ	Self energy [eV]	218
$\tau(\mathbf{k})$	Relaxation time [s]	17
$\tau^{(\alpha)}$	Out scattering time of the electrons from the α-valley [s]	(5.1)
τ_{ac}	Scattering time of carriers due to acoustical phonons [s]	181
τ_{coll}	Duration of a collision [s]	11
τ_d	Trapping time associated with a particle source [s]	(2.44)
τ_{diff}	Fall time induced by diffusion [s]	(3.12)
τ_D	Detrapping time [s]	(3.6)
$\tau_{\mathscr{E}}$	Energy relaxation time [s]	(4.10)
τ	Dielectric relaxation time [s]	89
τ_ε	Dielectric relaxation time [s]	89
τ_ε	Differential dielectric relaxation time [s]	(3.13)
τ_{fsc}	Space charge induced fall time [s]	(3.13)
τ_g	Carrier generation time [s]	89
τ_i	Relaxation time of the ith process [s]	(4.20)
	ith component of the momentum relaxation time [s]	150
τ_{imp}	Scattering time of carriers due to impurities [s]	181
τ_{int}	Relaxation time for intervalley scattering [s]	(4.21)
τ_L	Carrier lifetime [s]	89
τ_m	Momentum relaxation time [s]	120
τ_{min}	Minimum value of a characteristic time [s]	251
τ_{op}	Scattering time of carriers due to optical phonons [s]	179
τ_r	Rise time of the current signal [s]	(3.15)
τ_{RS}	Scattering time for resonance impurity scattering [s]	(2.213)
τ_{sc}	Scattering time for space charge-scattering [s]	(2.214)
τ_σ	Conductivity relaxation time [s]	(4.9)
τ_y	Relaxation time of y physical process [s]	(4.7)
τ_0	Constant value for the inverse of a scattering rate [s]	(2.64)
	Period of a characteristic cyclic motion [s]	(8.24)
τ^+	Mean time before trapping [s]	(3.8)
φ	Phase shift of the harmonic disturbance in the carrier concentration	(2.46)
$\Phi(z)$	Potential energy superimposed on $V(r)$ [eV]	(7.1)
ψ	Electron wavefunction [cm$^{-3/2}$]	(7.8)
$\Xi_{d,u}$	Dilatational and shear deformation potential for ellipsoidal energy surfaces [eV]	68
ω	Angular frequency [rad s^{-1}]	(1.2)
ω_c	Cyclotron angular frequency [rad s^{-1}]	193
$\omega_{o,op,Lo}$	Optical phonon angular frequency [rad s^{-1}]	127

List of Acronyms

Subject Index

O. Madelung

Introduction to Solid-State Theory

Translated from the German by B. C. Taylor
1978. 144 figures. XI, 486 pages. (Springer Series in Solid-State Sciences, Volume 2). ISBN 3-540-08516-5

"... it is still one of the most modern general texts, with the extensive bibliography serving the important function of establishing effective contact with more recent work. ... a few unifying concepts developed in depth and applied to a variety of basic situations, and then supplemented by a well-integrated access to the more detailed literature – will set the tone for future textbooks in this field."

Physics Today

K. Seeger

Semiconductor Physics

An Introduction
2nd corrected and updated edition. 1982. 288 figures.
XII, 462 pages. (Springer Series in Solid-State Sciences, Volume 40). ISBN 3-540-11421-1

Contents: Elementary Properties of Semiconductors. – Energy Band Structure. – Semiconductor Statistics. – Charge and Energy Transport in a Nondegenerate Electron Gas. – Carrier Diffusion Processes. – Scattering Processes in a Spherical One-Valley Model. – Charge Transport and Scattering Processes in the Many-Valley Model. – Carrier Transport in the Warped-Sphere Model. – Quantum Effects in Transport Phenomena. – Impact Ionization and Avalanche Breakdown. – Optical Absorption and Reflection. – Photoconductivity. – Light Generation by Semiconductors. – Properties of the Surface. – Miscellaneous Semiconductors. – Appendix: Physical Constants. – References. – Subject Index.

Springer-Verlag
Berlin
Heidelberg
New York
Tokyo

Two-Dimensional Systems, Heterostructures, and Superlattices

Proceedings of the International Winter School Mauterndorf, Austria, February 26 – March 2, 1984
Editors: G. Bauer, F. Kuchar, H. Heinrich
1984. 231 figures. IX, 293 pages. (Springer Series in Solid-State Sciences, Volume 53). ISBN 3-540-13584-7

Contents: Physics of Heterostructures and Inversion Layers. – Growth and Devices. – Multi Quantum Wells and Superlattices. – Doping Superlattices. – Quantum Hall Effect. – Index of Contributors.

Very Large Scale Integration (VLSI)

Fundamentals and Applications
Editor: **D.F.Barbe**
With contributions by numerous experts
2nd corrected and updated edition. 1982.
147 figures. XI, 302 pages. (Springer Series in
Electrophysics, Volume 5)
ISBN 3-540-11368-1

"... The book is well organized and provides a
virtually up to date (over 300 references, some
as recent as February 1980) overview of VLSI.
... Surprisingly, in a book with so many
authors, it is well written, contains few
misprints, and except for Chap. 4 is profusely
illustrated."

Applied Optics

Ion Implantation Techniques

Lectures given at the Ion Implantation School
in Connection with the Fourth International
Conference on Ion Implantation:
Equipment and Techniques. Berchtesgaden,
Federal Republic of Germany,
September 13–15, 1982
Editors: **H.Ryssel, H.Glawischnig**
1982. 245 figures. XII, 372 pages. (Springer
Series in Electrophysics, Volume 10)
ISBN 3-540-11878-0

Contents: Machine Aspects of Ion Implanta-
tion: Ion Implantation System Concepts. Ion
Sources. Faraday Cup Designs for Ion
Implantation. Safety and Ion Implanters. –
Ion Ranges in Solids: The Stopping and
Range of Ions in Solids. The Calculation of
Ion Ranges in Solids with Analytic Solutions.
Range Distributions. – Measuring Techniques
and Annealing: Electrical Measuring Tech-
niques. Wafer Mapping Techniques for
Characterization of Ion Implantation Process-
ing. Non-Electrical Measuring Techniques.
Annealing and Residual Damage. –
Appendix: Modern Ion Implantation Equip-
ment: Evolution and Performance of the
Nova NV-10 Predep™ Implanter. Ion Implan-
tation Equipment from Veeco. The Series
IIIA and IIIX Ion Implanters. Standard High-
Voltage Power Supplies for Ion Implantation.
The IONMICROPROBE A-DIDA 3000-30
for Dopant Depth Profiling and Impurity
Bulk Analysis. – List of Contributors. –
Subject Index.

Light Scattering in Solids I

Introductory Concepts
Editor: **M.Cardona**
2nd corrected and updated edition. 1983.
111 figures. XV, 363 pages. (Topics in
Applied Physics, Volume 8)
ISBN 3-540-11913-2

Contents: *M.Cardona:* Introduction. –
A.Pinczuk, E.Burstein: Fundamentals of
Inelastic Light Scattering in Semiconductors
and Insulators. – *R.M.Martin, L.M.Falicov:*
Resonant Raman Scattering. – *M.V.Klein:*
Electronic Raman Scattering. – *M.H.Brodsky:*
Raman Scattering in Amorphous Semicon-
ductors. – *A.S.Pine:* Brillouin Scattering in
Semiconductors. – *Y.-R.Shen:* Stimulated
Raman Scattering. – Overview. – Additional
References with Titles. – Subject Index. –
Contents of Light Scattering in Solids II, III
and IV.

Laser Processing and Diagnostics

Proceedings of an International Conference,
University of Linz, Austria, July 15–19, 1984
Editor: **D.Bäuerle**
1984. 399 figures. XI, 551 pages. (Springer
Series in Chemical Physics, Volume 39).
ISBN 3-540-13843-9

Contents: Laser – Solid Interactions: Funda-
mentals and Applications. – Photophysics and
Chemistry of Molecule – Surface Interactions.
– Photoassisted Chemical Processing. – Diag-
nostics of Laser Processing, Materials and
Devices. – Laser Diagnostics in Reactive
Gaseous Systems. – Subject Index. – Author
Index.

Springer-Verlag
Berlin
Heidelberg
New York
Tokyo